ADVANCED MOTION CONTROL AND SENSING FOR INTELLIGENT VEHICLES

ADVANCED MOTION CONTROL AND SENSING FOR INTELLIGENT VEHICLES

Dr. Li Li
Department of Systems and Industrial Engineering
University of Arizona
Tucson, AZ 85721

Dr. Fei-Yue Wang
Department of Systems and Industrial Engineering
University of Arizona
Tucson, AZ 85721
Institute of Automation
Chinese Academy of Sciences
Beijing, China 100080

Springer

Dr. Li Li & Dr. Fei-Yue Wang
Department of Systems and Industrial Engineering
University of Arizona
Tucson, AZ 85719

Advanced Motion Control and Sensing for Intelligent Vehicles

Library of Congress Control Number: 2006932384

ISBN 0-387-44407-6 e-ISBN 0-387-44409-2
ISBN 9780387444079 e-ISBN 9780387444093

Printed on acid-free paper.

9 8 7 6 5 4 3 2 1

springer.com

Table of Contents

Preface

As we enter a new millennium, automotive and associated industries are facing more challenges than ever due to the rapidly growing requirements for safe, time/energy-efficient, environment friendly, and comfortable traffic service. As a result, novel concepts and technologies have been continuously developed and applied into the automotive industry. In order to better serve the pressing needs of the professional communities interested in intelligent automotives and advanced transportation systems, we prepare this book to provide the latest information for engineers and researchers with a special focus on intelligent vehicle motion control and corresponding sensing technologies. Moreover, it intends to help senior undergraduate and beginning graduate students learn several cutting-edge developments of automotives/vehicles and transportation systems. This book also summarizes our understanding of the current trend and the likely future of intelligent vehicle motion control.

Our research on intelligent vehicles goes back to almost 20 years ago when I was a researcher in a project that built a mobile robot at the RPI/NASA Center of Intelligent Robotic Systems for Space Exploration. Since then, research and development of intelligent vehicles have experienced tremendous growth and enjoyed numerous successful applications in land, sea, and air, for scientific, civilian, or military missions. Our work on intelligent vehicles evolves from specific-application robotic vehicles to general-purpose intelligent vehicles too. In 1994, in collaboration with the National Institute of Standards and Technology (NIST), we constructed the SPIDER (Stewart Platform Instrumented Drive Environmental Robot) Vehicular System for in situ Martian/Lunar resource utilization and for a working prototype of a plant for producing oxygen from the Martian atmosphere at the UA/NASA Center for Space Engineering Research Center for Utilization of Local Planetary Resources. From 1996 to 1999, we modified a Caterpillar 98T wheel loader for autonomous digging and loading operations that can excavate efficiently both soils and rocks. In 2000, we completed our first fully autonomous driving passenger vehicle VISTA (Vehicles with Intelligent Systems for Transport Automation) and demonstrated its functions to the public in Highway 51 in Phoenix, Arizona. In 2002, we finished our design of automated proving grounds based on autonomous driving and wireless communication. In 2005, working together with Xi'an Jiaotong University and Shangdong Academy of Sciences, we developed the CASIC Mobile Platform for vASOS (vehicular Application Specific Operating Systems) that is capable of testing and evaluating various embedded electronic and software systems for driving safety, efficiency, and comfort.

This book draws from our recent works on intelligent vehicle research. Although many students and researchers have been involved in the development of this book, it is mainly based on Dr. Li Li's Ph.D. dissertation under my supervision from 2001 to 2005. It consists of nine chapters, which can be divided into three parts: introduction (Chapter 1), intelligent vehicle motion control (Chapter 2-7), and related vehicle sensing techniques (Chapter 8-9).

Chapter 1 mainly discusses the research objectives, tasks and backgrounds of intelligent vehicle technology. It aims to provide a summary of current achievements and trends. The relationships among intelligent vehicle, advanced transportation (highway) system and driver assistance are also discussed.

Chapter 2 carefully examines the tire/road friction phenomena and reviews some frequently used tire/road friction models. These topics provide important background for the individual vehicle motion control studied in Chapter 3-5, as understanding the interaction between the vehicle and the ground is essential to the study of vehicle driving performance. The subject of Chapter 3 is vehicle lateral dynamics and steering control. Several steering controller design approaches are analyzed in this chapter. Chapter 4 investigates the internal and external vehicle longitudinal dynamics and emphasizes vehicle tracking/breaking control. The decomposition and synthesis of vehicle lateral and longitudinal motion control are addressed in Chapter 5. Chapter 6 primarily studies vehicle vertical dynamics and examines vehicle suspension controllers. Finally, multiple-vehicle motion control is studied in Chapter 7, which presents additional technical challenges, especially in communication and coordination.

Chapters 8 and 9 review the existing approaches and several novel techniques applied in vehicle visual sensory and tire condition checking, respectively. Some research results in relevant research areas, i.e. material and device developments, are also outlined.

The goal of this book is to provide the state of the art survey of research in vehicle motion control and sensing, and to offer a multidisciplinary perspective for researchers who working in this field. Considering the broad set of the readers whom we would like to reach, we have omitted in several places the full technical details, although the related references have been provided. However, the key research findings and trends are addressed.

I would like to take this opportunity to thank all my colleagues and students who have contributed to our research projects in intelligent vehicles. First and most, I would like to express my deep gratitude to Professor George N. Saridis, my Ph.D. Advisor at the Rensselaer Polytechnic Institute 20 years ago, who had provided me first research opportunities in intelligent robotic systems, intelligent vehicles, and intelligent control for general intelligent systems. I also want to thank my former colleagues: Professors Paul J. Lever, Pitu B. Mirchandani and K. C. Kuo for their long time cooperation and support during the Caterpillar Project and VISTA Project. Several visiting scholars from China, Korea and Japan to my programs during 1996 to 2001 have also made significant contribution to our work, especially Professors Tang Nan, Guihe Qin, and Wenxue Liu for their effort in the VISTA Project. Last but not least, I want to thank my former students at the University of Arizona and the Chinese Academy of Sciences for their hard work and dedication to our intelligent vehicles projects, especially Drs. Xiaobo Shi, Hungman Kim, Yijia Xu, Yuetong Lin, Zhixue Wang, Shuming Tang and Mr. Deqian Chen, Hongyu Xie, Michael Williams, Javier Cortez, Pingzhong Li, Darko Babic, Chris Yao, Michael Do, Qunzhi Zhou, and many others.

Finally, the completion of this book would not be possible without the generous support from review and editorial staffs from the Springer. We are grateful to anonymous reviewers for their constructive comments and valuable suggestions. Patience and help from Editors Elaine Tham and Chris Simpson of the Springer are appreciated.

Many thanks are also due to various governmental agencies from both US and China for their continuous supporting our research over the years.

Fei-Yue Wang

Institute of Automation, Chinese Academy of Sciences, Beijing, China, and Department of Systems and Industrial Engineering, The University of Arizona, Tucson, Arizona, USA

Introduction

1.1 Intelligent Vehicles for Intelligent Transportation Systems

Over the last two decades, research and development of intelligent vehicles (IV) have experienced tremendous growth and enjoyed numerous successful applications in land, sea, and air, for scientific, civilian, or military missions. Interests in intelligent vehicles are motivated mainly by their great potentials for significant enhancements of driving safety and efficiency, and eventually notable improvements in the quality of life in modern societies. As a major component of Intelligent Transportation Systems (ITS), intelligent vehicles utilize sensing, communication, computing and control technologies to understand driving states and environments for the purpose of assisting vehicular operations, traffic control, service management, and many other activities. Furthermore, current progresses in several exciting and emerging techniques, especially pervasive computing, ad hoc networks, and intelligent spaces, will offer new thrusts to future development in this field. We are confident that smart vehicles driving on smart roads, or even driving in intelligent spaces, would be the "next wave" for research and development in intelligent vehicles.

Specifically, the concept of ground intelligent vehicles (smart cars) becomes publicly known since later 90s', although many related technologies can be dated back to early 80s or even 70s. As presented by Little in [13], "in the wake of the computer and information revolutions, motor vehicles are undergoing the most dramatic changes in their capabilities and how they interact with drivers since the early years of the century." "Recognizing the importance of smart vehicles and the potential for unintended consequences if human factors are not placed at the center of their design, DOT launched the Intelligent Vehicle Initiative (IVI) in 1997. This initiative aims to accelerate the development, availability, and use of integrated in-vehicle systems that help drivers of cars, trucks, and buses operate more safely and effectively." Soon after that, the 1998 Transportation Efficiency

Act for the 21st Century (TEA-21) authorized IVI as part of the USDOT's ITS program.

In the last decade, intelligent vehicles related researches were consistently impelled by enthusiasms of government, public, and the automotive/automotive associated products companies [1]-[25]. Now, researchers and customers believe that intelligent vehicles are potential answers to the quickly growing requirement for safe, time/energy-efficient, environment friendly, and comfortable traffic service. But there is no common conclusion on what an intelligent vehicle should be or what functions are essential of an intelligent vehicle. In other words, researchers still argue about what kind of vehicles can be called intelligent vehicles.

The US Department of Transportation now assumes that IVI should improve driving safety under at least three kinds of driving conditions: normal driving conditions, degraded driving conditions and imminent crash conditions. In order to achieve this goal, Federal Intelligent Transportation Systems program assumes that intelligent vehicles need to have three systems: collision avoidance systems, collision notification systems and driver assistance systems. More precisely, IVI addresses on the following eight problem (or function) areas: rear-end collision avoidance, lane change & merge collision avoidance, road departure collision avoidance, intersection collision avoidance, vision enhancement, vehicle stability, driver condition warning, and safety-impacting services.

Numerous achievements had been obtained and reported world-widely in the above areas. In [18], Bishop categorized the in-developing intelligent vehicles into three different levels according to how many functions they implement: a) the systems which provide an advisory/warning to the driver (collision warning systems); b) the systems which take partial control of the vehicle, either for steady-state driver assistance or as an emergency intervention to avoid a collision (collision avoidance); c) the systems which take full control of vehicle operation (vehicle automation). Generally, most researchers agree that a so called intelligent vehicle should be capable to implement function level a), because of the importance of collision avoidance. Dravidam and Tosunoglu in [27] estimated that 15% to 20% of the reported accidents involve rear-end collisions. The National Highway Traffic Safety Administration (NHTSA) estimates that about 88% of rear-end collisions in the United States are caused by driver inattention or by vehicles following too closely. Thus, collision warning systems is believed to be the fundamental function of an intelligent vehicle.

However, Bishop also pointed out that the so called collision warning system itself consists of several partly-overlapped sub-functions, i.e. leading vehicle collision warning, obstacle collision warning, rear impact warning, bicycle/pedestrian collision warning, etc. Till now, most experiment vehicles can only reach parts of these sub-functions. For instance, Fig.1.1 shows such an early intelligent driving approach designed by in Toyota's AHS project. But normally, they are still called intelligent vehicles to addresses the design objectives.

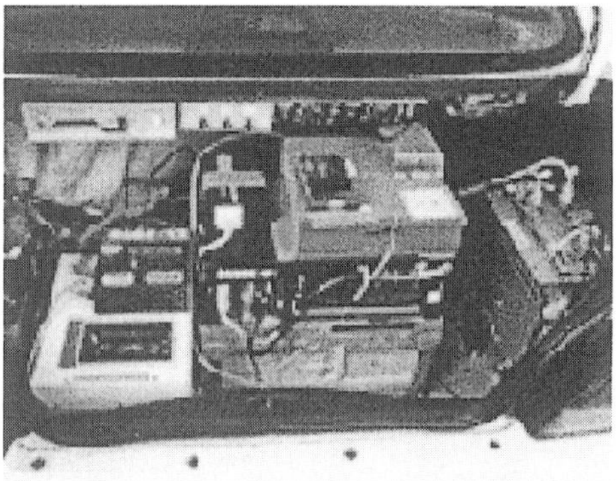

Fig.1.1 The trunk of Toyota's AHS vehicle contains early breadboard versions of computerized vehicle and steering controllers plus Global Positioning System based navigation, from [28].

Some researchers believed that cars won't drive themselves [21], [29]. Jones summarized their thoughts in [21] as follows:

"Though cars will soon be capable of doing much of what drivers do when guiding cars down a road, a car that operates without a driver's input may never see commercial production."

'New technologies are meant to complement rather than replace the driver,' said Daniel McGehee, director of the human factors research program at the University of Iowa's Public Policy Center in Iowa City. Giving the car total control, he said, raises a tangle of complex legal questions like, 'Who is responsible in the event of an accident - the driver or the carmaker?'

Other vehicle-sensing and human factors researchers, speaking under condition of anonymity, stressed the potential for crippling liability claims against auto manufacturers and makers of smart car systems.

In view of that deterrent, movie- and TV-inspired visions of a future in which cars ask only that their driver select the destination will remain confined to celluloid."

Their arguments are partly true, since fully autonomous commercial cars can not enter the city road in the near future because of varied reasons. But intelligent vehicles do not refer to intelligent commercial cars only. Although it is difficult to implement fully autonomous driving robot for ordinary use, more and more autonomous vehicles are build and use in several special situations. A typical example is DARPA Grand Challenge created in response to a Congressional and DoD mandate intended to accelerate research and development in autonomous ground vehicles that will help save American lives on the battlefield. Tens of groups from universities and companies all over the world energetically took part in this competition [30]-[33]. For example, Fig.1.2 shows an autonomous vehicle designed by Auburn University Group for DARPA Grand Challenge 2004. In 2005, there are several fully autonomous vehicles that had successfully passed the whole field test.

Fig.1.2 AVIDOR-2004 drives on the California Speedway QID track in Fontana, from [32]. The control system relies primarily on a differential GPS (Starfire) and a set of inertial sensors for navigating between the given set of waypoints. It uses infrared laser (LIDAR) and ultrasound sensing to provide the capability of obstacle avoidance and path following. (© [2004] IEEE)

Fully autonomous vehicles had also been designed and applied in mining ground, driving ground etc. [34]. Fig.1.3 shows another interesting commercial attempt built by Team of Technical University Braunschweig. This driving robot "Klaus" is a practical driving robot developed by German carmaker Volkswagen, which is driving a VW Multi-van on a test track. The sophisticated car control systems and sensor technology were used to identify the surroundings and calculate the desired driving behavior.

The concept and extensions of intelligent vehicles had being gradually and consistently changed within the last ten years. In some literals, not only some autonomous ground vehicles but also some under-water, air-cushion vehicles and aerial vehicles are named as intelligent vehicles, although the concept of IVI was proposed for ground transportation system only. On the other side, the borderline between intelligent vehicles and autonomous mobile robots are relatively vague too, since some big-size mobile robots are sometimes viewed as vehicles also. Notice that the strictly defined intelligent vehicle is yet to be developed; only the widely accepted intelligent ground vehicles are considered and discussed in the rest of this book. However, no one can deny that intelligent vehicles researchers inherit and extend many useful techniques and methodologies from the mobile robot research, since these techniques had yield satisfactory results in both applications fields [35]-[38].

Fig.1.3 "Klaus," the driving robot developed by German carmaker Volkswagen, drives a VW Multi-van on a test track, from http://www.vw-personal.de/.

1.2 Issues in Intelligent Vehicles Research and Development

As mentioned above, intelligent vehicles related research contains several multi-discipline objectives:

(1) safety is the most important objective that will be considered;
(2) time/energy-efficiency, which is related to traffic flow control;
(3) comfort, which needs to measure human driver's feeling.
(4) environment friendship (less noise, less pollution).

To optimize these objectives, an intelligent vehicle will execute the following tasks:

(1) fast road and vehicle information collection/processing, which include visual sensory (CCD/CMOS, Radar, Laser), position sensory (INS, GPS, MNS), engine sensory, and tire sensory, etc.;
(2) driving decision based on the obtained information, which include collision warning, navigation guidance, shortest road path planning, lane changing and lane merging assistance, etc.;
(3) driving control according to the generated decision, which include lateral motion control, longitudinal motion control, their compositions.

Based on the categories of tasks, intelligent vehicles related research can be divided into three areas: sensory, decision and control.

A modern vehicle is a highly complex system comprising a large number of mechanical, electronic, and electromechanical elements [39]-[44]. In order to describe the status of the vehicle, numerous sensors are developed in the last thirty years. For instance, Jones arranged the intelligent sensors into several independent groups in [21] as shown in Fig.1.4. In his plan, the input data is managed from myriad sensors and used for make split-second decisions that may involve taking control from the driver. When forward collision warning senses that a crash is imminent, data from body mass and position sensors in the cabin instantly adjust the amount of force with which air bags are deployed and seat belts are tightened.

If the intelligent sensors can be taken as human's eyes and ears, the intelligent center and the intelligent controller should be considered as human's brain and arms respectively. The intelligent center is usually a microcomputer that links all the sensors and control devices. It uses the information that is collected from the environment near the vehicle though intelligent sensors, and makes decisions of the vehicle motion based on pre-stored control algorithms. Apparently, they are the most important component of an intelligent vehicle.

Advanced decision/control algorithms may help the driver to find the optimal driving plan, translate the high-level driving demand into actual, usually complex, mechanical or electronic operations and execute them. It is also expected to be capable to detect any error driving behavior when the driver makes wrong actions, or any fault of sensors/actuators [55]-[64].

For example, when the intelligent center "see" a moving obstacle on the road in front of vehicle, it would invoke certain procedures to determine whether this obstacles will hinder the motion of it or not. If it decided that this obstacle would not bother its motion, it would continue its original motion. Otherwise it would either stop or steer this obstacle. Such simple control algorithms may collaborate with each other so as to implement certain complex actions. As a typical example, Fig.1.5 shows the flowchart of an intelligent vehicle's navigation algorithm that was proposed by Simon and Becker in [57]. It is clear that the complex obstacle avoidance function is divided and studied as several individual and relatively simpler functions. Due to cost consideration, most researcher focus on how to implement parts of these individual operations.

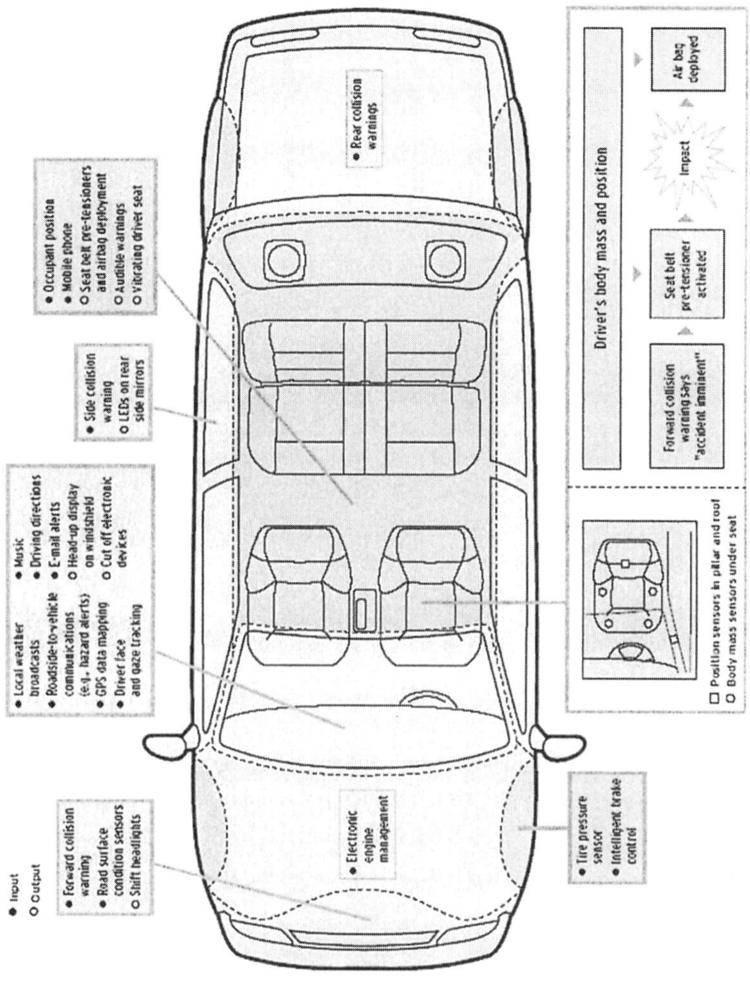

Fig.1.4 A vehicle intelligent sensor placement plan, from [21]. (© [2002] IEEE)

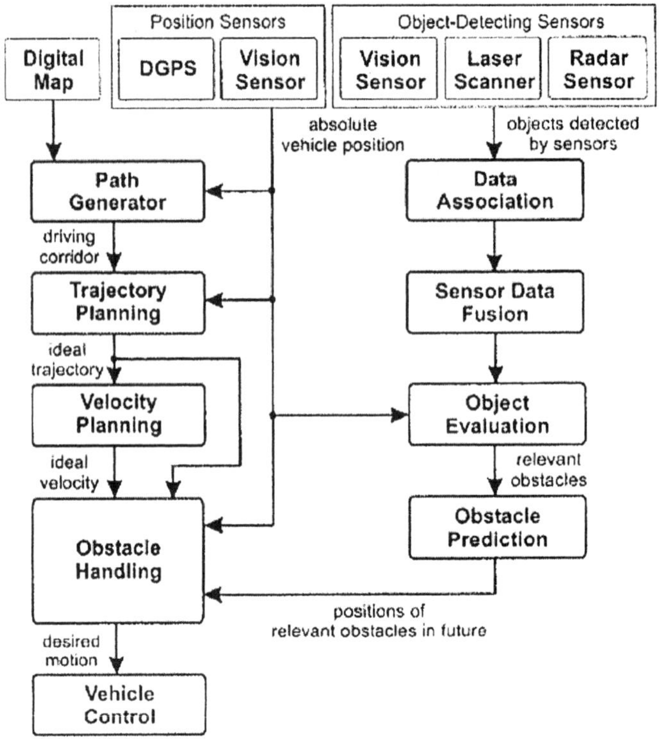

Fig.1.5 Intelligent vehicles navigation algorithm, from [57]. (© [1999] IEEE)

Besides, research on the software and hardware supporting these functions become increasingly interested recently. For example, increasing requirements of safe and comfortable driving have now led automotive manufacturers and suppliers to actively pursue development programs in the so called "by-wire" sub-systems [45]-[53]. These computer-controlled subsystems include steer-by-wire, brake-by-wire, drive-by-wire and etc., which are connected through in-vehicle computer networks. For instance, a steer-by-wire sys-tem can replace the traditional mechanical linkage between the steering wheel and the road wheel actuator with an electronic connection. Because it removes direct kinematical relationship between the steering and road wheels, it enables control algorithms to help enhance driver steering command.

Being more complex but also more powerful than "by-wire" subsystems, in-vehicle bus gained significant interests now. Usually, an in-vehicle bus refers to an electronic communication network that interconnects components inside an automobile. Due to the specialized requirements and constraints (cost, reliability and real-time characteristics, etc.), conventional computer networking protocols, i.e. Ethernet and TCP/I, cannot be used for in-vehicle buses. A typical instance for in-vehicle bus is OSEK-VDK. Now, more and more vehicle control components, i.e. Engine Control Modules (ECM), Transmission Control Modules (TCM), Anti-lock Brake System Modules (ABS), are linked with each other via in-vehicle bus, see Fig.1.6.

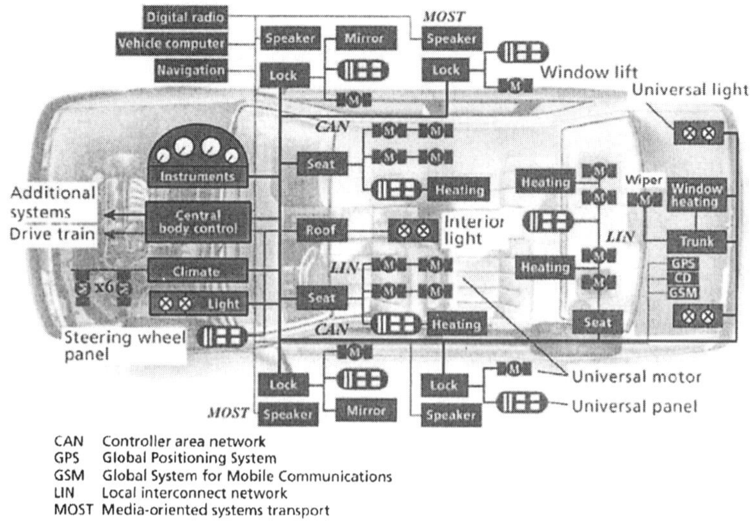

Fig.1.6 A diagram of in-vehicle buses, from [52]. (© [2002] IEEE)

It is widely expected that every smart vehicle components including windows, seats, wipers, fans, air condition controller, see Fig.1.7, can communicate with the intelligent center in the near future. Since they have little to do with vehicle motion control, the related contents are neglected in this book.

1.3 Book Organization

The rests of this book are divided into two related parts: advanced vehicle motion control and associated intelligent vehicle sensory. In the first part: tire/road friction, individual vehicle lateral/longitudinal/vertical/integrated motion control and multiple vehicles cooperative driving are discussed sequentially. In the second part, vehicle vision sensory and tire sensory are studied respectively.

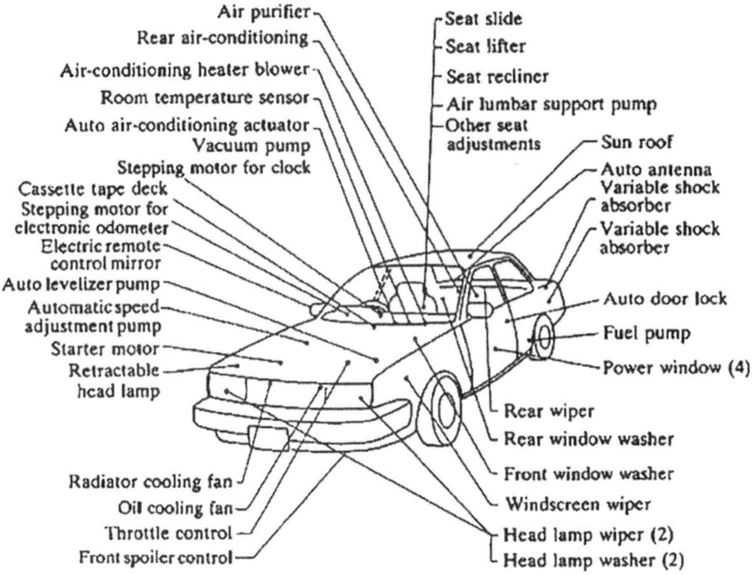

Fig.1.7 Electric motors in an EV or HEV, from [54]. (© [2004] IEEE)

The tires act as the interface between the vehicle and road. It does not only suspend other parts of the vehicle, driver/passengers and cargo, but also act as the function element for moving. The appropriate working states of the tires are essential to driving performance and safety [65]-[68]. Thus, *Chapter 2* carefully studies tire/road friction modeling and tire working status monitoring methods. It also provides the fundamental knowledge for the following *Chapter 3-6*.

As shown in Fig.1.8, advanced vehicle control systems help drivers by taking control of the steering, brakes and/or throttle to maneuver the vehicle in a safe state. Since a vehicle's movement can be decomposed into three dimensions: lateral, longitudinal and vertical, many associated studies only examines one dimensional motion control.

The main task of longitudinal control is vehicle following/tracking. It requires that an appropriate headway should be maintained between the lead vehicle and the controlled vehicle to avoid collision. The lateral control usually refers to vehicle steering control. Its prime task is path (road) following, or plainly, to keep the vehicle on the road. The vertical motion control mainly refers to vehicle suspension control. As shown in Fig.1.8, the engine is the power source of a vehicle and the powertrain subsystem transfers the linear motion of the engine's pistol into the rolling of the wheels and controls their rolling speeds [69]-[77]. The brake subsystem stops the rolling of the vehicle when the driver steps down the brakes; and the steering subsystem responses the driver's action and changes the rolling directions of the wheels. Vehicle longitudinal dynamics are really complex to analyze since the working conditions of engine, powertrain and braking systems all affects it. Besides, different vehicle shapes leads to different vehicle aerodynamic characteristic, which directly affects vehicle longitudinal driving behavior. Some important topics in vehicle longitudinal motion control, especially some newly developed intelligent control strategies, are examined in *Chapter 4*, since to discuss all the related issues would require a dedicated book.

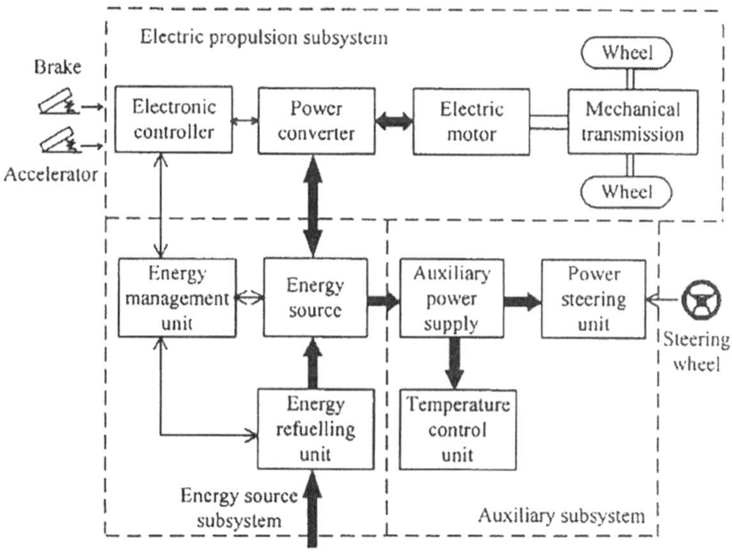

Fig.1.8 EV composition, from [69]. (© [2002] IEEE)

Comparing to longitudinal dynamics, vehicle lateral dynamics are relatively easy to control, because lateral dynamics mainly depend on steering subsystems control [78]-[82]. Therefore, *Chapter 3* gives a detailed systematical review for the recent advances in this direction.

The supporting frames and suspension parts are built upon tires and hold all the other vehicle components upon them. Active vehicle suspension controllers gained increasing interests, since they outperforms much better than passive and semi-active suspension controller and can significantly improve the driver and passenger's comfort. Because conventional suspension controllers are familiar to most readers, *Chapter 5* addresses some latest approaches in this direction.

Strictly speaking, a vehicle's lateral/longitudinal/vertical motion controls are not thoroughly independent, especially when the steering angle is big and/or the vehicle is running on a slope. Thus, *Chapter 6* studies integrated vehicle motion control, especially some representative trajectory planning methods.

In the last decade, advanced multiples vehicle motion control becomes a new hot direction in intelligent vehicle research areas. It is proven that proper cooperative driving of multiple vehicles can significantly reduces the probability of traffic accident and make the riding more smoothly. Therefore, *Chapter 7* gives a brief summary of current trends of three main directions in this area: vehicle platoon control, lane changing and merging, cooperative driving at intersections.

An intelligent vehicle usually equips varied sensors to gather environment information [83]-[85]. In some literals, sensors for vehicle motion control are classified into two types: infrastructure-based and infrastructure-independent. Examples of infrastructure-based systems include discrete magnetic reference markers and continuous magnetic tape. Infrastructure-independent methods use, for example, global positioning system (GPS), inertia position system (INS), or vision system for sensing. However, these infrastructure-independent systems still rely on infrastructure in the sense of reliable roadway markings in the former case, and a reliable and accurate roadway geographical information system (GIS) database.

Global positioning system (GPS), inertial navigation system (INS) and their combinations attract great interest in the last decade. The position and velocity of a vehicle can be directly measured by using global position systems. It had been proven in several literals including that sideslip angle, yaw rate, heading angle and position displacements can be indirectly estimated with cooperated inertial navigation system and global position systems. The newly developed fiber optic gyroscopes (FOP) are capable to measure sideslip angle, yaw rate, heading angle straightforward with high accuracy. But current FOPs require considerable installing and maintaining cost, which prevents their widely application in the near future.

Magnetic sensing is another promising technology that has been developed recently for the purposes of vehicle position measurement and guidance. By using either magnetic tape or magnetic markers, vehicle position displacement can be gotten as well some other useful information.

Position sensors and GIS are familiar to intelligent vehicles researcher and their operation theory had been well discussed in a great number of literals [109]-[146]. Therefore, no detailed discussions are made in this book. Besides, many conventional sensors, i.e. the rotary angle sensor mentioned in [147], are intentionally neglected and the corresponding measurements are assumed to be easily obtained.

Besides, vision sensors can also be employed to measure displacement. For example, the offset between the vehicle and road curb can be accurately obtained by using the laser sensor proposed in. However, measure performance of these two methods is more vulnerable to environment disturbances than that of the above two techniques.

The major usages of vision sensors are for vision-based road, vehicle/bicycle, pedestrian and traffic sign/lights detection and recognition. *Chapter 8* discusses the applications of different visual sensors for vehicles including CCD/CMOS camera, LiDAR, Laser sensors and FMCW radar [86]-[90]. The concentrations are given to vision and non-vision sensor fusion; vision information sharing via infrastructure-vehicle/inter-vehicle communication.

The comfort and safety of driving tightly depends on the good working states of the vehicle tires. However, it is not easy for people to estimate tire pressure, temperature and tire/road friction force etc. correctly, especially when driving on road. As a result, numerous novel tire sensors and inspection devices were developed in the last two decades to make driving more safe and comfortable. In addition, these approaches relates to resolving scrap tire pollution and highway noise reduction. *Chapter 9* reviews some newly developed sensors and inspecting devices for tire [91]. The prominence is given to SAW tire sensors due to their good properties.

1.4 Beyond Current Discussion

This book focuses on ground vehicle motion control and associated sensory techniques. Constrained by the length of this book, the following related research topics are not emphasized here.

1) EV and HEV

Originated tens of years ago, electric vehicle (EV) and hybrid electric vehicle (HEV) attracts increasing attentions and behaves as a promising solution to the coming Energy Crisis [92]. Countless corresponding results were reported in the last decade [69]-[77]. Because a thorough explanation and discussion for these topics requires a dedicated book due to its broad range, they are not studied in this book.

2) Tractor-Trailers

It is obvious that different vehicles may have varied dynamic behaviors because of varied vehicle's geometry size, loaded mass, engine type, tire type, and etc. Especially, a tractor-tailor's dynamic model is much more complicated than that of a passenger car [93]-[108]. Constrained by the length of the book, the discussions will mainly focus on passenger car and commercial trucks here.

3) Driver/Passenger Oriented Research

Driver/Passenger oriented research gains increasing interests in the last ten years. Conventional research focuses on how to make ride more comfortable. For example, designing advanced suspension and chair systems to avoid injury and implementing smart controller to adjust inside vehicle temperature.

In recent research, the following three topics are addressed [148]-[161]:

(1) Monitor drivers' behaviors so as to analyze driver's state, study, evaluate and even mimic his/her driving habits, etc., will be briefly discussed in *Chapter 8*;

(2) Advanced vehicle user interface design so as to more effectively ex-change information between driver or even passengers and intelligent vehicle;

With the rapid developments of intelligent assistance systems, more and more information needs to be shown for driver. As a result, how to arrange the display panel's placement and showing methods is believed to be another hot topic for the next twenty years. In [161], an interface was proposed to maximize the information representation by collapsing many of the separate dashboard controls, displays and systems into a single multi-function display (MFD), an acronyms often used in the aviation literature. A more challenging idea in [161] is to en-able end driver to switch representation of information as they drive in different situations (a city or a highway). However, static displays have its own advantages: information is always in the same place and format. Transferring to MFD will break that rule and introduce learning and usage trouble for the drivers. Before the final answer is obtained for this huge problem, countless efforts on driver ergonomics research are expected. And some known results for aviation assistance may be helpful for related research.

(3) man-in-loop hybrid vehicle control systems also received great attentions recently. It is believed that advanced driver assistance systems

which have the above two functions will be capable to ensure that the driver's reactions are appropriate and safe.

4) Intelligent Vehicles and Traffic Modeling/Monitoring/Control

It is obvious that intelligent vehicle motion control cannot be separated from advanced traffic control [162]-[185]. From the viewpoint of Automated Highway Systems, the general design policy of an intelligent vehicle is to make it run safely and smoothly at a good level of performance, and take advantage of cooperative elements so as to augment and enhance system performance. With guidance from the intelligent transportation systems, each vehicle acts the desired operations and the performance of vehicle platoon can then be improved. Thus, it is widely expected to function as the basic element to construct the whole transportation system and help it operate more effectively. Following the success of current research, intelligent vehicle is currently viewed as the "next wave" for ITS. As pointed out by Vahidi and Eskandarian in [26], due to financial and practical limitations, the short-term tendency has switched from AHS to IVI, since driver assist systems that can independently be implemented in today's generation of cars without the costly modifications in the infrastructure.

Now, the interaction between individual intelligent vehicle and automated highway system remains as a hot topic in intelligent vehicle related research. An automated highway system may consist of thousands of intelligent vehicles. All these intelligent vehicles are linked through certain communication network within the automated highway system. This information sharing technique is believed of capability of improving driving performance.

There is no unique framework to analyze the interrelationship between each individual intelligent vehicles and the whole intelligent transportation system. As pointed out in [26], its basic idea is to substitute a group of man-made driving decisions and actions with more systematic and precise machine tasks to achieve regulated traffic flow, and thus improve highways' safety and capacities. In some other literals, different layers were proposed for different levels of automation and a detailed account of the operations belonging to each layer was explained explicitly. Such an outlook paves the path to the long-term goal of advanced highways and is also beneficial for short term applications.

A typical example was given by Toy et. al in [61] for California Partners for Advanced Transit and Highways (PATH) hierarchical control architecture as Fig.1.9. This framework is constructed by five hierarchical layers: network, link, coordination, regulation, and physical. The first two layers are roadside control systems, and the latter three reside on each vehicle. The purpose of the hierarchical control structure is to partition the complex problem of controlling an AHS and all the vehicles into several control problems which can be separately designed. Each hierarchical layer provides a reference model to its adjacent control layers. Each hierarchical layer is described below.

One network-layer controller exists for the entire automated highway network. It is responsible for assigning a specific route to the vehicles based on the vehicle destinations. The network layer controller minimizes the travel time of vehicles by choosing optimal vehicle routes. Control is exerted by specifying activities at highway junctions to the link-layer controller.

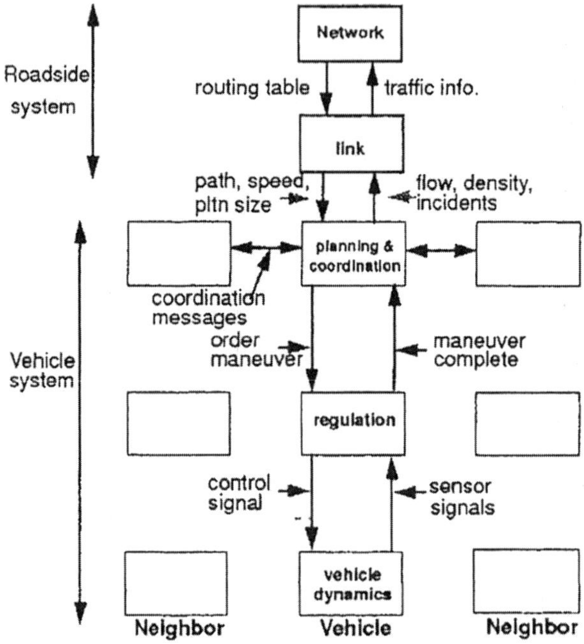

Fig.1.9 California partners for advanced transit and highways (PATH) hierarchical control architecture, from [61]. (© [2002] IEEE)

An AHS network is divided into links, or sections, that can vary from hundreds of meters to a few kilometers. A single link-layer controller controls several links. The link layer does not identify individual vehicles, but rather specifies velocities, platoon size, and proportions of activities for a particular vehicle destination or type on each link. Roadside sensors provide density information for the different types of vehicles on each link.

Control commands from the link layer are passed to the coordination layer. The coordination layer determines what maneuvers to perform, manages inter-vehicle communication, and coordinates the movement of the vehicle with neighboring cars. The choice of maneuvers and when to execute them depend on safety, the vehicle's route, commands from the link layer, and local traffic conditions. The vehicle speed transmitted by the link layer to the coordination layer is passed directly to the regulation-layer controllers and is not directly utilized by the coordination layer. The proportion of activities broadcasted to each vehicle's coordination layer determines a probability for maneuver initiation (e.g., merge, split, or change lane). At present, each vehicle's probability of maneuver execution is independent of its local state.

The regulation layer receives the commands from the coordination layer and executes the chosen maneuvers. It is a set of continuous-time feedback-based controllers. The lowest hierarchical level is the physical layer, which pertains to the vehicle's actuators and sensors. It receives steering, throttle, and brake actuator commands from the regulation layer and returns information such as vehicle speed, acceleration, and engine state.

Another framework was presented by Martin et. al in [14]-[15]. As shown in Fig.1.10, it covers the full range of IVHS technologies associated with Advanced Vehicle Control System (AVCS), which is normally understood in the IVHS framework. The four application fields include information systems applied to single vehicles, inter-vehicle (vehicle-to-vehicle interaction), vehicle-to-road/vehicle-to-driver interactions. These fields were combined with the overall objectives of the SSVS as: a cooperative driving system and a control-configured vehicle system. The first system includes radar systems and vehicle-to-vehicle communication is expected to increase safety and efficiency of driving. This Advanced Vehicle Control System (AVCS) will provide automatic longitudinal and lateral control, including lane changing and merging capabilities. The system will result in more efficient road capacity use and increased safety.

The second system consists of ultra-little vehicles with high performance that can be driven either independently of each other or linked by mechanical coupling or vehicle-to-vehicle communication. The system decreases congestion by increasing road capacity use. It involves the development and use of special vehicles that actually occupy less space and can be linked in closely spaced formations.

By comparing the above two frameworks, it is obvious to find that they both emphasize the importance of communication between intelligent vehicle and automated highway system. Now, it is widely believed that the communication based intelligent vehicle control is capable to redirect traffic flow for unexpected situations, and avoid potential hazards. It is widely expected that the vehicle networks will emerge as a hot research topic in the coming future.

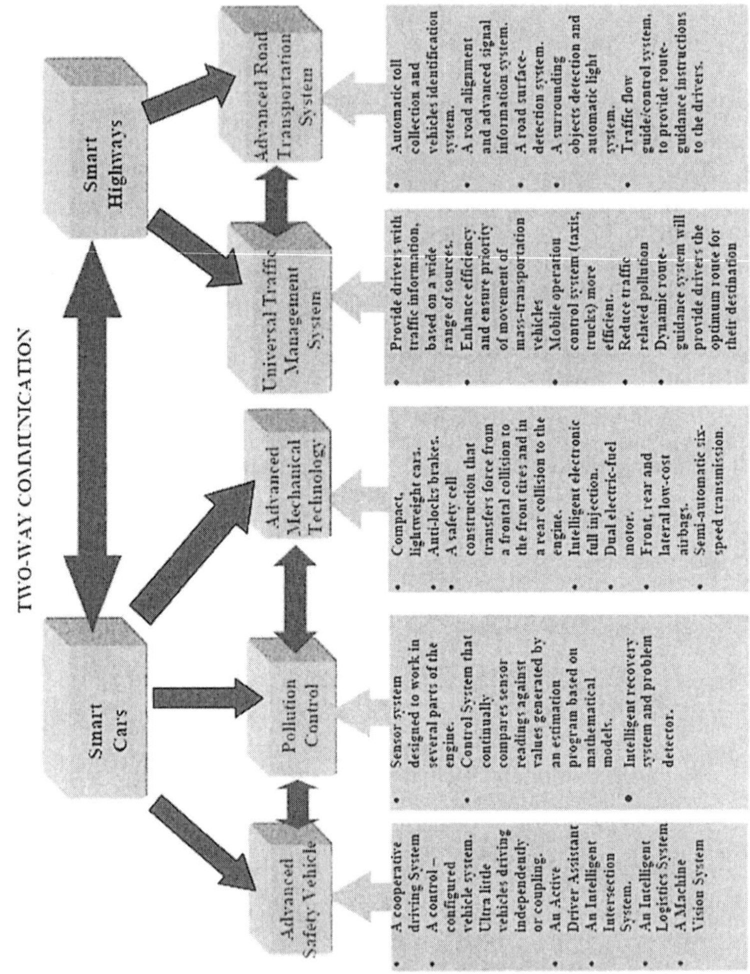

Fig.1.10 Intelligent vehicles – intelligent highway systems, from [15].

5) Vehicle-Roadside, Inter-Vehicle and Intra-Vehicle Communication

Vehicle-roadside, inter-vehicle and intra-vehicle information sharing is another attractive direction of intelligent vehicle research currently. Consequently, an important problem that needs to solve is setting up communication protocols so that products from different manufacturers can communicate with each other. Since no single company or institution can provide an intelligent vehicle with all the functions, interoperability among varied sensors and actuators offers a promising challenge for the new century [186]-[194]. Some detailed discussions are presented in Chapter 7-8.

6) Vehicle Dynamics Simulation and Vehicle-Traffic Simulation

Integrated vehicle dynamics simulation received fast boosting interests now, since it may reduce the design and manufactory cost and also provide platforms to researchers and engineers who cannot carry out testing for their own motion controllers on the real prototype vehicles [195]-[202].

Furthermore, integration of individual vehicle motion simulation/control and traffic flow simulation/control also attracted significant considerations now. Increasing researchers believed that this approach can provide us a better solution to the current traffic problems.

1.5 References

1. D. F. Moore, The friction of pneumatic tyres, Elsevier Scientific Publishing Co., New York, 1975.
2. U. Kiencke, "A view of automotive control systems," IEEE Control Systems Magazine, vol. 8, no. 4, pp. 11-19, 1988.
3. K. Chen and B. A. Galler, "An overview of intelligent vehicle-highway systems (IVHS) activities in North America," Proceedings of the 5th Jerusalem Conference on Information Technology, 'Next Decade in Information Technology', pp. 694-701, 1990.
4. S. E. Shladover, C. A. Desoer, and J. K. Hedrick, et. al, "Automated vehicle control developments in the PATH program," IEEE Transactions on Vehicular Technology, vol. 40, no. 1, pp. 114-130, 1991.
5. P. Varaiya, "Smart cars on smart roads: problems of control," IEEE Transactions on Automatic Control, vol. 38, no. 2, pp. 195-207, 1993.
6. T. Tsumura, "AGV in Japan-recent trends of advanced research, development, and industrial applications," Proceedings of the IEEE/RSJ/GI International Conference on Intelligent Robots and Systems, 'Advanced Robotic Systems and the Real World', vol. 3, pp. 1477-1484, 1994.
7. R. French, Y. Noguchi, and K. Sakamoto, "International competitiveness in IVHS: Europe, Japan, and the United States," Proceedings of Vehicle Navigation and Information Systems Conference, pp. 525–530, 1994.
8. A. Puri and P. Varaiya, "Driving safely in smart cars," Proceedings of the American Control Conference, vol. 5, pp. 3597-3599, 1995.
9. R. Larsen, "AVCS: an overview of current applications and technology," Proceedings of Intelligent Vehicles Symposium, pp. 152-157, 1995.

10. S. E. Shladover, "Review of the state of development of advanced vehicle control systems," Vehicle Systems Dynamics, vol. 24, pp. 551-595, 1995.
11. R. Whelan, Smart Highways, Smart Cars, Artech House, 1995.
12. S. Hahn, "Automation of driving functions - future development, benefits and pitfalls," Proceedings of IEEE Intelligent Vehicles Symposium, pp. 309-312, 1996.
13. S. Tsugawa, M. Aoki, and A. Hosaka, et. al, "A survey of present IVHS activities in Japan," Proceedings of the 13th International Federation of Automatic Control World Congress, pp. 147–152, 1996.
14. C. Little, The Intelligent Vehicle Initiative: advancing 'Human-Centered' smart vehicles, Federal Highway Administration, Public Roads, U.S. Department of Transportation, vol. 61, no. 2, pp. 18, 1997.
15. A. Martin, H. Marini, and S. Tosunoglu, "Intelligent vehicle/highway system: a survey -Part 1," Proceedings of Florida Conference on Recent Advances in Robotics, 1999.
16. A. Martin, H. Marini, and S. Tosunoglu, "Intelligent vehicle/highway system: design of a blind spot detector using proximity sensors - Part 2," Proceedings of Florida Conference on Recent Advances in Robotics, 1999.
17. S. N. Huang and W. Ren, "Autonomous intelligent vehicle and its performance in automated traffic systems," International Journal of Control, vol. 72, no. 18, pp. 1665-1688, 1999.
18. R. Bishop, "A survey of intelligent vehicle applications worldwide," Proceedings of IEEE Intelligent Vehicles Symposium, pp. 25-30, 2000.
19. R. Bishop, "Intelligent vehicle applications worldwide," IEEE Intelligent Systems and Their Applications, vol. 15, no. 1, pp.78-81, 2000.
20. U. Kiencke and L. Nielsen, Automotive Control Systems: for Engine, Driveline and Vehicle, Springer-Verlag, 2000.
21. W. D. Jones, "Building safer cars," IEEE Spectrum, vol. 39, no. 1, pp. 82-85, 2002.
22. M. Weber and J. Weisbrod, "Requirements engineering in automotive development: experiences and challenges," IEEE Software, vol. 20, no. 1, pp. 16-24, 2003.
23. D. Honnery and P. Moriarty, "Future vehicles: an introduction," International Journal of Vehicle Design, vol. 35, no. 1/2, pp. 1-8, 2004.
24. J. Sutherland, K. Gunter, and D. Allen, et. al, "A global perspective on the environmental challenges facing the automotive industry: state-of-the-art and directions for the future," International Journal of Vehicle Design, vol. 35, no. 1/2, pp. 86-110, 2004.
25. L. Li, J. Song, and F.-Y. Wang, et. al, "New developments and research trends for intelligent vehicles," IEEE Intelligent Systems, vol. 20, no. 4, pp. 10-14, 2005.
26. A. Vahidi and A. Eskandarian, "Research advances in intelligent collision avoidance and adaptive cruise control," IEEE Transactions on Intelligent Transportation Systems, vol. 4, no. 3, pp. 143-153, 2003.

27. U. Dravidam and S. Tosunoglu, "A survey on automobile collision avoidance system," Proceedings of Florida Conference on Recent Advances in Robotics, 1999.
28. S. Ashley, "Smart cars and automated highways," Mechanical Engineering Magazine, 1998.
29. W. D. Jones, "Keeping cars from crashing," IEEE Spectrum, vol. 38, no. 9, pp. 40-45, 2001.
30. Q. Chen, U. Ozguner, and K. Redmill, "Ohio State University at the 2004 DARPA Grand Challenge: developing a completely autonomous vehicle," IEEE Intelligent Systems, vol. 19, no. 5, pp. 8-11, 2004.
31. A. Bacha, C. Reinholtz, and A. Wicks, et. al, "The DARPA Grand Challenge: overview of the Virginia Tech vehicle and experience," Proceedings of International IEEE Conference on Intelligent Transportation Systems, pp. 481-486, 2004.
32. R. Behringer, S. Sundareswaran, and B. Gregory, et. al, "The DARPA grand challenge - development of an autonomous vehicle," Proceedings of IEEE Intelligent Vehicles Symposium, pp. 226-231, 2004.
33. U. Ozguner, K. A. Redmill, and A. Broggi, "Team TerraMax and the DARPA grand challenge: a general overview," Proceedings of IEEE Intelligent Vehicles Symposium, pp. 232-237, 2004.
34. P. Saeedi, P. D. Lawrence, and D. G. Lowe, et. al, "An autonomous excavator with vision-based track-slippage control," IEEE Transactions on Control Systems Technology, vol. 13, no. 1, pp. 67-84, 2005.
35. Y. Ohshima, et al. "Control system for automatic automobile driving," Proceedings of IFAC Tokyo Symposium on Systems Engineering for Control System Design, pp. 347-357, 1965.
36. D. Kuan, G. Phipps, and A.-C. Hsueh, "Autonomous robotic vehicle road following," IEEE Transactions on Pattern Analysis and Machine Intelligence, vol. 10, no. 5, pp. 648-658, 1988.
37. D. P. Miller, "Evaluation of vision systems for teleoperated land vehicles," IEEE Control Systems Magazine, vol. 8, no. 3, pp. 37-41, 1988.
38. C.-H. Ku and W.-H. Tsai, "Obstacle avoidance in person following for vision-based autonomous land vehicle guidance using vehicle location estimation and quadratic pattern classifier," IEEE Transactions on Industrial Electronics, vol. 48, no. 1, pp. 205-215, 2001.
39. J. Fenton, Handbook of Vehicle Design Analysis, Mechanical Engineering Publications, London, 1996.
40. R. Bosch, Automotive Handbook, 4th ed., Robert Bentley Publisher, 1997.
41. R. K. Jurgen ed., Automotive Electronics Handbook, McGraw-Hill Professional, 1999.
42. J.-W. Wong, The Theory of Ground Vehicle, 3rd ed., John Wiley & Sons, 2001.
43. W. Ribbens, Understanding Automotive Electronics, Newnes, 2002.

44. [P. Lugner and M. Plochl, "Modelling in vehicle dynamics of automobiles," ZAMM - Journal of Applied Mathematics and Mechanics, vol. 84, no. 4, pp. 219-236, 2004.
45. R. K. Jurgen, "Global '90 cars: electronics-aided," IEEE Spectrum, vol. 26, no. 12, pp. 45-49, 1989.
46. E. G. Chowanietz, "Automobile electronics in the 1990s. Part 1: Power-train electronics," Electronics & Communication Engineering Journal, vol. 7, no. 1, pp. 23-36, 1995.
47. E. G. Chowanietz, "Automobile electronics in the 1990s. Part 2: Chassis electronics," Electronics & Communication Engineering Journal, vol. 7, vol. 2, pp. 53-58, 1995.
48. M. Tomizuka and K. J. Hedrick, "Adcanced control methods for automotive applications," Vehicle System Dynamics, vol. 24, pp. 449-468, 1995.
49. V. Claesson, S. Poledna, and J. Soderberg, "The XBW model for dependable real-time systems," Proceedings of International Conference on Parallel and Distributed Systems, pp. 130-138, 1998.
50. J.G. Kassakian, "Automotive electrical systems: the power electronics market of the future," Proceedings of Applied Power Electronics Conference and Exposition, pp. 3-9, 2000.
51. H.-P. Schoner and P. Hille, "Automotive power electronics: new challenges for power electronic," Proceedings of IEEE Annual Power Electronics Specialists Conference, vol. 1, pp. 6-11, 2000.
52. G. Leen and D. Heffernan, "Expanding automotive electronic systems," IEEE Computer, vol. 35, no. 1, pp. 88-93, 2002.
53. S. Zetterstrom, Electromechanical steering, suspension, drive and brake modules, Proceedings of IEEE Vehicular Technology Conference, vol. 3, pp: 1856-1863, 2002.
54. D. Diallo, M. E. H. Benbouzid, and A. Makouf, "A fault-tolerant control architecture for induction motor drives in automotive applications," IEEE Transactions on Vehicular Technology, vol. 53, no. 6, pp. 1847-1855, 2004.
55. T. Jochem, D. Pomerleau, and B. Kumar, et. al, "PANS: a portable navigation platform," Proceedings of Intelligent Vehicles Symposium, pp. 107-112, 1995.
56. I. Sohnitz and K. Schwarze, "Control of an autonomous vehicle: design and first practical results," Proceedings of IEEE/IEEJ/JSAI International Conference on Intelligent Transportation Systems, pp. 448-452, 1999.
57. A. Simon and J. C. Becker, "Vehicle guidance for an autonomous vehicle," Proceedings of IEEE/IEEJ/JSAI International Conference on Intelligent Transportation Systems, pp. 429-434, 1999.
58. A. Simon, I. Sohnitz, and J. Becker, et. al, "Navigation and control of an autonomous vehicle," Proceedings of the 9th IFAC Symposium on Control in Transportation Systems, 2000.

59. M. Cellario, "Human-centered intelligent vehicles: toward multimodal interface integration," IEEE Intelligent Systems, vol. 16, no. 4, pp. 78-81, 2001.
60. L. Li and F.-Y. Wang, "Vehicle trajectory generation for optimal driving guidance," Proceedings of the IEEE International Conference on Intelligent Transportation Systems, pp. 231-235, 2002.
61. C. Toy, K. Leung, and L. Alvarez, et. al, "Emergency vehicle maneuvers and control laws for automated highway systems," IEEE Transactions on Intelligent Transportation Systems, vol. 3, no. 2, pp. 109-119, 2002.
62. N.-N. Zheng, S. Tang, and H. Cheng, et. al, "Toward intelligent driver-assistance and safety warning system," IEEE Intelligent Systems, vol. 19, no. 2, pp. 8-11, 2004.
63. A. Pick and D. Cole, "Neuromuscular dynamics and the vehicle steering task," Vehicle System Dynamics, Supplement, vol. 41, pp. 182-191, 2004.
64. M. A. Morrisey and D. C. Grabowski, "State motor vehicle laws and older drivers," Health Economics, vol. 14, no. 4, pp. 407-419, 2005.
65. J. Bernard and C. L. Clover, "Tire modeling for low-speed and high-speed calculations," SAE #950311, 1995.
66. C. L. Clover and J. E. Bernard, "Longitudinal tire dynamics," Vehicle System Dynamics, vol. 29, pp: 231-259, 1998.
67. J. Lacombe, "Tire model for simulations of vehicle motion on high and low friction road surfaces," Proceedings of Winter Simulation Conference, vol. 1, pp: 1025-1034, 2000.
68. D. M. Bevly, R. Sheridan, J. C. Gerdes, "Integrating INS sensors with GPS velocity measurements for continuous estimation of vehicle side-slip and tire cornering stiffness," Proceedings of American Control Conference, vol. 1, pp: 25-30, 2001.
69. C. C. Chan, "The state of the art of electric and hybrid vehicles," Proceedings of the IEEE, vol. 90, no. 2, pp. 245-246, 2002.
70. I. Husain, Electric and Hybrid Vehicles: Design Fundamentals, CRC Press, 2003.
71. M. C. Wehrey, "What's new with hybrid electric vehicles," IEEE Power and Energy Magazine, vol. 2, vol. 6, pp. 34-39, 2004.
72. J. T. Pukrushpan, A. G. Stefanopoulou, and H. Peng, Control of Fuel Cell Power Systems - Principles, Modeling, Analysis and Feedback Design, Springer 2004.
73. G. L. Plett, "High-performance battery-pack power estimation using a dynamic cell model," IEEE Transactions on Vehicular Technology, vol. 53, no. 5, pp. 1586-1593, 2004.
74. R. W. Johnson, J. L. Evans, and P. Jacobsen, et. al, "The changing automotive environment: high-temperature electronics," IEEE Transactions on Electronics Packaging Manufacturing, Part C, vol. 27, no. 3, pp. 164-176, 2004.

75. I. J. Albert, E. Kahrimanovic, and A. Emadi, "Diesel sport utility vehicles with hybrid electric drive trains," IEEE Transactions on Vehicular Technology, vol. 53, no. 4, pp. 1247-1256, 2004.
76. S. Barsali, C. Miulli, and A. Possenti, "A control strategy to minimize fuel consumption of series hybrid electric vehicles," IEEE Transactions on Energy Conversion, vol. 19, no. 1, pp. 187-195, 2004.
77. Y. Hori, "Future vehicle driven by electricity and control-research on four-wheel-motored 'UOT Electric March II'," IEEE Transactions on Industrial Electronics, vol. 51, no. 5, pp. 954-962, 2004.
78. R. Hayama, K. Nishizaki, and S. Nakano, et. al, "The vehicle stability control responsibility improvement using steer-by-wire," Proceedings of the IEEE Intelligent Vehicles Symposium, pp. 596-601, 2000.
79. P. Yih, J. Ryu, and J. C. Gerdes, "Modification of vehicle handling characteristics via steer-by-wire," Proceedings of American Control Conference, vol. 3, pp. 2578-2583, 2003.
80. R. McCann, L. R. Pujara, and J. Lieh, "Influence of motor drive parameters on the robust stability of electric power steering systems," Power Electronics in Transportation, pp. 103-108, 1998.
81. C. Hatipoglu, U. Ozguner, and K. A. Redmill, "Automated lane change controller design," IEEE Transactions on Intelligent Transportation Systems, vol. 4, no. 1, pp. 13-22, 2003.
82. H. Fritz, "Autonomous lateral road vehicle guidance using neural network controllers," Proceedings of European Control Conference, vol. 3, pp. 285–290, 1995.
83. Lawrence A. Klein, Sensor Technologies and Data Requirements for ITS, Artech House, Boston, 2001.
84. M. E. Russell, C. A. Drubin, and A. S. Marinilli, "Integrated automotive sensors," IEEE Transactions on Microwave Theory and Techniques, vol. 50, no. 3, pp. 674-677, 2002.
85. S. M. Donecker, T. A. Lasky, and B. Ravani, "A mechatronic sensing system for vehicle guidance and control," IEEE/ASME Transactions on Mechatronics, vol. 8, no. 4, pp. 500-510, 2003.
86. S. Tsugawa, "Vision-based vehicles in Japan: machine vision systems and driving control systems," IEEE Transactions on Industrial Electronics, vol. 41, no. 4, pp. 398-405, 1994.
87. A. Broggi, M. Bertozzi, and A. Fascioli, "ARGO and the MilleMiglia in Automatico Tour," IEEE Intelligent Systems, vol. 14, no. 1, pp. 55-64, 1999.
88. M. Bertozzi, A. Broggi, and A. Fascioli, "Development and test of an intelligent vehicle prototype," Proceedings of World Congress on Intelligent Transportation System, 2000.
89. E. D. Dickmanns, "The development of machine vision for road vehicles in the last decade," Proceedings of IEEE Intelligent Vehicle Symposium, vol. 1, pp. 268-281, 2002.

90. A. Broggi, M. Cellario, and P. Lombardi, et. al, "An evolutionary approach to visual sensing for vehicle navigation," IEEE Transactions on Industrial Electronics, vol. 50, no. 1, pp. 18-29, 2003.
91. A. Pohl, R. Steindl, and L. Reindl, "The 'intelligent tire' utilizing passive SAW sensors measurement of tire friction," IEEE Transactions on Instrumentation and Measurement, vol. 48, no. 6, pp. 1041-1046, 1999.
92. C. Sulzberger, "An early road warrior: electric vehicles in the early years of the automobile," IEEE Power and Energy Magazine, vol. 2, no. 3, pp. 66-71, 2004.
93. S.-G. Kong and B. Kosko, "Adaptive fuzzy systems for backing up a truck-and-trailer," IEEE Transactions on Neural Networks, vol. 3, no. 2, pp. 211-223, 1992.
94. H. Ichihashi and M. Tokunaga, "Neuro-fuzzy optimal control of backing up a trailer truck," Proceedings of IEEE International Conference on Neural Networks, pp. 306-311, 1993.
95. K. Tanaka and M. Sano, "A robust stabilization problem of fuzzy control systems and its application to backing up control of a truck-trailer," IEEE Transactions on Fuzzy Systems, vol. 2, no. 2, pp. 119-134, 1994.
96. A. W. Divelbiss and J. T. Wen, "Trajectory tracking control of a car-trailer system," IEEE Transactions on Control Systems Technology, vol. 5, no. 3, pp. 269-278, 1997.
97. K. Tanaka, T. Taniguchi, and H. O. Wang, "Trajectory control of an articulated vehicle with triple trailers," Proceedings of IEEE International Conference on Control Applications, vol. 2, pp. 1673-1678, 1999.
98. P. Hingwe, A. K. Packard, and M. Tomizuka, "Linear parameter varying controller for automated lane guidance experimental study on tractor semi-trailers," Proceedings of American Control Conference, vol. 3, pp. 2038-2042, 2000.
99. B.-S. Chen, C.-S. Wu, and H.-J. Uang, "A minimax tracking design for wheeled vehicles with trailer based on adaptive fuzzy elimination scheme," IEEE Transactions on Control Systems Technology, vol. 8, no. 3, pp. 418-434, 2000.
100. Y. Nakamura, H. Ezaki, and Y. Tan, et. al, "Design of steering mechanism and control of nonholonomic trailer systems," IEEE Transactions on Robotics and Automation, vol. 17, no. 3, pp. 367-374, 2001.
101. C. Altafini, A. Speranzon, and B. Wahlberg, "A feedback control scheme for reversing a truck and trailer vehicle," IEEE Transactions on Robotics and Automation, vol. 17, no. 6, pp. 915-922, 2001.
102. K. Tanaka, S. Hori, and H. O. Wang, "Multiobjective control of a vehicle with triple trailers," IEEE/ASME Transactions on Mechatronics, vol. 7, no. 3, pp. 357-368, 2002.
103. P. Hingwe, H.-S. Tan, and A. K. Packard, et. al, "Linear parameter varying controller for automated lane guidance: experimental study on tractor-trailers," IEEE Transactions on Control Systems Technology, vol. 10, no. 6, pp. 793-806, 2002.

104.D. Zobel, "Trajectory segmentation for the autonomous control of backward motion for truck and trailer," IEEE Transactions on Intelligent Transportation Systems, vol. 4, no. 2, pp. 59-66, 2003.
105.C. Chen and M. Tomizuka, "Lateral control of commercial heavy vehicles," Vehicle System Dynamics, vol. 33, no. 6, pp. 391-420, 2000.
106.F. Cuesta, F. Gomez-Bravo, and A. Ollero, "Parking maneuvers of industrial-like electrical vehicles with and without trailer," IEEE Transactions on Industrial Electronics, vol. 51, no. 2, pp. 257-269, 2004.
107.A. Astolfi, P. Bolzern, and A. Locatelli, "Path-tracking of a tractor-trailer vehicle along rectilinear and circular paths: a Lyapunov-based approach," IEEE Transactions on Robotics and Automation, vol. 20, no. 1, pp. 154-160, 2004.
108.J. Stephen, "Lateral dynamics of motorcycles towing single-wheeled trailers," Vehicle System Dynamics, vol. 43, no. 8, pp. 581-599, 2005.
109.Y. Narita, S. Katahara,and M. Aoki, "Lateral position detection using side looking line sensor cameras," Proceedings of IEEE Intelligent Vehicles Symposium, pp. 271-275, 2003.
110.J. Wang, S. Schroedl, and K. Mezger, et. al, "Lane keeping based on location technology," IEEE Transactions on Intelligent Transportation Systems, vol. 6, no. 3, pp. 351-356, 2005.
111.C. Drane and C. Rizos, Positioning Systems in Intelligent Transportation Systems, Artech House, 1998.
112.C. F. Donnell, Inertial Navigation, Analysis and Design, McGraw-Hill, 1964.
113.A. Lawrence, Modern Inertial Technology, Springer-Verlag, 1993.
114.A. Lawrence, Modern Inertial Technology: Navigation, Guidance and Control, Springer-Verlag, 1998.
115.A. B. Chatfield, Fundamentals of High Accuracy Inertial Navigation, American Instistute of Aeronautics, 1997.
116.A. D. King, "Inertial navigation - forty years of evolution," GEC Rreview, vol. 13, no. 3, pp. 140-149, 1998.
117.J. L. Weston and D. H. Titterton, "Modern inertial navigation technology and its application," Electronics & Communication Engineering Journal, vol. 12, no. 2, pp. 49-64, 2000.
118.N. Barbour and G. Schmidt, "Inertial sensor technology trends," IEEE Sensors Journal, vol. 1, no. 4, pp. 332-339, 2001.
119.E. D. Kaplan, Understanding GPS, Principles and Applications, Artech House, MA, 1996.
120.J. V. Sickle, GPS for Land Surveyors, Ann Arbor Press, 1996.
121.B. W. Parkinson and J. J. Spilker, ed., Global Positioning System: Theory and Applications, Volume I, Amer Inst of Aeronautics, 1996.
122.B. W. Parkinson and J. J. Spilker, ed., Global Positioning System: Theory and Applications, Volume II, Amer Inst of Aeronautics, 1996.
123.Y. Zhao, Vehicle Location and Navigation Systems, Artech House, 1997.

124. G. T. French, Understanding the GPS: An Introduction to the Global Positioning System, Baker GeoResearch, 1997.
125. L. C. Larijani, GPS for Everyone: How the Global Positioning System Can Work for You, Amer Interface Corp., 1998.
126. S. Dye and F. Baylin, The GPS Manual: Principles & Applications, Baylin/Gale Productions, 1998.
127. K. Monahan and D. Douglass, GPS Instant Navigation: A Practical Guide from Basics to Advanced Techniques, Fine Edge Productions, 1998.
128. J. A. Farrel and M. Barth, The Global Positioning System & Inertial Navigation, McGraw-Hill, 1999.
129. N. J. Hotchkiss, A Comprehensive Guide to Land Navigation with GPS, Alexis Pub., 1999.
130. J. B.-Y. Tsui, Fundamentals of Global Positioning System Receivers: A Software Approach, Wiley Series, 2000.
131. M. S. Grewal, R. W. Lawrence, and P. A. Angus, Global Positioning Systems, Inertial Navigation, and Integration, John Wiley & Sons, Inc., NY, 2001.
132. B. Hofmann-Wellenhof, H. Lichtenegger, and J. Collins, GPS: Theory and Practice, Springer Verlag, 2001.
133. S. Kamewaka and S. Uemura, "A magnetic guidance method for automated guided vehicles," IEEE Transactions on Magnetics, vol. 23, no. 5, pp. 2416-2418, 1987.
134. D. Polvani, "Magnetic guidance of autonomous vehicles - part 1," Oceans, vol. 18, pp. 1407-1412, 1986.
135. D. Polvani, "Magnetic guidance of autonomous vehicles - part 2," Proceedings of International Symposium on Unmanned Untethered Submersible Technology, vol. 5, pp. 257-264, 1987.
136. W. Zhang, R. E. Parsons, and T.West, "An intelligent roadway reference system for vehicle lateral guidance/control," Proceedings of American Control Conference, pp. 281–286, 1990.
137. D. Hopstock and L. Wald, "Verification of field model for magnetic pavement marking tape," IEEE Transaction on Magnetics, vol. 32, no. 5, pp. 5088–5090, 1996.
138. J.-C. Hsu and M. Tomizuka, "Analyses of vision-based lateral control for automated highway system," Vehicle System Dynamics, vol. 30, pp. 345-373, 1998.
139. H.-S. Tan, J. Guldner, and S. Patwardhan, et. al, "Development of an automated steering vehicle based on roadway magnets - a case study of mechatronic system design," IEEE/ASME Transaction on Mechatronics, vol. 4, no. 3, pp. 258–272, 1999.
140. K.-S. Park, C.-S. Kim, and Y.-J. Lee, et. al, " Development of an unmanned vehicle driving system by MR sensor," Proceedings of Annual Conference of IEEE Industrial Electronics Society, pp. 842-847, 2000.
141. C.-Y. Chan, B. Bougler, and D. Nelson, et. al, "Characterization of magnetic tape and magnetic markers as a position sensing system for vehicle

guidance and control," Proceedings of American Control Conference, vol. 1, pp. 95-99, 2000.
142. C.-Y. Chan, "Magnetic sensing as a position reference system for ground vehicle control," IEEE Transactions on Instrumentation and Measurement, vol. 51, no. 1, pp. 43-52, 2002.
143. M. Kennedy, The Global Positioning System and GIS: An Introduction, Ann Arbor Pr Inc., 1988.
144. M. Birkin, G. Clarke, and M. Clarke, et. al, Intelligent GIS: Location Decisions and Strategic Planning, John Wiley & Sons, 1996.
145. K. Steede-Terry, Integrating GIS and the Global Positioning System, ESRI, 2000.
146. I. Heywood, S. Cornelius, and S. Carver, An Introduction to Geographical Information Systems, Prentice Hall College Division, 2000.
147. A. M. Pawlak, S. Schultz, and V. Gangla, "Rotary actuators with multipole ring magnets," IEEE Transactions on Industry Applications, vol. 31, no. 6, pp. 1306-1314, 1995.
148. Y. I. Noy, Ergonomics and Safety of Intelligent Driver Interfaces (Human Factors in Transportation), Lawrence Erlbaum Associates, 1996.
149. D. J. LeBlanc, R. D. Ervin, and G. E. Johnson, et. al, "CAPC: an implementation of a road-departure warning system," IEEE Control Systems Magazine, vol. 16, no. 6, pp. 61-71, 1996.
150. D. J. LeBlanc, R. D. Ervin, and G. E. Johnson, et. al, "Warning and intervention system to prevent road-departure accidents," Vehicle System Dynamics, vol. 25. Suppl., pp. 383-396, 1996.
151. T. Pilutti and A. G. Ulsoy, "Decision making for road departure warning systems," Proceedings of American Control Conference, pp. 1838-1842, 1998.
152. T. Pilutti and A. G. Ulsoy, "Identification of driver state for lane-keeping tasks," IEEE Transactions on Systems, Man and Cybernetics, Part A, vol. 29, no. 5, pp. 486-502, 1999.
153. L. K. Chen and A. G. Ulsoy, "Driver model uncertainty," Proceedings of American Control Conference, pp. 714-718, 1999.
154. S.-N. Huang, S. C. Chan, and W. Ren, "Mixture of automatically- and manually-controlled vehicles in intelligent transport systems," Journal of Intelligent and Robotic Systems, vol. 24, pp. 175–205, 1999.
155. J. Tanaka, S. Ishida, and H. Kawagoe, et. al, "Workload of using a driver assistance system," Proceedings of IEEE Intelligent Transportation Systems, pp. 382-386, 2000.
156. M. Cellario, "Human-centered intelligent vehicles: toward multimodal interface integration," IEEE Intelligent Systems, vol. 16, no. 4, pp. 78-81, 2001.
157. L.-K. Chen and A. G. Ulsoy, "Identification of a driver steering model, and model uncertainty, from driving simulator data," ASME Journal of Dynamic Systems, Measurement, and Control, vol. 123, pp. 623-629, 2001.

158. J. McCall, S. Mallick, and M. M. Trivedi, "Real-time driver affect analysis and tele-viewing system," Proceedings of IEEE Intelligent Vehicles Symposium, pp. 372-377, 2003.
159. J. C. McCall, O. Achler, and M. M. Trivedi, "Design of an instrumented vehicle test bed for developing a human centered driver support system," Proceedings of Intelligent Vehicles Symposium, pp. 483-488, 2004.
160. Y. Lin, P. Tang, and W.J. Zhang, et. al, "Artificial neural network modelling of driver handling behaviour in a driver-vehicle-environment system," International Journal of Vehicle Design, vol. 37, no. 1, pp. 24-45, 2005.
161. B. B. Champoux, "A mode of interaction for driver vehicle interface (DVI)," Proceedings of IEEE Intelligent Vehicles Symposium, pp. 795–800m, 2005.
162. P. A. Ioannou, A. Kanaris, and F.-S. Ho, "Spacing and capacity evaluations for different AHS concepts," Technical Report, University of Southern California, Los Angeles, 1995.
163. J. K. Hedrick, M. Tomizuka, and P. Varaiya, "Control issues in automated highway systems," IEEE Control Systems Magazine, vol. 14, no. 6, pp. 21-32, 1994.
164. D. N. Godbole and J. Lygeros, "Tools for safety-throughput analysis of automated highway systems," Proceedings of American Control Conference, vol. 3, pp. 2031-2035, 1997.
165. P. A. Ioannou, Automated highway systems, Plenum Press, New York, 1997.
166. C. J. Khisty, Transportation Engineering: An Introduction, Prentice Hall, 1998.
167. B. McQueen and J. McQueen, Intelligent Transportation Systems Architectures, Artech House, 1999.
168. L. Alvarez, R. Horowitz and P. Li. "Traffic flow control in automated highway systems," IFAC Journal of Control Engineering Practice, vol. 7, no. 9, pp. 1071-1078, 1999.
169. S. Tsugawa, "An overview on control algorithms for automated highway systems," Proceedings of IEEE Intelligent Transportation Systems Conference, pp. 234-239, 1999.
170. S. Huang, W. Ren, and S. C. Chan, "Design and performance evaluation of mixed manual and automated control traffic," IEEE Transactions on Systems, Man and Cybernetics, Part A, vol. 30, no. 6, pp. 661-673, 2000.
171. R. Parsons and W. B. Zhang, "Program on advanced technology for the highway - lateral guidance system requirements definition," Proceedings of International Conference on Application of Advanced Technology in Transportation Engineering, pp. 257–280, 1989.
172. W. Zhang, S. Shladover, and R. Hall, et. al, "A functional definition of automated highway systems," TRB#940988, 1994.
173. S. Ashley, "Smart cars and automated highways," Mechanical Engineering Magazine, 1998.

174. A. Gollu and P. Varaiya, "SmartAHS: A simulation framework for automated vehicles and highway systems," Mathematical and Computer Modeling, vol. 27, no. 9-11, pp. 103-128, 1998.
175. P. Menig and C. Coverdill, "Transportation recorders on commercial vehicles," Proceedings of International Symposium on Transportation Recorders, 1999.
176. J. Lygeros, D. N. Godbole, and M. Broucke, "A fault tolerant control architecture for automated highway systems," IEEE Transactions on Control Systems Technology, vol. 8, no. 2, pp. 205-219, 2000.
177. S. Ghosh and T. S. Lee, Intelligent Transportation Systems: New Principles and Architectures, CRC Press, 2000.
178. J. Sussman, Introduction to Transportation Systems, Artech House, 2000.
179. F.-Y. Wang, P. B. Mirchandani, and Z. Wang, "The VISTA project and its applications," IEEE Intelligent Systems, vol. 17, no. 6, pp. 72-75, 2002.
180. F.-Y. Wang, S. Tang, and Y. Sui, et. al, "Toward intelligent transportation systems for the 2008 Olympics," IEEE Intelligent Systems, vol. 18, no. 6, pp. 8-11, 2003.
181. V. N. Lukanin, A. P. Buslaev, and A. V. Novikov, et. al, "Traffic flows modelling and the evaluation of energy-ecological parameters. Part I.," International Journal of Vehicle Design, vol. 32, no. 3/4, pp. 381-399, 2003.
182. V. N. Lukanin, A. P. Buslaev, and A. V. Novikov, et. al, "Traffic flows modelling and evaluation of energy-ecological parameters. Part II," International Journal of Vehicle Design, vol. 32, no. 3/4, pp. 400-421, 2003.
183. P. Mirchandani and F.-Y. Wang, "RHODES to intelligent transportation systems," IEEE Intelligent Systems, vol. 20, no. 1, pp. 10-15, 2005.
184. J. M. Sussman, Perspectives on Intelligent Transportation Systems (ITS), Springer, New York, 2005.
185. F.-Y. Wang, "Agent-Based Control for Networked Traffic Management Systems," IEEE Intelligent Systems, vol. 20, no. 5, pp. 92-96, 2005.
186. A. Cochran, "AHS communications overview," Proceedings of IEEE Intelligent transportation Systems Conference, 1997.
187. G. Leen, D. Heffernan, and A. Dunne, "Digital networks in the automotive vehicle," IEE Journal of Computer and Control Engineering, vol. 10, no. 6, pp. 257-266, 1999.
188. M. Krug and A. V. Schedl, "New demands for invehicle networks," Proceedings of EUROMICRO Conference, 'New Frontiers of Information Technology', pp. 601-605, 1997.
189. A. Zahir, "OIL-OSEK implementation language," Proceedings of IEE Seminar on OSEK/VDX Open Systems in Automotive Networks, pp. 8/1-8/3, 1998.

190. D. John, "OSEK/VDX history and structure," Proceedings of IEE Seminar OSEK/VDX Open Systems in Automotive Networks, pp. 2/1-214, 1998.
191. G. Leen, D. Heffernan, and A. Dunne, "Digital networks in the automotive vehicle," Computing & Control Engineering Journal, vol. 10, no. 6, pp. 257-266, 1999.
192. T. Muller and S. Furst, "Flexible system architecture for rapid algorithmic prototyping of multi-sensorial driver assistance systems," Proceedings of IEEE Intelligent Transportation Systems, pp. 303-308, 2001.
193. P. C. Richardson, A. Elkateeb, and L. Sieh, "An adaptive real-time intravehicle network protocol for intelligent vehicle systems," IEEE Transactions on Vehicular Technology, vol. 53, no. 5, pp. 1594-1606, 2004.
194. L. Wang, Z. Wu, and M. Zhao, "Worst-case response time analysis for OSEK/VDX compliant real-time distributed control systems," Proceedings of Annual International Computer Software and Applications Conference, vol. 1, pp. 148-153, 2004.
195. G. Rill, Simulation von Kraftfahrzeugen, Vieweg, 1994.
196. K. B. Wipke, M. R. Cuddy, and S. D. Burch, "ADVISOR 2.1: a user-friendly advanced powertrain simulation using a combined backward/forward approach," IEEE Transactions on Vehicular Technology, vol. 48, no, 6, pp. 1751-1761, 1999.
197. D.-S. Yun, J.-H. Choi, and G.-Y. Lee, et. al, "The concept for the integration of tele-operated unmanned vehicle and driving simulator," Proceedings of IEEE International Symposium on Industrial Electronics, vol. 3, pp. 1419-1424, 2001.
198. J. Andreasson, "VehicleDynamics library, Proceedings of the 3rd International Modelica Conference, pp. 11-18, 2003.
199. PowerTrain Library, http://www.dynasim.se
200. OPAL-RT, http://www.opal-rt.com
201. Modelica Association, http://www.modelica.org
202. ADAMS, Mechanical Dynamics Inc., http://www.adams.com/

Advanced Tire Friction Modeling and Monitoring

2.1 Introduction

A proper tire friction model is essential to describe overall vehicle dynamics for simulation, analysis, or control purposes, since the motion of a ground vehicle is primarily determined by the friction forces transferred from roads via tires. Thus, analysis of tire/road friction can provide us an in-sight understand of vehicle dynamics and help us to improve ride performance [1]-[8].

The main task of tire/road friction modeling and monitoring is to determine the relationship between friction forces and the longitudinal/lateral slip ratios and steering angles. However, tire/road friction phenomenon is hard to analyze for at least the following three reasons.

Firstly, tire/road friction force is affected by several different factors including tire/road surface conditions, tire pressure, vehicle load, and steering angle, etc. It is now believed to come from three major aspects: deformation, adhesion, and tearing/wear, but none of which has been thoroughly explained [1]-[6].

Movement of a tire slider on rough surfaces results in the deformation of rubber called asperities. The load of a vehicle causes these asperities to penetrate the rubber; see Fig.2.1. The road asperities exert pulsating forces on the sliding rubber block, which leads to energy dissipation in the rubber via its internal friction. When the rubber drapes over the asperities, it yields a resistance force as a whole. Deformation friction provides most of the friction force, especially when the road is wet.

Adhesion is a property of rubber that causes it to stick to other materials. It appears to come from two factors: one is related to the molecular bonds between tread rubber and road; the other related to the shearing of the rubber just below the surface layer. The nature of the first factor is not complete clarified [1]-[4]. However, it is believed that these forces are highly dependent on sliding velocity and contact interface temperature.

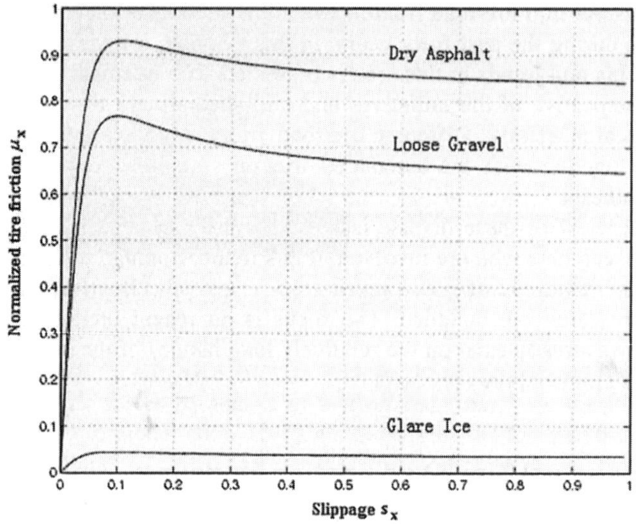

(a)

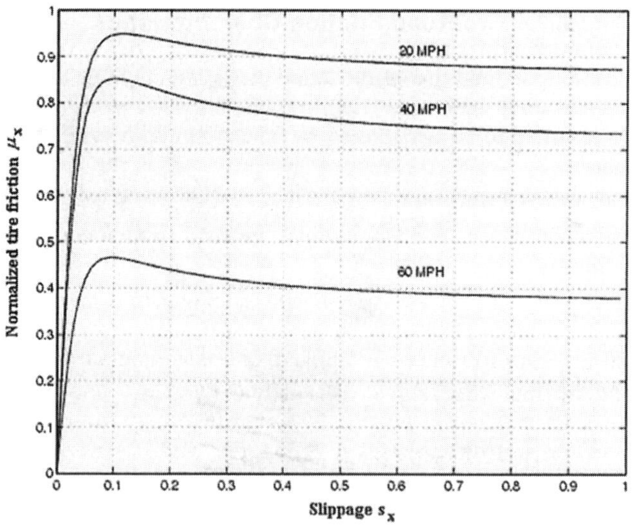

(b)

Fig.2.3 Typical tire/road friction profiles for: (a) vehicle running on different road surface conditions with velocity 20m/h, (b) vehicle s running on dry asphalt road with varied velocities, from [9].

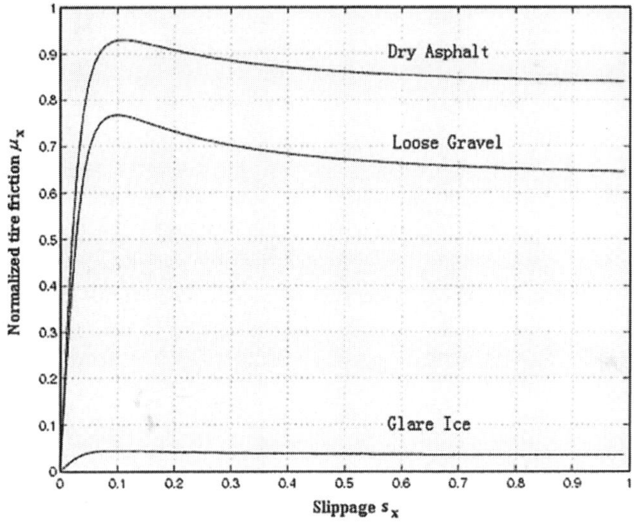

(a)

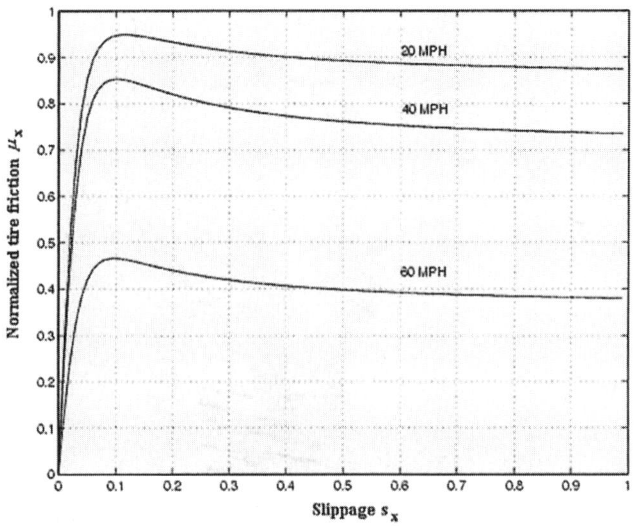

(b)

Fig.2.3 Typical tire/road friction profiles for: (a) vehicle running on different road surface conditions with velocity 20m/h, (b) vehicle s running on dry asphalt road with varied velocities, from [9].

interest was put into tire/road friction modeling to conquer the above three difficulties during the past forty years. In this Chapter, a review of recent developments and trends in this area is presented with attempts to provide a broad perspective of the initiatives and multidisciplinary techniques for related research efforts. Different tire/road friction models are carefully examined together with the associated friction situation monitoring and control synthesis.

This overview of these diverse issues aims to provide useful perspectives for researchers who are involved in this field. Although it is impossible to cover hundreds of publications in this area, the key findings and trends of research are included. The focus is on recent literature, since good reviews already exist on the relatively long history of the subject [5], [7], [18], [32], [43], [100]-[102], [118].

2.2 Longitudinal Tire/Road Friction Modeling

2.2.1 Longitudinal Tire/Road Friction Characteristics

Literatures for longitudinal tire/road friction modeling could be dated back to late 1980s and early 1990s [9]-[12]. In most proposed models, the friction force is assumed to be determined by wheel slip with regard to some other parameters.

Suppose F_x is used to denote longitudinal friction force and F_{xn} to denote the normalized force. The longitudinal tire/road friction coefficient is normally defined as

$$\mu_x = \frac{F_x}{F_{xn}} \tag{2.1}$$

and the longitudinal wheel slip s_x is usually defined as

$$s_x = \begin{cases} 1 - \dfrac{R\omega}{v_x}, if\ v_x > R\omega, v \neq 0, braking \\ 1 - \dfrac{v_x}{R\omega}, if\ v_x < R\omega, \omega \neq 0, driving \end{cases} \tag{2.2}$$

where R is the effective radius of the wheel, ω represents the angular velocity, v_x represents the longitudinal velocity.

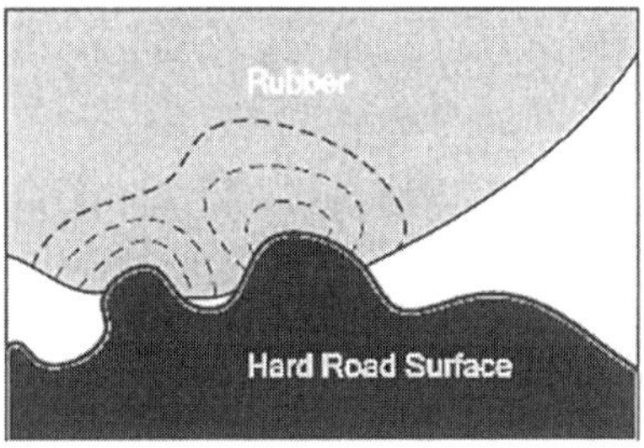

Fig.2.1 Diagram of the road asperities exert pulsating forces on the sliding rubber block.

In addition to adhesive friction and deformation friction, rubber produces traction forces by means of tearing and wear, too, see Fig.2.2. An appropriate tire model should be able to describe the resulted behaviors of all these factors.

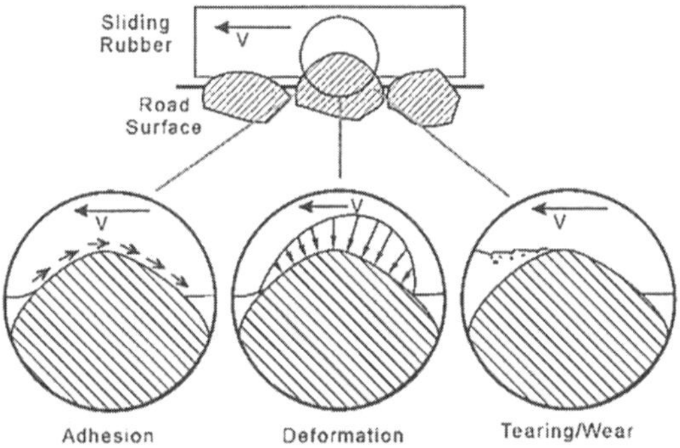

Fig.2.2 Diagram of friction force generated by adhesion and tearing and wear.

Secondly, the nonlinear and dynamic properties of tire/road friction, i.e. the viscous and hysteresis phenomena, are difficult to describe and analyze; while most empirical tire/ road friction formulas are hard to explain by physical laws. Thus, it requires a manageable tire model, in which measurement data can be translated into tire properties and vice versa.

Thirdly, an appropriate tire/road friction model should be easy to employ in vehicle control systems, i.e. anti-lock brake systems. It should be able to express specific vehicle driving behaviors in terms of some tire characteristic parameters, which finally produces insight into the desired tire properties that are given to tire manufacturers and vehicle designers.

Motivated by the development of high performance anti-lock brake systems (ABS), traction control, and steering systems, a significant research

Advanced Tire Friction Modeling and Monitoring

2.1 Introduction

A proper tire friction model is essential to describe overall vehicle dynamics for simulation, analysis, or control purposes, since the motion of a ground vehicle is primarily determined by the friction forces transferred from roads via tires. Thus, analysis of tire/road friction can provide us an in-sight understand of vehicle dynamics and help us to improve ride performance [1]-[8].

The main task of tire/road friction modeling and monitoring is to determine the relationship between friction forces and the longitudinal/lateral slip ratios and steering angles. However, tire/road friction phenomenon is hard to analyze for at least the following three reasons.

Firstly, tire/road friction force is affected by several different factors including tire/road surface conditions, tire pressure, vehicle load, and steering angle, etc. It is now believed to come from three major aspects: deformation, adhesion, and tearing/wear, but none of which has been thoroughly explained [1]-[6].

Movement of a tire slider on rough surfaces results in the deformation of rubber called asperities. The load of a vehicle causes these asperities to penetrate the rubber; see Fig.2.1. The road asperities exert pulsating forces on the sliding rubber block, which leads to energy dissipation in the rubber via its internal friction. When the rubber drapes over the asperities, it yields a resistance force as a whole. Deformation friction provides most of the friction force, especially when the road is wet.

Adhesion is a property of rubber that causes it to stick to other materials. It appears to come from two factors: one is related to the molecular bonds between tread rubber and road; the other related to the shearing of the rubber just below the surface layer. The nature of the first factor is not complete clarified [1]-[4]. However, it is believed that these forces are highly dependent on sliding velocity and contact interface temperature.

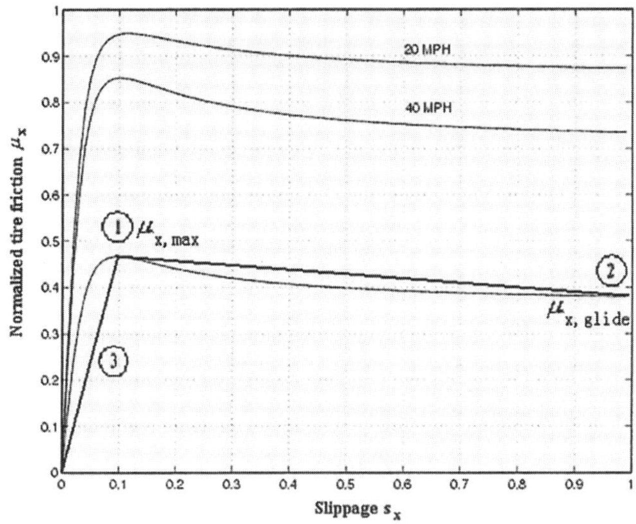

Fig.2.4 Diagram of steady friction and sliding.

The two plots in Fig.2, which were obtained by Harned et. al. 1987, demonstrate the typical variation trends between μ_x and s_x under different road conditions (a) and different vehicle velocities (b). It is obvious that μ_x normally increase when vehicle velocity decrease or the road surface become rougher. As shown in Fig.2.4, with given conditions, μ_x is a nonlinear function of s_x with a distinct maximum (maximum of friction $\mu_{x,\max}$, point 1 in Fig.2.4). The variation relationship of μ_x and s can be clearly distinguished into two parts: the steady rising part of $\mu_x - s_x$ graph (3 in Fig.2.4) and the local sliding part in which μ_x gradually decrease to $\mu_{x,glide}$ (2 in Fig.2.4). Usually, $\mu_{x,glide}$ is notably smaller than $\mu_{x,\max}$.

2.2.2 Representative Longitudinal Tire/Road Friction Models

From the theoretical aspects, different longitudinal tire/road friction models can be classified into two types: empirical (semi-empirical) models and analytical models. The empirical (semi-empirical) models are based on curve fitting techniques, which usually formulate the $\mu_x - s_x$ curve into a complex function of μ_x mainly in terms of s_x.

Empirical (semi-empirical) models can accurately catch the steady state characteristics of tire/road friction phenomena [10]-[12]. However, they cannot describe several significant dynamic behaviors such as hysteresis. That's why they were also called static models in some literals. Moreover, they lack physical interpretations, and cannot directly reflect the effects of some special factors such as humidity of the road or tire pressure. Thus, some analytical models, especially the brush model in [13]-[14], [19]-[34], became popular recently.

Most analytical models use differential equations to describe tire/road friction properties. For instance, the main idea of the brush model is to view the tire as a collection of elastic bristles that touches the road plane and deflects in the direction parallel to the road surface. The bristles' compliance is added into a set of forces and torques acting on the tire, which finally leads to a set of differential equations. It has been proven that the brush model can be used to interpret dynamic tire/road phenomena. Therefore, it is also called dynamic model in some literals.

Constrained by the length, this Chapter only discusses some representative longitudinal tire/road friction models as below.

1) Piecewise Linear Model

Piecewise linear model is a very simple model, where the $\mu_x - s_x$ relationship is roughly approximated with a piecewise linear function that passes the original point, $\mu_{x,\max}$ and $\mu_{x,glide}$; see Fig.2.5.

Although it cannot accurately capture the nonlinear dynamic characteristics of tire/road friction, this model still received considerable attentions due to its simplicity. A typical usage of this model will be discussed in Chapter 4.

2) Burckhardt Model

It is a frequently mentioned tire/road friction model that was proposed in [15]. Originally, it is written as

$$\mu_x = (C_1(1 - e^{-C_2|s_x|}) - C_3|s_x|)e^{-C_4 v_x} \tag{2.3}$$

where C_i, $i = 1,\ldots,4$ are parameters that can be determined by experiment data. The item $e^{-C_4 v_x}$ was added to represent different friction force according to speed.

It was shown that this model can approximately describe the $\mu_x - s_x$ curve with appropriately assigned parameters C_i. But the parameter identification cost is relatively high. Thus, several revised formulas that come out of Eq.(2.3) were proposed. For example, it was modified in [16] as

$$\mu_x = C_1 e^{-C_2|s_x|} \cdot |s_x|^{(C_3|s_x|+C_4)} \cdot e^{-C_5 v_x} \tag{2.4}$$

of which the logarithm linear formation can be written as

$$\ln \mu_x = C_1 - C_2|s_x| + (C_3|s_x| + C_4) \cdot \ln|s_x| - C_5 v_x \tag{2.5}$$

It's clear that model (2.5) is much simpler to be identified than the original model, since all the parameters C_i are formatted as linear coefficients. Moreover, it was shown that model (2.5) can yield similar approximation properties as (2.4).

3) Rill Model

The Rill model proposed in [17] is another semi-empirical model. It is based on steady-state force/torque characteristics of a tire together with a simple transient tire deflection model. It calculates the slip in steady-state and a corresponding tire force with a curve fit using initial inclination at $s_x = 0$, location/magnitude of $\mu_{x,\max}$ and $\mu_{x,glide}$ as parameters. The nonlinear dependence of vertical load is handled by an interpolation between a set of the parameters for predefined load cases.

4) Magic Formula

In 1980, Pacejka, in conjunction with Volvo, developed an empirical formula in which the properties shown above could be described in closed form [10]-[12].

It is originally written as

$$\mu_x = C_1 \sin(C_2 \tan^{-1}(C_3 s_x - C_4(C_3 s_x - \tan^{-1}(C_3 s_x)))) \tag{2.6}$$

where C_i, $i = 1,\ldots,4$ are determined by experiments [109].

This model receives successive modifications in the last two decades and now becomes the most important semi-empirical tire friction model. It can be applied to denote aspects such as camber, vertical load and transient behaviors now. The level of details is controlled by User Modes (UM).

This "magic formula" had been widely used in tire/road friction simulation and control system design [10]-[12], [18], and [106]-[107]. However, it is over-parameterized and thus difficult to analyze. That is partially why the above model is called "magic formula".

5) Dahl Model

Dahl Model is the first well known analytical tire/road friction model. In the late 1970s, Dahl developed a comparatively simple model, which is indeed a generalization of Coulomb friction. But it produces a smooth transi tion around zero velocity. The frictional hysteresis during pre-sliding is approximated by a generalized first order equation of the position depending only on the sign of the velocity [19]-[20].

Specifically, Dahl proposed the following equation:

$$\frac{dF}{dt} = \sigma_0 \left(1 - \operatorname{sgn}(v_r) \cdot \frac{F}{F_s} \right)^{\delta_d} v_r \tag{2.7}$$

where σ_0 denotes the initial stiffness of the contact at velocity reversal and δ_d denotes a model parameter determining the shape of the hysteresis. v_r is the relative moving speed.

6) Bliman-Sorine Model

This Dahl model captures many important phenomena of friction including zero slip displacement and hysteresis. The most instructive contribution of Dahl is that he modeled the stress-strain curve by a differential equation. However, it simultaneously discards some basic merit of static models. For example, it does not reflect the relationship between the velocity and the friction and neglects the stiction, too. Thus, Dahl's dynamic modeling idea soon led to the birth of several other models, such as Bliman-Sorine model [21]-[23].

Bliman-Sorine model actually consists of two first-order models in parallel:

$$\begin{cases} \dfrac{dz'}{dt} = v_r - \sigma_0' \dfrac{|v_r|}{F_s - F_c} z' \\ \dfrac{dz''}{dt} = v_r - \sigma_0'' \dfrac{|v_r|}{F_s - F_c} z'' \\ F_x = \sigma_0' z' + \sigma_0'' z'' \end{cases} \tag{2.8}$$

where $0 < \sigma_0'' < \sigma_0'$ are two initial stiffness of the contacts and indicate that one first order model varies faster than the other one. F_s is the highest steady state friction and F_c is the steady state friction force.

Bliman and Sorine showed that the model behaves as a classical model with Coulomb friction and stiction force if σ_0' is assigned to be infinitive. And the friction force from the slower model is subtracted from the faster model, which gives the resulting friction force.

7) LuGre Model

Inspired by the skills used in Dahl model, Canudas de Wit et. al. proposed a novel friction model which incorporated the idea of introducing an averaged characteristic pre-sliding displacement [24]-[25]. Since this model was developed at the universities of Lund and Grenoble, it is then called the LuGre model.

The LuGre model combines the merits of Dahl model with arbitrary steady-state characteristics, i.e. the Stribeck curve. However, the interpretation of the internal state has been set upon the bristle model instead of the average random behavior model of the asperities. In this explanation, surfaces are considered to be very irregular at the microscopic level and two surfaces therefore make contact at a number of asperities. When tangential force is applied, the bristles will deflect like springs which gives rise to the friction force; see Fig.2.5(a). By defining a variable z to represent the average deflection of the asperities as shown in Fig.2.5(b), LuGre-model for tire/road friction model is written as

$$\begin{cases} \dfrac{dz}{dt} = v_r - \sigma_0 \dfrac{|v_r|}{g(v_r)} z \\ v_r = v_x - r\omega \\ F_x = (\sigma_0 z + \sigma_1 \dfrac{dz}{dt} + \sigma_2 v_r) F_{xn} \end{cases} \tag{2.9}$$

where σ_0 is the stiffness, σ_1 is a damping coefficient, and σ_2 is a proportional coefficient to the relative velocity to account for viscous friction.

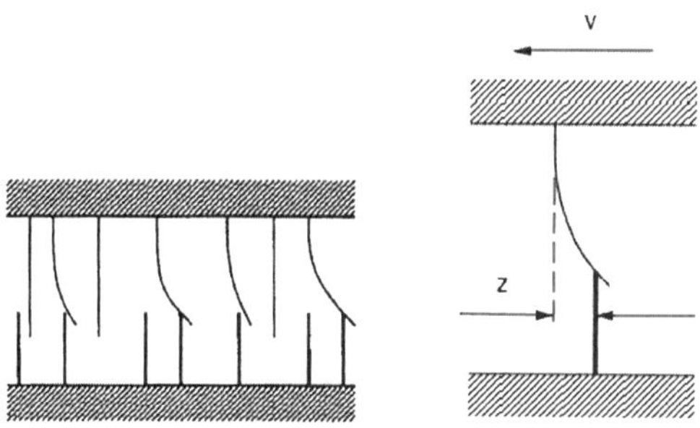

Fig.2.5 Bristle model: (left) the friction interface between two surfaces is thought of as a contact between bristles. The bristles on the lower part are shown as being rigid, (right) the average deflection of the asperities [24]. (© [1995] IEEE)

The function $g(\cdot)$ is positive and depends on many factors such as material properties and temperature. For typical bearing friction, $g(v)$ will decrease monotonically from $g(v(t=0))$ so as to model the Stribeck effect when v increase. Usually, it is written as $g(v_r) = \mu_c + (\mu_x - \mu_c)e^{-|v_r/v_s|^{1/2}}$, where μ_c is the normalized Coulomb friction. v_s denotes the Stribeck relative velocity. Normally it is assumed that $\mu_c \le \mu_s \in [0,1]$.

The properties of LuGre model had been well explored in the Canudas de Wit et. al [24], Olsson [26] and Barabanov and Ortega [27]. Only two important properties that are associated with tire/road friction analysis are discussed here. The detailed discussion about other dynamical model behaviors like pre-sliding displacement, frictional lag and varying break-away force could be found in [24]-[25].

The first property, which reveals the intuitive properties of LuGre model, is that the deflection z should be finite. And this boundary property could be described as follows:

Boundary Property: assume $0 < g(v) \le a$. If $|z(0)| \le a$ then $|z(t)| \le a$, $\forall t \ge 0$.

Brief Proof: Let us consider Lyapunov function $V = z^2/2$, then it has the time derivative of V evaluated along the solution (7.3) should be

$$\frac{dV}{dt} = \frac{dV}{dz}\cdot\frac{dz}{dt} = z\left(v - \frac{|v|}{g(v)}z\right) = -\|v\|\|z\|\left(\frac{|v|}{g(v)} - \operatorname{sgn}(v)\operatorname{sgn}(z)\right) \le 0$$

Based on the Lyapunov stability theory, the conclusion follows directly.

Dissipativity Property is another item that attracts our interest. Intuitively, it may be expected that friction will dissipate energy. However, the concept of micro viscous damping, which is introduced into the pre-sliding friction, does not have obvious physical meanings. Olsson reported in his Ph.D. thesis that this term is introduced into the friction equation to avoid un-damped oscillation in the bristle deflection. It seems that this damping parameter also affects the dissipativity properties of the friction model. But its actual effect has not been experimentally proven yet.

From the pure math viewpoint, Barabanov and Ortega showed that LuGre model is dissipative if and only if:

$$\sigma_1 < \overline{\sigma}_1 \frac{\operatorname{sgn}(v)\left[F_c + (F_s - F_c)\exp\left(-\left|\frac{v}{v_s}\right|^{\sigma_v}\right)\right]}{v}$$

where $\overline{\sigma}_1$ is a pre-chosen constant parameter. This indicates that LuGre model is highly desirable because the damping coefficient is velocity dependent.

From another viewpoint, Canudas de Wit narrated the dissipativity property in terms of the mapping relationship between v and z as follows:

Dissipativity Property: The map $\varphi : v \to z$ as defined by (6.3) is dissipative with respect to Lyapunov function $V = z^2 / 2$.

Brief Proof: Notice

$$zv = z \cdot \frac{dz}{dt} + \frac{|v|}{g(v)} z^2 \geq z \cdot \frac{dz}{dt}$$

and

$$\int_0^t z(\tau)v(\tau)d\tau \geq \int_0^t z(\tau) \cdot \frac{dz(\tau)}{dt} d\tau \geq V(t) - V(0)$$

The conclusion follows directly.

Since LuGre model is still a first order model, it is easier to analyze than most second-order models. Its properties had been well explored in [27]-[38], of which the most important two are that the deflection z should be finite and the system (8) is dissipative. Deur and Canudas de Wit independently showed that LuGre model can be physically explained using distributed tire models in [33]-[37]. In [92], it is further shown that a slightly modified LuGre model can incorporate disturbance caused by unsteady road surfaces and other reasons. Because of its simplicity and convenient combination of pre-sliding and sliding into one equation set, LuGre model becomes a popular model for friction compensation and estimation soon after its initial presentation.

As shown by Gafvert 1997 [28], for rate dependency, Bliman-Sorine model can only reproduce rate independent transient phenomena after a sign change of velocity. This means that Bliman-Sorine model cannot reproduce unidirectional phenomena such as the Stribeck effect and frictional lag. It should be pointed out that the Stribeck effect mentioned in Bliman and Sorine's papers is not exactly the one described by Stribeck.

Since Bliman-Sorine model has two states, it directly leads to problems with initial conditions, comparing to single state LuGre model. As a result, it can behave unpredictably with respect to stiction force and presliding displacement for small motions around the stick region, if the states do not reach their asymptotic values before each sign change of the velocity. These problems do not occur for the LuGre model since it only has one state. It was revealed that the LuGre model performs better, especially in modeling break-away force and frictional lag.

Moreover, drawbacks of the proposed identification procedure for Bliman-Sorine model add significant difficulty to achieve good experimental data to plot the curve, and that the nonlinear map does not always have a solution. If the friction force drop in the transition from stiction to Coulombic friction is too sharp, no solution exists. Hence, the Bliman-Sorine model will be poorly adapted to such systems.

The mechanism of this model and the physical meaning of the introduced variable z could be further revealed by its corresponding distributed formulation. In such distributed formulations, the contact patch area is discretized to a series of elements, and the microscopic deformation effects are studied in detail. Suppose the proposed first order LuGre-model is used to characterize the elastic and Coulomb friction forces at each point of the contact patch, see Fig.2.5.

Define $g(v_r) = \mu_c + (\mu_s - \mu_c)e^{-|v_r/v_s|^{1/2}}$, it could have

$$\frac{d\delta z}{dt}(\varsigma,t) = v_r - \sigma_0 \frac{|v_r|}{g(v_r)}\delta z \tag{2.10}$$

and

$$F = \int_0^L \delta F d\varsigma = \int_0^L (\sigma_0 \delta z + \sigma_1 \frac{d\delta z}{dt} + \sigma_2 v)\delta F_n d\varsigma \tag{2.11}$$

Here δF is the differential friction force, $\delta F_n = F/L$ is the differential normal force, and δz is the differential internal friction state.

Obviously, this distributed model assumes that:

(1) the normal force is uniformly distributed. It is possible to include different normal force distribution if necessary. But the model then will become too difficult to handle;

(2) the contact velocity of each differential state element is equal to v_r.

Under these assumptions, the above PDE can be approximated by a set of n ordinary differential equations via a spatial discretization. To this end, let's divide the contact patch into N equally spaced discrete points, to each associate the "discrete" average displacement δz_i, i.e. $\delta z_i = \delta z(iL/N,t)$, $\forall i = 0,1,\ldots,N-1$. Suppose the space/time scalar $\delta z(\varsigma,t)$ is approximated by an N dimensional time vector $\delta z = [\delta z_0 \quad \delta z_1 \quad \ldots \quad \delta z_{N-1}]^T$. For the simplicity of notation, let's further assume $\delta z_i(t) = \delta z_i$. Thus, it could have

$$\frac{d\delta z}{d\varsigma} = \begin{bmatrix} \frac{d\delta z_0}{d\varsigma} & \frac{d\delta z_1}{d\varsigma} & \cdots & \frac{d\delta z_{N-2}}{d\varsigma} & \frac{d\delta z_{N-1}}{d\varsigma} \end{bmatrix}$$
$$= \begin{bmatrix} \frac{\delta z_1 - \delta z_1}{L/N} & \frac{\delta z_2 - \delta z_1}{L/N} & \cdots & \frac{\delta z_{N-1} - \delta z_{N-2}}{L/N} & 0 \end{bmatrix} \quad (2.12)$$

Hence for each i th item, it has

$$\delta \dot{z}_i = -\frac{\delta z_{i+1} - \delta z_i}{L/N} r\omega + v_r - \sigma_0 \frac{|v_r|}{g(v_r)} \delta z_i \quad (2.13)$$

Use $F_{n,i} = F_N / L$, $\forall i$ and $\Delta\varsigma = L/N$, F can be approximated as

$$F = \sum_{i=0}^{N-1} \Delta F_i = \sum_{i=0}^{N-1} [(\sigma_0 \delta z_i + \sigma_1 \delta \dot{z}_i) \Delta F_{n,i} \Delta\varsigma + \sigma_2 v_r F_n]$$
$$= \frac{F_n}{N} \sum_{i=0}^{N-1} (\sigma_0 \delta z_i + \sigma_1 \delta \dot{z}_i) + \sigma_2 v_r F_n \quad (2.14)$$

To simplify the model, a virtual variable z is introduced in the lumped model, which represents the average deformation of asperities. Similarly, let's introduce a variable $\bar{z}$ in the distributed model to represents the mean value of all the δz_i as blow

$$\bar{z} = \frac{1}{N} \sum_{i=0}^{N-1} \delta z_i \quad (2.15)$$

Then, it leads to

$$\frac{d\bar{z}}{dt} = -\frac{1}{L} \sum_{i=0}^{N-1} (\delta z_{i+1} - \delta z_i) r\omega + v_r - \sigma_0 \frac{|v_r|}{g(v_r)} \delta \bar{z} \quad (2.16)$$

Noticing that $\sum_{i=0}^{N-1} (\delta z_{i+1} - \delta z_i) = \delta z_0$ has $\delta z_0 = 0$, it have

$$\frac{d\bar{z}}{dt} = v_r - \sigma_0 \frac{|v_r|}{g(v_r)} \bar{z} \quad (2.17)$$

and

$$F = (\sigma_0 \bar{z} + \sigma_1 \frac{d\bar{z}}{dt} + \sigma_2 v_r) F_n \quad (2.18)$$

Compare to (2.9) and (2.17)-(2.18), it indicates that the lumped model can be used as a suitable approximation of the distributed one.

2.3 Lateral Tire/Road Friction Modeling

2.3.1 Lateral Tire/Road Friction Models

Lateral vehicle dynamics has been studied since the 1950s [36]-[37]. It is found that for a given tire/road friction condition and a constant load, lateral tire force will initially increase with slip angle and then saturate, see Fig.2.6. For very small slip angles, the force is proportional to slip angle and the proportionality constant is called the cornering stiffness. In general, the tire force increases with the tire–road friction coefficient for a given cornering stiffness, except at very small slip angles.

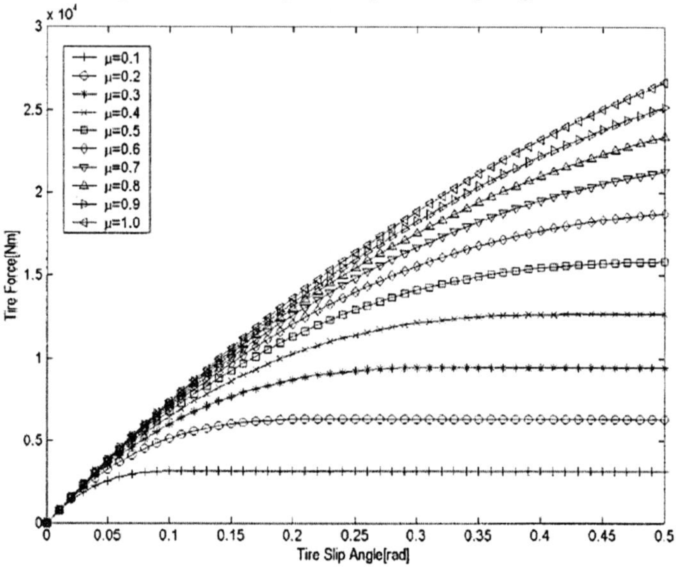

Fig.2.6 Lateral tire force versus lateral slip angle [54]. (© [2002] IEEE)

In the last two decades, some lateral tire/road friction models were proposed to handle lateral tire/road friction independently to longitudinal friction. This method greatly simplifies the steering controller design problems. Let's discuss the front steering vehicle for instance. Suppose the vehicle is moving on a flat road surface, thus the front steering model can be formulated as Fig.2.7 [37]-[39].

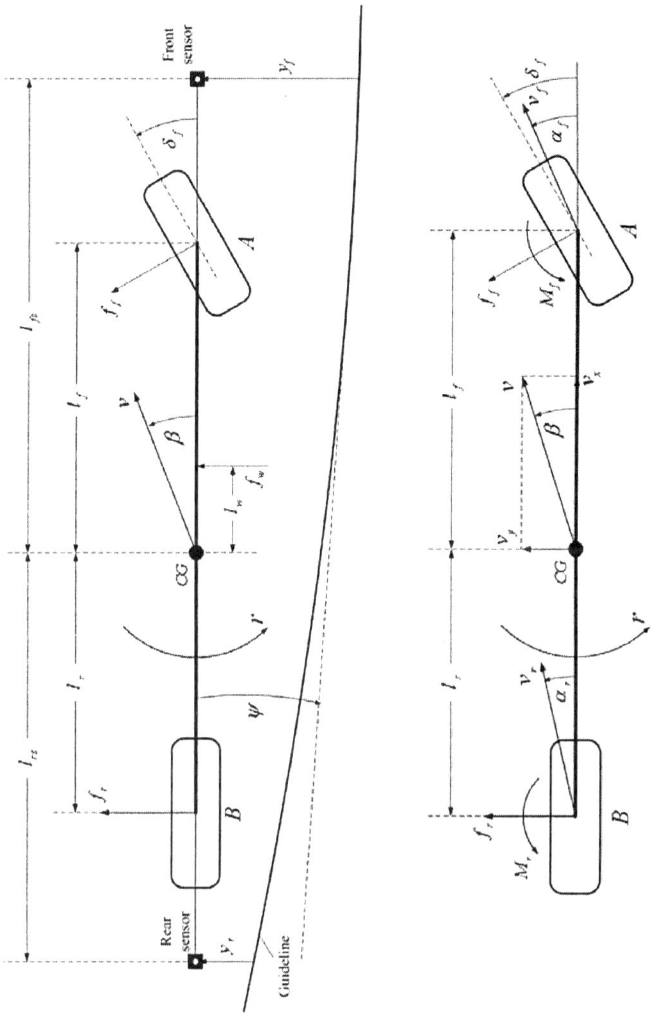

Fig.2.7 "Bicycle Model" for front steering vehicles: (left) external dynamic model, (right) internal friction model.

Here reference point ***CG*** is chosen to represent the center of gravity for vehicle body, where vehicle velocity v is defined. Symbol ***A*** and ***B*** denote the positions of front and rear tire/road interfaces respectively. Heading angle ψ is the angle from the guideline to the longitudinal axis of vehicle body ***AB***. Slide-slip angle β is the angle from the longitudinal axis of vehicle body to the direction of the vehicle velocity. δ_f is the front tires steering angle. δ_r is the rear tires steering angle. Yaw rate is denoted as r. f_f and f_r are the front and rear tire forces which are perpendicular to the directions of tire movements, respectively. f_w is the wind forces acting on the aerodynamic center of the side surface and l_w denotes the distance between ***CG*** and aero-dynamical center of the side surface. l_{fs} and l_{rs} denotes the distances from the front and rear sensor "looking at" points to ***CG***, respectively. y_f and y_r represent the displacements from the front and rear "looking at" points to the guideline.

1) Linear Proportional Model

In some literals, lateral force is taken as linear proportional to the slip angle, in which the proportionality constant is called the cornering stiffness. Usually it is written as

$$\begin{cases} F_{yf} = C'_f \alpha_f \\ F_{yr} = C'_r \alpha_r \end{cases} \tag{2.19}$$

where F_{yf} and F_{yr} are the front and rear tire forces which are perpendicular to vehicle moving direction. C'_f and C'_r are the cornering stiffness coefficients. α_f and α_r are the slip angles of front and rear tires respectively that are shown in Fig.2.8(b).

Although it is not so accurate, this model gets used in some online estimation cases due to its simplicity. An example will be discussed in *Chapter 4*.

2) Nonlinear Proportional Model

To capture the saturation property of lateral tire/road friction, several nonlinear models were proposed. For instance, the nonlinear proportional model in [113] was chosen as

$$F_{yf} = \begin{bmatrix} C''_f & \dfrac{C''^2_f}{\mu} & \dfrac{C''^3_f}{\mu^2} \end{bmatrix} \cdot \begin{bmatrix} |\tan\alpha_f| & \dfrac{-|\tan\alpha_f|^2}{3F_z} & \dfrac{-|\tan\alpha_f|^3}{27F_z^2} \end{bmatrix}^T \tag{2.20}$$

It was proven in several reports, i.e. [50], that such nonlinear proportional models can provide more accurate descriptions for the lateral tire/road friction phenomena and are still easy to identification.

3) Magic Formula

In [10]-[12], Bakker, Pacejka et. al. developed a famous lateral tire/road friction model - "magic formula". Different from the above proportional models, "magic formula" assumes that front tire force f_f does not only depend on front slip angle α_f but also vehicle slide slip angle β, steering angle δ_f and read slip angle α_r; and so does rear tire force f_r.

Normally, it is written as

$$f_f = D_f \sin\left\{C_f \tan^{-1}\left(B_f\left[1 - E_f\right]\alpha_f + E_f \tan^{-1}\left(B_f \alpha_f\right)\right)\right\} \tag{2.21}$$

and

$$f_r = D_r \sin\left\{C_r \tan^{-1}\left(B_r\left[1 - E_r\right]\alpha_r + E_r \tan^{-1}\left(B_r \alpha_r\right)\right)\right\} \tag{2.22}$$

together with

$$\begin{cases} \alpha_f = \beta + \tan^{-1}\left(\dfrac{l_f}{v} \cdot r \cos\beta\right) - \delta_f \\ \alpha_r = \beta - \tan^{-1}\left(\dfrac{l_f}{v} \cdot r \cos\beta\right) \end{cases} \tag{2.23}$$

where the coefficients B_j, C_j, D_j and E_j ($j = f, r$) in the models can be calculated in practice.

Magic formula outperforms the above two kinds of models in modeling accuracy, since it reflect the effect of yaw rate r on tire force distribution. Moreover, it can be incorporated into steering model to directly describe road surface condition.

Since lateral slide skipping forces f_f and f_r are relatively difficult to measure, recent approaches further establish some lateral tire/road friction models that do not directly describe lateral friction force.

2.3.2 Bicycle Model

In 1956, Segel presented a vehicle model with three degrees of freedom to describe lateral movements including roll and yaw. If roll movement is neglected, a simple model known as "bicycle model" can be obtained. Now, this model is the most important model that has been used for vehicle lateral motion studies.

Assuming that vehicle has a constant velocity, front steering model can be described by the differential equations

$$\frac{d}{dt}\begin{pmatrix}\beta\\ r\end{pmatrix}=\begin{pmatrix}\dfrac{f_f+f_r}{mv}-r\\ \dfrac{l_f f_f-l_r f_r}{I_z}\cos\beta\end{pmatrix} \tag{2.24}$$

In [42], Ono, Hosoe and Tuan et. al analyzed the bifurcation phenomena in (2.21)-(2.24). Approximating the nonlinearities with

$$\cos\beta\cong 1,\ \alpha_f\cong\beta+\frac{l_f}{v}\cdot r+\delta_f,\ \alpha_r\cong\beta-\frac{l_f}{v}\cdot r \tag{2.25}$$

they obtained the Jacobian matrix for $\frac{d}{dt}\begin{pmatrix}\beta\\ r\end{pmatrix}=F(\beta,r,\delta_f)$ at equilibrium point χ_0 as

$$A_{\chi_0}=\begin{bmatrix}-\dfrac{c_f^*+c_r^*}{mv} & -1-\dfrac{l_fc_f^*-l_rc_r^*}{mv^2}\\ -\dfrac{l_fc_f^*-l_rc_r^*}{I_z} & -\dfrac{l_f^2c_f^*+l_r^2c_r^*}{I_zv}\end{bmatrix} \tag{2.26}$$

where c_f^* and c_r^* are the tangents to slopes of front and rear side force characteristics at equilibrium point χ_0 respectively.

The bifurcation situation around point χ_0 has been checked in [42]. It was showed that for most vehicles, the linearized model is stable only at a narrow neighborhood of this equilibrium point. They also showed that bifurcation diagrams of the simplified vehicle model (2.27), when Δ is a nonlinear function which agrees (2.28) with the characteristics of the rear cornering force on a low friction road. The equilibrium points are such satisfying

$$\begin{cases}\dfrac{d}{dt}\begin{pmatrix}\beta\\ r\end{pmatrix}=\begin{bmatrix}-\dfrac{c_f^*+c_r^*}{mv} & -1-\dfrac{l_fc_f^*-l_rc_r^*}{mv^2}\\ -\dfrac{l_fc_f^*-l_rc_r^*}{I_z} & -\dfrac{l_f^2c_f^*+l_r^2c_r^*}{I_zv}\end{bmatrix}\begin{pmatrix}\beta\\ r\end{pmatrix}-W\begin{bmatrix}\dfrac{c_r^*}{mv}\\ -\dfrac{l_rc_r^*}{I_z}\end{bmatrix}\Delta\alpha_r+\begin{bmatrix}\dfrac{c_f^*}{mv}\\ \dfrac{l_fc_f^*}{I_z}\end{bmatrix}\delta_f\\ \alpha_r=C\begin{pmatrix}\beta\\ r\end{pmatrix}\end{cases} \tag{2.27}$$

and

$$F_r=-C_r(1+W\Delta)\alpha_r,\ -1\le\Delta\le 1 \tag{2.28}$$

where Δ is a function of α_r, representing the nonlinear part of F_r. Since Δ is considered as uncertainty term of the plant, it can be normalized to [-1, 1].

Here, the saturation characteristics of rear tires are modeled by a linear function with uncertainty terms of a special structure. The nonlinearity of the saturation is included in the part of uncertainty since it changes considerably due to road environments, whose rigorous identification is usually very difficult.

The characteristic equation of matrix A_{χ_0} is

$$s^2 + ps + q = 0 \tag{2.29}$$

with

$$\begin{cases} p = \dfrac{c_f^* + c_r^*}{mv} + \dfrac{l_f^2 c_f^* + l_r^2 c_r^*}{I_z v} \\ q = \dfrac{(l_f + l_r)^2 c_f^* c_r^*}{m I_z v^2} - \dfrac{l_f c_f^* - l_r c_r^*}{I_z} \end{cases} \tag{2.30}$$

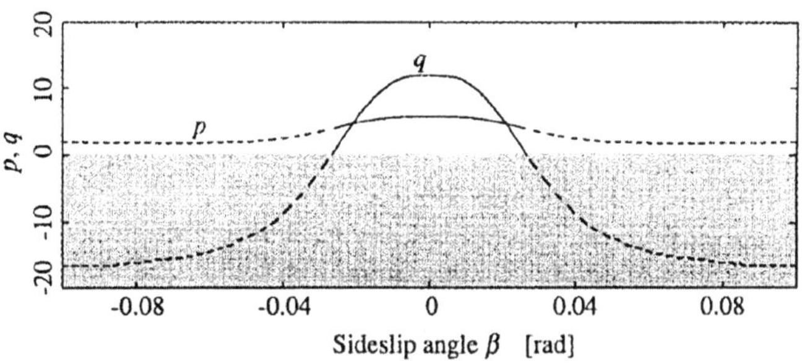

Fig.2.8 Coefficients of characteristic Eq.(2.29), from [42]. (© [1998] IEEE)

In Fig.2.8, the solid and dashed lines express stable and unstable equilibrium points, respectively. With increasing sideslip angle, becomes to be negative that makes the equilibrium point unstable. Note that the case of $q < 0$ corresponds to

$$c_r^* < \frac{mv^2 l_f c_f^*}{mv^2 l_r + (l_f + l_r)^2 c_f^*} \tag{2.31}$$

Thus, vehicle instability can be well interpreted in terms of saturation of rear side force characteristics concerned with road environments, i.e., vehicle instability is occurred when the slope of the rear side force c_r^* is bounded by the right-hand side of (2.31). Full steering vehicles outperform front steering vehicles in handling and stability significantly.

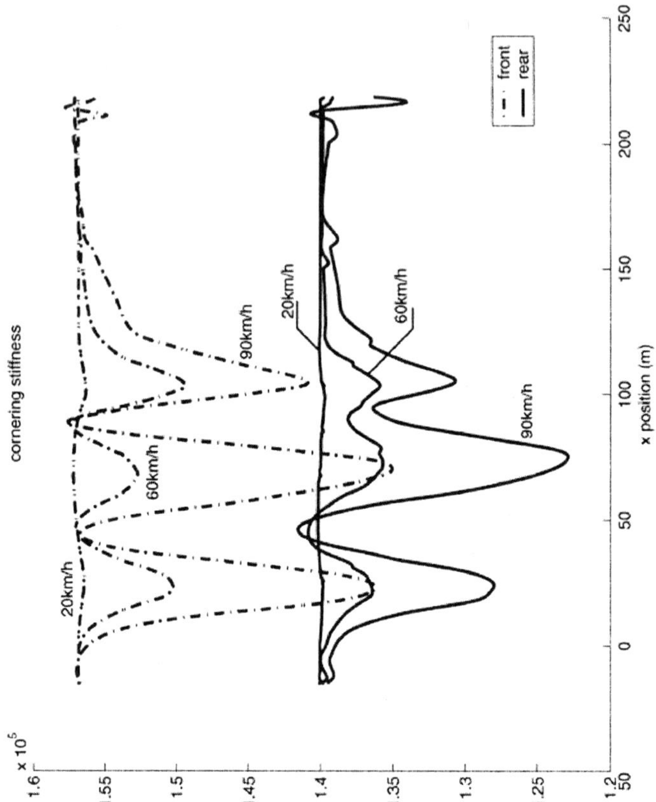

Fig.2.9 Variation of rear and front cornering stiffness at speed 20, 60, and 90 km/h, from [43]. (© [2004] IEEE)

If cornering stiffness c_f^* and c_r^* is taken to be constant, the linear model for front steering vehicle can be written as

$$\begin{bmatrix} \dot{\beta} \\ \dot{r} \\ \dot{\psi} \\ \dot{y}_f \\ \dot{y}_r \end{bmatrix} = \begin{bmatrix} a_{11} & a_{12} & 0 & 0 & 0 \\ a_{21} & a_{22} & 0 & 0 & 0 \\ 0 & 1 & 0 & 0 & 0 \\ v & l_{fs} & v & 0 & 0 \\ -v & -l_{rs} & v & 0 & 0 \end{bmatrix} \begin{bmatrix} \beta \\ r \\ \psi \\ y_f \\ y_r \end{bmatrix} + \begin{bmatrix} b_{11} & 0 & d_1 \\ b_{21} & 0 & d_2 \\ 0 & -v & 0 \\ 0 & -vl_{fs} & 0 \\ 0 & vl_{rs} & 0 \end{bmatrix} \begin{bmatrix} \delta_f \\ \rho_{ref} \\ f_w \end{bmatrix} \tag{2.32}$$

where

$$a_{11} = -(c_r + c_f)/\tilde{m}v, \; a_{12} = -1 + (c_r l_r - c_f l_f)/\tilde{m}v^2$$

$$a_{21} = (c_r l_r - c_f l_f)/\tilde{I}_z, \; a_{22} = -(c_r l_r^2 + c_f l_f^2)/\tilde{I}_z v$$

$$b_{11} = c_f/\tilde{m}v, \; b_{21} = c_f l_f/\tilde{I}_z, \; d_1 = 1/mv, \; d_2 = l_w/I_z$$

Here $\tilde{m} = m/\lambda$ and $\tilde{I}_z = I_z/\lambda$ are the normalized mass and inertia, respectively. And λ is a controllable parameter to model road adhesion factor with $\lambda = 1$ for dry road and $\lambda = 0.5$ for wet road.

"Bicycle model" can also be viewed as a certain lateral tire/ road friction model, in which the friction situation is described by two stiffness coefficients c_f and c_r. Based on vehicle movement information, c_f and c_r can be estimated and monitored online.

Cornering stiffness varied with several factors. One factor is that it increases with tire pressure. When the car turns, the mass transfer onto the external wheels increases tire pressure, which can lead to notable variations in cornering stiffness. As shown in Fig.2.9, Stephant, Charara and Meizel showed in [43] that the variations caused by this factor are normally less than 10% and still tolerable for most robust steering controllers.

Besides the above models, there were some other lateral tire models proposed, i.e. [47]. But researchers gradually realized that integrated models which consider both longitudinal and lateral tire/road friction can describe the friction phenomena more accurately. Thus, more and more efforts are put into integrated tire/road friction modeling.

2.4 Integrated Tire/Road Friction Modeling

2.4.1 Integrated Tire/Road Friction Characteristics

Let's use F_y to denote lateral friction force and F_{yn} the normalized force. The lateral tire/road friction coefficient could be defined as

$$\mu_y = \frac{F_y}{F_{yn}} \tag{2.33}$$

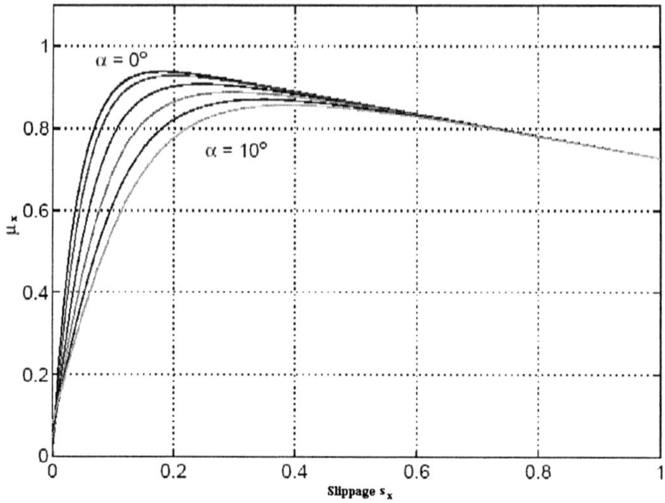

(a)

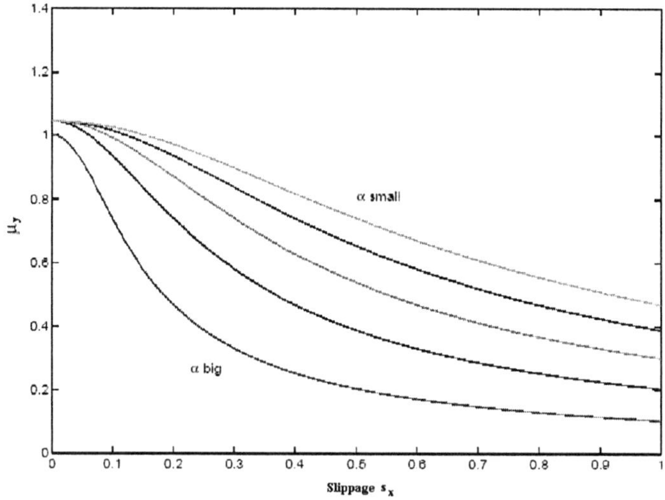

(b)

Fig.2.10 Typical tire side slip/friction curves: (a) $\mu_x - s_x$, (b) $\mu_y - s_x$, from [48]. (© [2003] IEEE)

Fig.2.10 shows the dependence relationships of $\mu_x - s_x$ and $\mu_y - s_x$. It is clear that lateral force F_y notably depends on side slip angle α. The larger the side slip angle is, the smaller the lateral force gets. This property attracts constant interests and is now employed in ABS design. An example of it will be discussed in *Chapter 4*.

2.4.2 Empirical and Semi-Empirical Integrated Models

There were several different empirical and semi-empirical integrated models proposed in the last ten years [49]-[55]. Usually, these models first decompose the whole friction force into two orthogonal parts, longitudinal and lateral force. Then, longitudinal and lateral forces are represented as a function in terms of μ_x / μ_y (or s_x / s_y) and slip angle α.

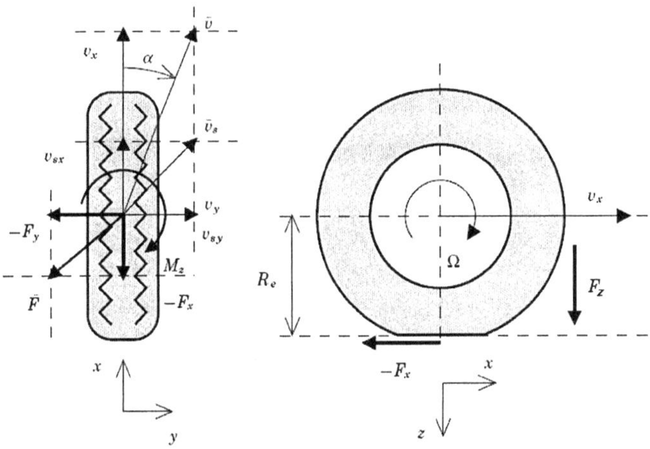

Fig.2.11 Kinematics of a tire during braking and cornering, (left) top view, (right) side view, partly from [63].

For instance, in the Kiencke model proposed in [49], see Fig.2.11, the longitudinal force is written as

$$F_x = \mu \frac{F_z}{g}[g_x \cos(\alpha) - k_{\mu y} g_y \sin(\alpha)] \tag{2.34}$$

and the lateral force is written as

$$F_y = \mu \frac{F_z}{g}[g_x \sin(\alpha) + k_{\mu y} g_y \cos(\alpha)] \tag{2.35}$$

where α denotes the slip angle α_f or α_r. F_z is the vertical pressure force generated by vehicle load. μ is the common friction coefficient which is associated with s_x. g_l / g and $k_{\mu l} g_l / g$ represent the longitudinal and lateral (transversal) decomposition coefficients, respectively.

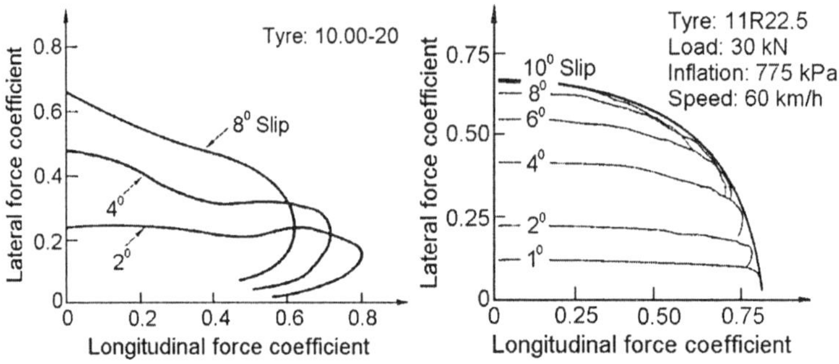

Fig.2.12 Braking and cornering, (left) bias tire from [61] and (right) radial tire from [62].

To approximate the nonlinear properties of tire/road friction, recent empirical models become increasingly complex, i.e. the models proposed in [54]. In some recently developed models, the forces generated by deformation and adhesion are considered separately [63]. This provides the possibility to analysis non-orthogonal distribution phenomena of tire force; Fig.2.12(left). The orthogonal distribution of tire force indicates a quart eclipse curve as the envelope curve shown in Fig.2.12(right); however, it is clear that the real curve does not fit a quart eclipse, let alone those bias tires.

2.4.3 Analytical Integrated Models

Comparing to empirical integrated models, analytical ones received much less consideration [56]-[60], among which those brush models are the most important approaches. In order to explain tire/road friction phenomena, when vehicle simultaneously makes lateral steering and longitudinal motion, Claeys et. al. in [57] and Deur et. al in [58] extended the above 2D LuGre-type model into 3D LuGre models.

For example, in the new model proposed in [57], wheel velocity is featured and projected into two orthogonal components, which are denoted as v_x and v_y. The microscopic bristle deflections are represented by two components z_x and z_y respectively.

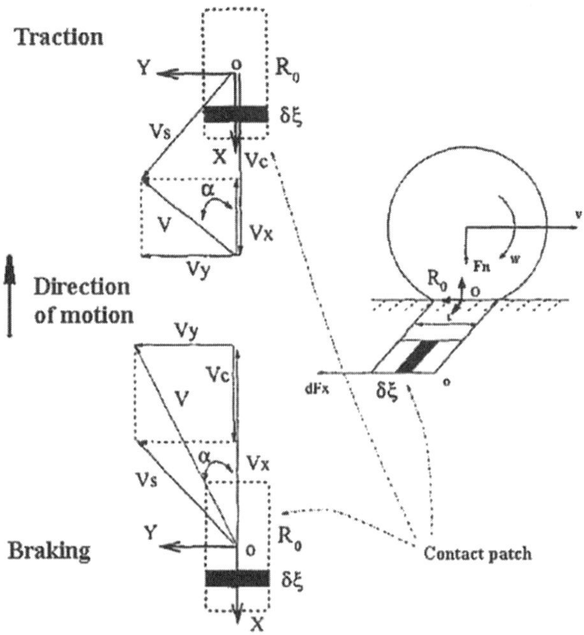

Fig.2.13 Convention for the slip definition for both braking and traction cases, from [57]. (© [2001] IEEE)

For rigid tire belt, the extend distributed model is given by

$$\begin{cases} \dfrac{d\delta z_x}{dt}(\varsigma,t) = v_{xr} - \sigma_0 \dfrac{|v_{xr}|}{g(v_{xr})}\delta z_x \\ \dfrac{d\delta z_y}{dt}(\varsigma,t) = v_{yr} - \sigma_0 \dfrac{|v_{yr}|}{g(v_{yr})}\delta z_y \\ F_x = \int_0^L \delta F_x d\varsigma = \int_0^L (\sigma_{x0}\delta z_x + \sigma_{x1}\dfrac{d\delta z_x}{dt} + \sigma_{x2}v_x)\delta F_n d\varsigma \\ F_y = \int_0^L \delta F_y d\varsigma = \int_0^L (\sigma_{y0}\delta z_y + \sigma_{y1}\dfrac{d\delta z_y}{dt} + \sigma_{y2}v_y)\delta F_n d\varsigma \end{cases} \tag{2.36}$$

where

$$g(v_{xr}) = \mu_{xc} + (\mu_{xs} - \mu_{xc})e^{-|v_{xr}/v_{xs}|^{1/2}}$$

$$g(v_{yr}) = \mu_{xc} + (\mu_{xs} - \mu_{xc})e^{-|v_{yr}/v_{ys}|^{1/2}}$$

All the symbols above have similar meanings to what are given for 2D LuGre model expect foot notes x and y denote longitudinal and lateral directions respectively, see Fig.2.13.

v_{xs} and v_{ys} re x and y coordinate Stribeck relative velocity, respectively. v_{xr} and v_{yr} denote x and y coordinate relative velocities for tire/road contact-point. It is assumed $\mu_{xc} \le \mu_{xs} \in [0,1]$, $\mu_{yc} \le \mu_{ys} \in [0,1]$.

One apparent shortcoming of this proposed 3D LuGre model is its distributed formulation. Several convenient conclusions given for 2D LuGre Model cannot be directly employed. Thus, there were several discussions to approximate the lumped 3D LuGre model in [54]-[56]. However, all these approaches still need further discussions.

2.5 Tire/Road Friction Monitors

2.5.1 Scheme for Tire/Road Friction Monitors

A general scheme of tire/road friction monitor is shown in Fig.2.14. It is obvious that correct information of vehicle position, velocity, wheel speed, and steering angle need to be measured before an acceptable estimation of the frictions is obtained.

Since a vehicle is a highly complex system that constitutes of varied mechanical, electronic and electromechanical elements, numerous sensors were designed and applied to measure the movement information. Detailed discussions on tire monitoring devices and sensors can be found in *Chapter 9*. But not all vehicle characteristics are directly measured due to high cost or some other reasons. Instead, several special observers are used to reconstruct the needed information. In literals, these observers were also called virtual sensors [43].

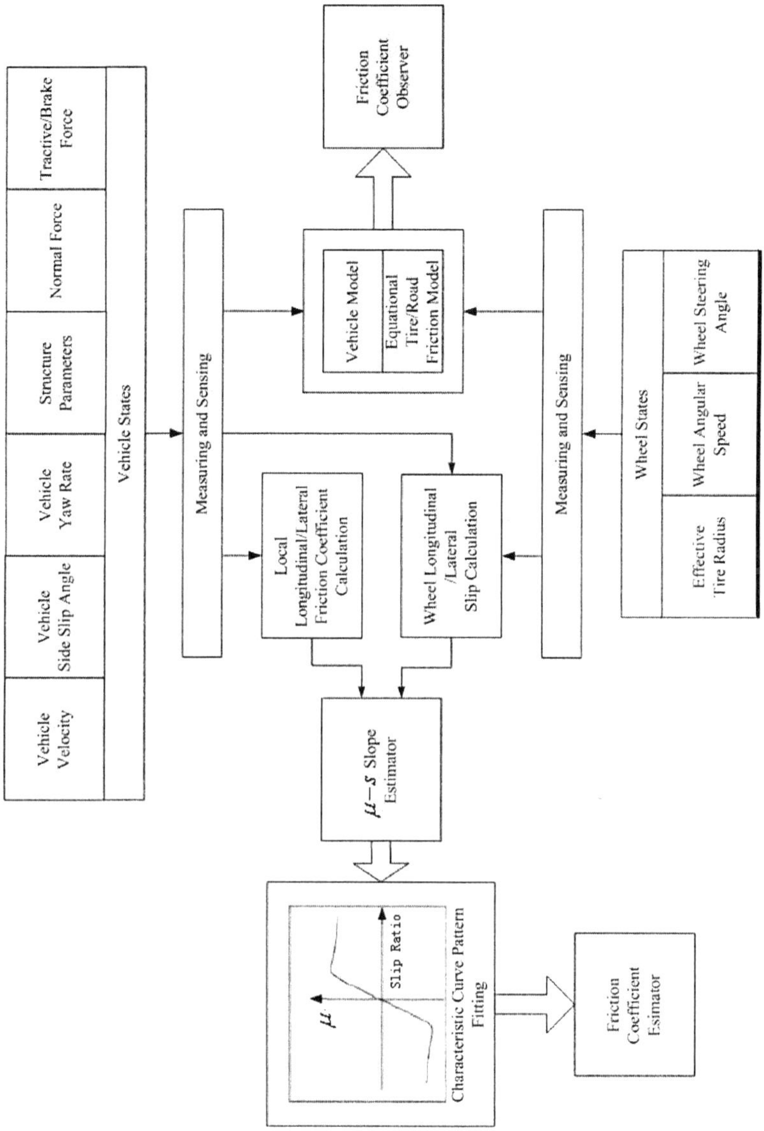

Fig.2.14 Scheme of tire/road friction monitor.

2.5.2 Identification of Empirical Longitudinal Friction Models

Online identification for empirical longitudinal tire/road friction models had been studied since early 1990s [74]-[77].

In [78]-[79], Gustafsson proposed a linear regression model for tire/road friction status monitor, which emphasized the slip slope of the friction only. As shown in Fig.2.15, he considered the slippage s_x to be a linear function of μ_x as

$$s_x = \frac{1}{k}\mu_x + \delta \tag{2.37}$$

where $1/k$ is used to represent the slip slope and δ is used to compensate the offset.

Apparently this much simplified model neglects all the dynamic features of tire/road friction. But it is very easy to put into use. Gustafsson tested this filter on a Volvo 850, and proved that the alarm time after a change is in the order of a second. However, as shown in Fig.2.15, the estimation error is a little large.

There were some other more complex identification models. For instance, in [80], the empirical identification model is chosen as

$$\mu_x = \mu_0' \frac{s_x}{c_1 s_x^2 + c_2 s_x + 1} \tag{2.38}$$

where μ_0', c_1 and c_2 are parameters that need to be identified. And some sample identification results using (2.38) are shown in Fig.2.16.

In [81], Muller, Uchanski, Hedrick proposed the following semi-empirical model for identification

$$\mu_x = 3\gamma s_x - \frac{3\gamma^2}{\mu_0} s_x^2 \frac{s_x}{|s_x|} + \frac{\gamma^3}{\mu_0^2} s_x^3 \tag{2.39}$$

where $\gamma = 4a^2 bk / 3F_z$ is a pre-defined coefficient, a and b are the length and width of the rectangle contact area. They all need to be estimated as well as μ_0.

It was claimed that these nonlinear models can approach the tire/road curve more accurately. Besides them, there are many other empirical/semi-empirical nonlinear estimation models [82]-[83]. However, none of them can be directly used to describe the dynamic properties of tire/road friction.

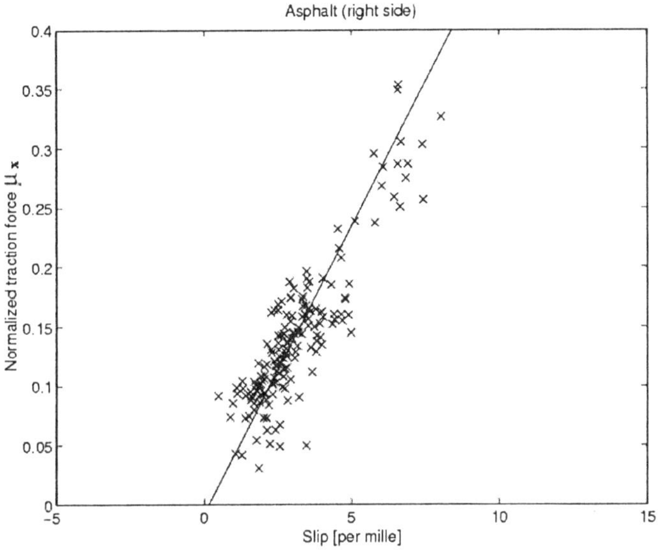

(a)

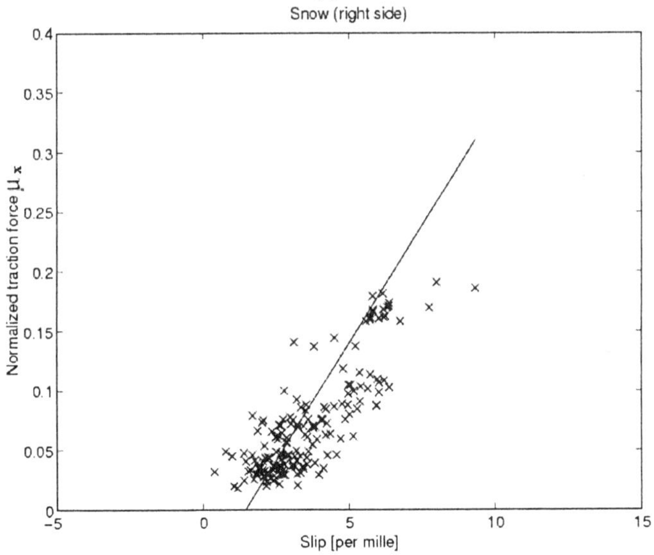

(b)

Fig.2.15 Samples of u and s, computed from ABS signals on the right side of the car, one dry asphalt (a) and hard snow (b). Crosses denote measurements and the solid line is a straight-line approximation, from [79]. (© [1998] IEEE)

2.4.3 Observers of Analytical Longitudinal Friction Models

Different from empirical models, analytical longitudinal tire/ road friction models are usually described by differential equations, which thus require observers to monitor friction changes [84]-[90].

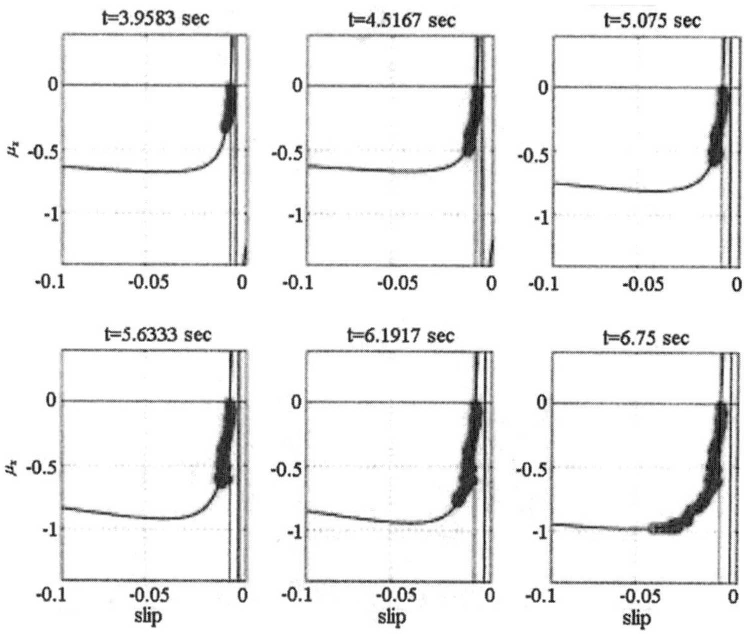

Fig.2.16 Measured μ_x versus slip data at several time instants during a hard braking maneuver (circles) and their least squares slip curves using tire model of Eq.(2.38) (solid line). $\mu_{x\max}$ is taken from the fitted slip curve tends to depend on the most extreme μ_x attained, from [80].

Canudas de Wit et. al. developed an observer for 2D LuGre model in [86], in which the uncertainty was modeled by a new parameter θ as

$$\begin{cases} \dfrac{dz}{dt} = v_r - \theta\sigma_0 \dfrac{|v_r|}{g(v_r)} z \\ v_r = v_x - r\omega \\ F_x = (\sigma_0 z + \sigma_1 \dfrac{dz}{dt} + \sigma_2 v_r) F_{xn} \end{cases} \tag{2.40}$$

where different road adhesion coefficients can be represented by different θ. Thus, to monitor tire/road friction is changed to estimate θ.

The vehicle dynamics can be written as:

$$\begin{cases} F_x = m\dot{v}_x / 4 \\ J\dot{\omega} = -rF_x + u \end{cases} \tag{2.41}$$

where r is the inertia and radius of the wheel and u denotes the accelerating or braking torque.

Combining (2.40) and (2.41), Canudas de Wit et. al. introduced the following transformation as

$$\begin{cases} \eta = rmv + J\omega \\ \chi = J\omega + rF_n\sigma_1 z \end{cases} \tag{2.42}$$

and the output was chosen as $\xi = \frac{1}{J}[\varsigma_2 - rF_n\sigma_1\varsigma_3] = \omega$.

Thus, the longitudinal vehicle dynamic model is written as

$$\begin{cases} \dot{\varsigma} = \widetilde{A}\varsigma + \widetilde{B}[\theta\varphi(\xi,u,\varsigma)] + \widetilde{R}u + \widetilde{E}\xi \\ \dot{\theta} = 0 \\ \xi = \widetilde{C}\varsigma \end{cases} \tag{2.43}$$

where $\varsigma = [\eta \quad \chi \quad z]^T$ are the transformed state variables. $\widetilde{A}$, $\widetilde{B}$, $\widetilde{C}$, $\widetilde{E}$ are the corresponding matrix.

Notice there exist a function $q(\cdot)$ for LuGre model that

$$\left|\varphi(\hat{\xi},u,\hat{\varsigma})\right| \le q(\hat{\varsigma}) \le q_{\max}$$

the following observer was proposed in [81]

$$\begin{cases} \dot{\hat{\varsigma}} = \widetilde{A}\hat{\varsigma} + \widetilde{B}\left[\hat{\theta}\varphi(\hat{\xi},u,\hat{\varsigma})\right] + \widetilde{E}u + \widetilde{K}(\xi - \hat{\xi}) + \widetilde{B}\eta \\ \hat{\theta} = \gamma\varphi(\hat{\varsigma})\hat{\xi} \\ \hat{\xi} = \widetilde{C}\hat{\varsigma} \end{cases} \tag{2.44}$$

where $\eta = 2\hat{\theta}_{\max}(q_{\max} + q(\|\hat{\varsigma}\|^2))\,\mathrm{sgn}(\hat{\xi})$.

It was proven in [86]-[88] that θ can be correctly estimated using (2.44) if the wheel angular velocity ω can be measured. A typical simulation result is shown in Fig.2.17, when θ varied from 1 to 4 to mimic different road surface conditions.

It was further shown in [89] that this method can be extended to build a fault detection observer with respect to disturbance caused by unsteady road surface or unavoidable vehicle vertical vibration.

In [89], the LuGre model (2.43) was extended as (2.45) below to describe the unavoidable disturbance introduced by vehicle vibration or road roughness. The disturbance effect is addressed onto road profile parameter θ by introducing a new variable ε to represent the disturbance

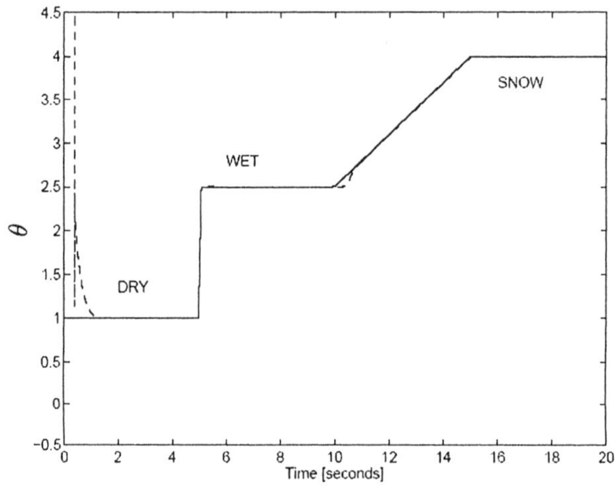

Fig.2.17 Estimation result of varied θ, from [86]. (© [1999] IEEE)

$$\begin{cases} \dot{\varsigma} = \tilde{A}\varsigma + \tilde{B}[\theta\varphi(\xi,u,\varsigma)] + \tilde{R}u + \tilde{E}\xi \\ \dot{\theta} = \varepsilon \\ \xi = \tilde{C}\varsigma \end{cases} \tag{2.45}$$

where it satisfies

1) ε is boundary, i.e., $|\varepsilon| \le \tau$, $\tau > 0$;

2) θ is boundary, i.e., $|\theta - \bar{\theta}| = \left|\theta_0 + \int_0^t \varepsilon d\tau - \bar{\theta}\right| \le \Delta\theta_{\max}$, $\Delta\theta_{\max} > 0$,

$t > 0$. Here $\bar{\theta}$ denotes the expectation of θ, i.e. $\bar{\theta} = E[\theta]$; θ_0 represents the initial state of θ; and $\Delta\theta_{\max}$ denotes the maximum offset of θ.

Corresponding to (2.44), the observer was designed as

$$\begin{cases} \dot{\hat{\varsigma}} = A\hat{\varsigma} + \widetilde{B}\,\hat{\theta}\varphi(\hat{\varsigma}) - L(\xi - \hat{\xi}) + k\widetilde{B}\,\mathrm{sgn}(\xi - \hat{\xi}) + \widetilde{R}u + \widetilde{E}\xi \\ \dot{\hat{\theta}} = \gamma_1 \varphi(\hat{\varsigma})(\xi - \hat{\xi}) \\ \xi = \widetilde{C}\varsigma \end{cases} \tag{2.46}$$

where $k = \max_x \{x : 2\theta_{\max} |\varphi(x)|\}$. L is an appropriately chosen observer matrix.

It was proven in [89] that the observer error state system designed above is strong practical stable with respect to a clear bound

$$\bar{d} = \sqrt{\frac{\lambda_{\max}(P) \cdot 4|\theta_e|_{\max} \tau + 4|\theta_e|^2_{\max}}{\lambda_{\min}(P) \cdot \lambda_{\min}(Q)} \cdot \frac{\gamma_2}{\gamma_1}} \tag{2.47}$$

when there exit two positive symmetric matrices P and Q satisfy the following equations

$$-Q = [A + LC]^T P + P[A + LC], \; PB = \gamma_2 C^T \tag{2.48}$$

where γ_2 is a positive weighted number. And γ_1 is a positive real number that is introduced to adjust the variation rate of $\hat{\theta}$. Notice that

$$\left|\xi - \hat{\xi}\right| = \left|\frac{1}{r}[(\varsigma_2 - \hat{\varsigma}_2) - (\varsigma_3 - \hat{\varsigma}_3)]\right| \le \frac{1}{r}(|\varsigma_2 - \hat{\varsigma}_2| + |\varsigma_3 - \hat{\varsigma}_3|) \le \frac{2}{r}\bar{d} \tag{2.49}$$

two fault detection rules can be naturally formulated as:

(1) trigger the fault alarm should if $\left|\xi - \hat{\xi}\right| = |\omega - \hat{\omega}| > \frac{2}{r}\bar{d}$;

(2) trigger the fault alarm should if $\left|\hat{\theta} - \bar{\theta}\right| > \Delta\theta_{\max}$.

In order to test the fault detection ability of this model, two simulation tests were carried out as shown in Fig.2.18 and Fig.2.19. In the first simulation test, the value of θ is assumed to jump from 1 to 3 at $t = 5s$. It is easy to find that $\xi - \hat{\xi} = \omega - \hat{\omega}$ exceeds the chosen threshold $2\bar{d}/r = 0.495$ immediately when fault occurs. And θ reaches new value 3 soon after the fault occurs, too. In the second simulation test, the value of θ is assumed to shift from 1 to 3 during time span $t = 5s$ to $t = 10s$. It is easy to find that $\xi - \hat{\xi} = \omega - \hat{\omega}$ does not exceed the threshold $2\bar{d}/r = 0.495$. But the statistic characteristics of the error output changes. While the estimation of road profile coefficient $\hat{\theta}$ keeps tracking θ all the time. It's obvious that the fault alarm will be trigger in time for two cases.

2.4.4 Identification of Empirical Lateral Friction Models

Besides, longitudinal tire/road friction identification research, There were several discussions on empirical lateral tire/road friction identification [54], [91]-[99].

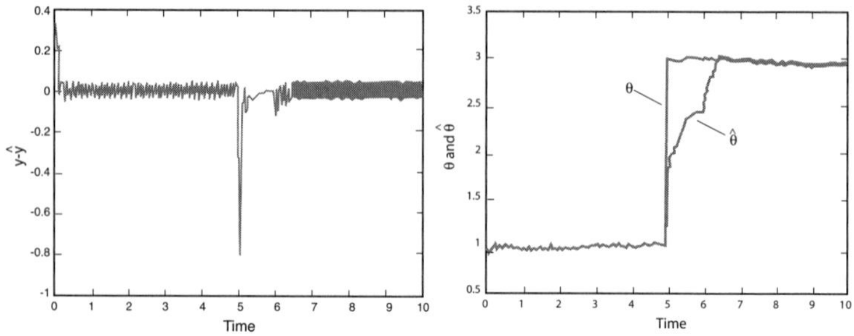

Fig.2.18 Variation of $\xi - \hat{\xi} = \omega - \hat{\omega}$ (left), θ and $\hat{\theta}$ (right) with the jump error occurs at $t = 5$, from [89].

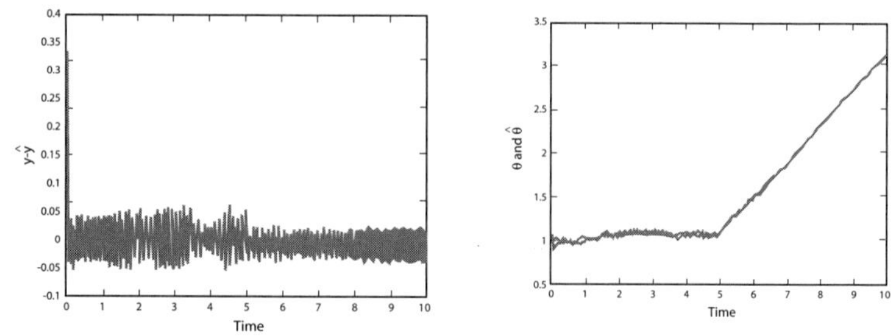

Fig.2.19 Variation of $\xi - \hat{\xi} = \omega - \hat{\omega}$ (left), θ and $\hat{\theta}$ (right) with the shift error begins at $t = 5$, from [89].

and the side slip angles are tracked using GPS signals as

$$\begin{cases} \phi = \tan^{-1}(v_y / v_x) \\ \alpha_f = \phi - \delta_f \\ \alpha_r = \phi \end{cases} \tag{2.52}$$

In [94], Sienel proposed a recursive derivation formula of the lateral friction coefficients in terms of previous measurement for rotation and lateral acceleration, where α_f is tracked as

$$\dot{\alpha}_f = \dot{\delta}_f - \frac{\alpha_f}{v} + r \tag{2.53}$$

At sampling time i, $C'_{f,i}$ is estimated as

$$C'_{f,i} = \begin{cases} m \cdot \dfrac{l_r}{l_f + l_r} \cdot \dfrac{\dot{\alpha}_{f,i}}{\dot{\delta}_f - \dfrac{\alpha_f}{v} + r}, for \left| \dot{\delta}_{f,i} - \dfrac{\alpha_{f,i}}{v_i} + r_i \right| > \varepsilon \\ C'_{f,i-1}, else \end{cases} \tag{2.54}$$

Apparently, it requires to measure yaw rate r additionally with respect to the above method.

In [95], the cornering stiffness coefficients are estimated by

$$\begin{cases} C'_f = C'_{f0}(1 - \dfrac{\dot{v}}{\mu_f g}) \\ C'_r = C'_{r0}(1 - \dfrac{\dot{v}}{\mu_r g}) \end{cases} \tag{2.55}$$

where μ_f and μ_r are the common adhesion coefficients. C'_{f0} and C'_{r0} are the bound value.

There were also some integrated estimation models that had been proposed. In [96]-[97], Ray used extended Kalman-Bucy filtering and Bayesian hypothesis selection to estimate motion, tire forces and tire/road friction coefficient on asphalt surface. It requires no *a priori* knowledge of friction coefficients and does not require a tire force model. It first estimates tire friction based on measurement of vehicle motion and then determines friction coefficient directly. Similar methods are reported in [98]-[99].

2.6 Summary

The recent trend of research on developments of tire/road friction modeling and related control techniques is reviewed in this Chapter. It sequentially scans the previous research efforts in longitudinal, lateral and integrated tire/road friction modeling and discusses their applications in recent vehicle traction, steering and braking controllers designs.

However, some important contents leave untouched in this Chapter due to length limits. For instance,

(1) those tire/road friction force estimation methods based on torque converter/ powertrain models are not discussed here [112]-[113]. For instance, the force axis of slip curve was estimated from an observer based on model of torque converter in [113];

(2) Some novel tire/road modeling methods, which use neural networks or some other techniques to "learn" the nonlinear characteristics of tire/road friction phenomena, are not mentioned in here [114]. Moreover, the algorithms for tire/road friction model parameter identification [115]-[116], such as the identification algorithm using genetic algorithms (GA) that was proposed in [116], are not discussed here;

(3) In this Chapter, it is assumed that the vehicles are moving a plat surface and their vertical position changes are neglected. However, this simplification may yield notable errors in some cases. The methods that were developed to deal with such situation, i.e. [117], are not examined;

(4) In this Chapter, the tire/road friction force is indirectly measured via resulted vehicle motion. This is to use tire as a "virtual" sensor as what was summarized in [118]. Recently, some novel tire sensors, i.e. surface acoustic wave (SAW) sensors [119]-[120], were proposed to directly measure tire deformation and temperature [121], and therefore determine tire/road friction force. How to employ these "real" sensors in tire/road friction modeling, monitoring and control issues is expected to become a hot topic of this field in the near future [122].

2.7 Reference

1. D. F. Moore, The friction of pneumatic tyres, Elsevier Scientific Publishing Co., New York, 1975.
2. T. French, Tyre Technology, Ed. Adam Hilger, Bristol, 1989.
3. D. Dowson, History of tribology, Longman Ltd., London, 1998.
4. B. Feeny, A. Guran, and N. Hinrichs, et. al, "A historical review on dry friction and stick-slip phenomena," ASME Applied Mechanical Reviews, vol. 51, no. 5, pp. 321-341, 1998.
5. J. Y. Wong, Theory of ground vehicles, John Wiley & Sons, Inc., New York, 1993.
6. U. Kiencke and L. Nielsen, Automotive Control System for Engine, Driveline, and Vehicle, Springer. 2000.
7. W. Hirschberg, G. Rill, and H. Weinfurter, "User-appropriate tyre-modelling for vehicle dynamics in standard and limit situations," Vehicle System Dynamics, vol. 38, no. 2, pp. 103-125, 2002.
8. Y. P. Chang, M. El-Gindy, and D. A. Streit, "Literature survey of transient dynamic response tyre models," International Journal of Vehicle Design, vol. 34, no. 4, pp. 354-386, 2004.

9. J. Harned, L. Johnston, and G. Scharpf, "Measurement of tire brake force characteristics as related to wheel slip (anti-block) control system design," SAE Transactions, vol. 78, SAE #690214, pp: 909-925, 1969.
10. E. Bakker, L. Nyborg, and H. B. Pacejka, "Tire modeling for use in vehicle dynamic studies," Society of Automotive Engineers, SAE #870421, 1987.
11. E. Bakker, H. B. Pacejka, and L. Lidner, "A new tire model with an application in vehicle dynamics studies," Proceedings of International Congress and Exposition, SAE #890087, 1989.
12. H. B. Pacejka and R. S. Sharp, "Shear force development by pneumatic tires in steady state conditions: a review of modeling aspects," Vehicle System Dynamics, vol. 20, pp: 121-176, 1991.
13. P. W. A. Zegelaar and H. B. Pacejka, "The in-plane dynamics of tyres on uneven roads," Vehicle System Dynamics Supplement, vol. 25, pp. 714-730, 1996.
14. E. Denti and D. Fanteria, "Models of wheel contact dynamics: an analytical study on the in-plane transient response of a brush model," Vehicle System Dynamics, vol. 32, pp.199-225, 2000.
15. M. Burckhardt, Fahrwerktechnik: Radschlupfregelsysteme, Vogel-Verlag, Germany, 1993.
16. L. Alvarez and J. Yi, "Adaptive emergency braking control in automated highway systems," Proceedings of IEEE Conference on Decision and Control, vol. 4, pp: 3740-3745, 1999.
17. G. Rill, Simulation von Kraftfahrzeugen, Vieweg, 1994.
18. H. B. Pacejka and E. Bakker, "The magic formula tyre model," Proceedings of 1st International Colloquium on Tyre Models for Vehicle Dynamics Analysis, pp: 1-18, 1991.
19. P. Dahl, "A solid friction model," Technical Report TOR-0158(3107-18)-1, The Aerospace Corporation, El Segundo, CA, 1976.
20. P. Dahl, "Measurement of solid friction parameters of ball bearings," Proceedings of 6th Annual Symposium on Incremental Motion, Control Systems and Devices, pp: 49-60, 1977.
21. P.-A. Bliman and M. Sorine, "Friction modeling by hysteresis operators: application to Dahl, stiction and stribeck effects," Proceedings of the Conference 'Models of Hysteresis', 1991.
22. P.-A. Bliman and M. Sorine, "A system-theoretic approach of systems with hysteresis: application to friction modeling and compensation," Proceedings of European Control Conference, 1993.
23. P.-A. Bliman and M. Sorine, "Easy-to-use realistic dry friction models for automatic control," Proceedings of European Control Conference, 1994.
24. C. Canudas de Wit, H. Olsson, and K. J. Astrom, et al, "A new model for control of systems with friction," IEEE Transactions on Automatic Control, vol. 40, no. 3, pp: 419-425, 1995.

25. C. Canudas de Wit and P. Tsiotras, "Dynamic tire friction models for vehicle traction control," Proceedings of IEEE Conference on Decision and Control, vol. 4, pp: 3746-3751, 1999.
26. H. Olsson, "Control systems with friction," PhD thesis, Department of Automatic Control, Lund Institute of Technology, Lund, Sweden, 1996.
27. N. Barabanov and R. Ortega, "Necessary and sufficient conditions for passivity of the LuGre friction model," IEEE Transactions on Automatic Control, vol. 45, no. 4, pp: 830–832, 2000.
28. M. Gafvert, "Comparisons of two dynamic friction models," Proceedings of IEEE International Conference on Control Applications, pp: 386-391, 1997.
29. H. Olsson, K.J. Astrom, and C. Canudas de Wit, et. al, "Friction models and friction compensation," European Journal of Control, vol. 4, no. (3), 1998.
30. J. Deur, "Modeling and analysis of longitudinal tire dynamics based on the LuGre friction model," Proceedings of 3rd IFAC Workshop Advances in Automotive Control, pp: 101-106, 2001.
31. J. Deur, "A brush-type dynamic tire friction model for non-uniform normal pressure distribution," CD-ROM Proceedings of 15th Triennial IFAC World Congress, 2002.
32. C. Canudas-de-Wit, P. Tsiotras, and E. Velenis, "Dynamic friction models for longitudinal road/tire interaction: theoretical advances," 21st IASTED Conference on Modelling, Identification and Control, pp: 48-53, 2002.
33. C. Canudas-de-Wit, P. Tsiotras, and E. Velenis, "Dynamic friction models for longitudinal road/tire interaction: experimental results," 21st IASTED Conference on Modelling, Identification and Control, 2002.
34. C. Canudas de Wit, P. Tsiotras, and E. Velenis, et. al, "Dynamic friction models for road/tire longitudinal interaction," Vehicle System Dynamics, vol. 39, no. 3, pp. 189-226, 2003.
35. J. Yi, L. Alvarez, and X. Claeys, et. al, "Emergency braking control with an observer-based dynamic tire road friction model and wheel angular velocity measurement," Vehicle System Dynamics, vol. 39, no. 2, pp. 81-97, 2003.
36. M. Segel, "Theoretical prediction and experimental substantiation of the response of the automobile to steering control," Proceedings of Automobile division of the institute of mechanical engineers, vol. 7, pp: 310–330, 1956.
37. J. Kasselmann and T. Keranen, "Adaptive steering," Bendix Technical Journal, vol. 2, pp. 26-35, 1969.
38. J. Ackermann, "Robust car steering by yaw rate control," Proceedings of the 29th IEEE Conference on Decision and Control, pp: 2033-2034, 1990.
39. J. Ackermann and W. Dareberg, "Automatic track control of a city bus," IFAC Theory Report on Benchmark Problems for Control Systems Design, 1990.

40. B. Breuer, V. Eichorn, and J. Roth, "Measurement of tyre/road friction ahead of car and inside the tyre," Proceedings of International Symposium on Advanced Vehicle Control, 1992.
41. Q. Qu and Y. Liu, "On lateral dynamics of vehicles based on nonlinear characteristics of tires," Vehicle System Dynamics, vol. 34, pp. 131-141, 2000.
42. E. Ono, S. Hosoe, and D. Tuan, et. al, "Bifurcation in vehicle dynamics and robust front wheel steering control," IEEE Transactions on Control Systems Technology, vol. 6, no. 3, pp: 412-420, 1998.
43. J. Stephant, A. Charara, and D. Meizel, "Virtual sensor: application to vehicle sideslip angle and transversal forces, " IEEE Transactions on Industrial Electronics, vol. 51, no. 2, pp: 278-289, 2004.
44. Y. Furukawa, N. Yuhara, and S. Sano, et. al, "A review of four-wheel steering studies from the viewpoint of vehicle dynamics and control," Vehicle System Dynamics, no. 18, pp: 151-186, 1989.
45. P. Raksincharoensak, M. Nagai, and H. Mouri, "Investigation of automatic path tracking control using four-wheel steering vehicle," Proceedings of the IEEE International Vehicle Electronics Conference, pp: 73-77, 2001.
46. J. Ackermann and W. Sienel, "Robust yaw damping of cars with front and rear wheel steering," IEEE Transactions on Control Systems Technology, vol. 1, no. 1, pp: 15-20, 1993.
47. H. Peng and M. Tomizuka, "Lateral control of front-wheel-steering rubber-tire vehicles," PATH Research Report UCB-ITS-PRR-90-5, 1990.
48. T. A. Johansen, I. Petersen, J. Kalkkuhl, J. Ludemann, Gain-scheduled wheel slip control in automotive brake systems, IEEE Transactions on Control Systems Technology, vol. 11, pp. 799-811, 2003.
49. J. P. Maurice, M. Berzeri, and H. B. Pacejka, "Pragmatic tyre model for short wavelength side slip variations," Vehicle System Dynamics, vol. 31, pp: 65–94, 1999.
50. J. Stephant, A. Charara, and D. Meizel, "Force model comparison on the wheel-ground contact for vehicle dynamics," IEEE Intelligent Vehicle Symposium, vol. 2, pp: 589-593, 2002.
51. G. Mastinu and E. A. Pairana, "A semi-analytical tyre model for steady and transient state simulations, Proceedings of 1st International Colloquium on Tyre Models for Vehicle Dynamics Analysis, pp: 58-81, 1991.
52. H. B. Pacejka and I. Besseling, "Magic formula tyre model with transient properties," Proceedings of 2nd International Colloquium on Tyre Models for Vehicle Dynamics Analysis, 1997.
53. M. Gafvert and J. Svendenius, "Construction of semi-empirical tire models for combined slip," Technical Report ISRN LUTFD2/TFRT7606SE, Department of Automatic Control, Lund Institute of Technology, Sweden, 2003.
54. J.-O. Hahn, R. Rajamani, and L. Alexander, "GPS-based real-time identification of tire-road friction coefficient," IEEE Transactions on Control Systems Technology, vol. 10, no. 3, pp: 331-343, 2002.

55. D. J. Schuring, W. Pelz, and M. G. Pottinger, "A model for combined tire cornering and braking forces," Investigations and Analysis in Vehicle Dynamics and Simulation, SAE International, SAE #960180, pp: 61–83, 1996.
56. G. Gim and P. E. Nikravesh, "An analytical model of pneumatic tyres for vehicle dynamics simulations, Part 2: Comprehensive slips," International Journal of Vehicle Design, vol. 12, no. 1, pp: 19–39, 1991.
57. X. Claeys, J. Yi, and L. Alvarez, et. al, "A dynamic tire/road friction model for 3D vehicle control and simulation," Proceedings of IEEE Intelligent Transportation Systems Conference, pp: 483-488, 2001.
58. J. Deur, J. Asgari, and D. Hrovat, "A 3D brush-type dynamic tire friction model," Vehicle System Dynamics, vol. 42, no. 3, pp. 133-173, 2004.
59. S. Velenis, C. Canudas de Wit, and P. Tsiotras, "Extension of the LuGre Dynamic Friction Model to 2D Motion," Internal report, Laboratoire d'Automatique de Grenoble, France, 2001.
60. J. Martinez, J. Avila and C. Canudas, "A new bicycle vehicle model with dynamic contact friction," First IFAC Symposium on Automotive Control, 2004.
61. T. L. Ford and F. S. Charles, "Heavy duty truck tire engineering," SAE #880001, 1988.
62. E. Gohring, E. C. Von Glasner, and H. C. Pflug, "Contribution to the force transmission behavior of commercial vehicle tires," SAE #912692, 1991.
63. M. Gafvert, "Topics in modeling, control, and implementation in automotive systems," Ph.D. Dissertation, ISSN 0280-5316, ISRN LUTFD2/TFRT-1066-SE, 2003.
64. G. Mavros, H. Rahnejat and P. King, "Investigation of steady-state tyre force and moment generation under combined longitudinal and lateral slip conditions," Vehicle System Dynamics, vol. 41, pp. 351-360, 2004.
65. A. Lawrence, Modern Inertial Technology, Springer Verlag, 1993.
66. F. Napolitano, T. Gaiffe, and Y. Cottreau, et. al. "PHINS: the first inertial navigation system based on fiber optic gyroscopes," Proceedings of St Petersbourg International Conference on Navigation Systems, 2002.
67. R. Usui and A. Ohno, "Recent progress of fiber optic gyroscope and application at JAE," Optical Fiber Sensors Conference Technical Digest, vol.1, pp: 11-14, 2002.
68. E. Nebot, S. Sukkarieh, and H. Durrant-Whyte, "Inertial navigation aided with GPS information," Proceedings of the Fourth Annual Conference on Mechatronics and Machine Vision in Practice, pp: 169-174, 1997.
69. F. X. Cao, D. K. Yang, and A. G. Xu, et. al, "Low cost SINS/GPS integration for land vehicle navigation," Proceedings of the IEEE 5th International Conference on Intelligent Transportation Systems, pp: 910-913, 2002.

70. C.-Y. Chan, "Magnetic sensing as a position reference system for ground vehicle control," IEEE Transactions on Instrumentation and Measurement, 51 (1), pp: 43-52, 2002.
71. J. I. Hernandez and C.-Y. Kuo, "Steering control of automated vehicles using absolute positioning GPS and magnetic markers," IEEE Transactions on Vehicular Technology, vol. 52, no. 1, pp: 150-161, 2003.
72. S. M. Donecker, T. A. Lasky, and B. Ravani, "A mechatronic sensing system for vehicle guidance and control," IEEE/ASME Transactions on Mechatronics, vol. 8, no. 4, pp: 500-510, 2003.
73. W. S. Wijesoma, K. R. S. Kodagoda, and A. P. Balasuriya, et. al, "Road edge and lane boundary detection using laser and vision," Proceedings of 2001 IEEE/RSJ International Conference on Intelligent Robots and Systems, vol. 3, pp: 1440-1445, 2001.
74. U. Eichhorn and J. Roth, "Prediction and monitoring of tyre/road friction," XXIV FISITA Congress, London, 1992.
75. B. Breuer, U. Eichhorn, and J. Roth, "Measurement of tyre/road friction ahead of the car and inside the tyre," Proceedings of International Symposium on Advanced Vehicle Control, pp: 347–353, 1992.
76. S. Germann, M. Wurtenberger, and A. Daiss, "Monitoring of the friction coefficient between tyre and road surface," Proceedings of the Third IEEE Conference on Control Applications, vol. 1, pp: 613-618, 1994.
77. C. Liu and H. Peng, "Road friction coefficient estimation for vehicle path prediction," Vehicle System Dynamics, vol. 25 Supplement, pp. 413-425, 1996.
78. F. Gustafsson, "Slip-based tire-road friction estimation," Automatica, vol. 33, no. 6, pp: 1087-1099, 1997.
79. F. Gustafsson, "Monitoring tire-road friction using the wheel slip," IEEE Control Systems Magazine, vol. 18, no. 4, pp: 42-49, 1998.
80. S. Muller, M. Uchanski, K. Hedrick, "Estimation of the maximum tire-road friction coefficient," ASME Journal of Dynamic Systems, Measurement, and Control, vol. 125, no. 4, pp: 607-617, 2003.
81. Kiencke, U. and Daiss, A., "Estimation of tyre friction for enhanced ABS systems," Proceedings of International Symposium on Advanced Vehicle Control, 1994.
82. W. Hwang and B.-S. Song, "Road condition monitoring system using tire-road friction estimation," Proceedings of International Symposium on Advanced Vehicle Control, pp: 437–442, 2000.
83. J. Wang, L. Alexander, R. Rajamani, "Friction estimation on highway vehicles using longitudinal measurements," ASME Journal of Dynamic Systems, Measurement, and Control, vol. 126, no. 2, pp: 265-275, 2004.
84. H. Nishira, T. Kawabe, and S. Shin, "Road friction estimation using adaptive observer with periodical σ-modification," Proceedings of IEEE International Conference on Control Applications, vol. 1, pp: 662-667, 1999.

85. K. Yi, K. Hedrick, and S. C. Lee, "Estimation of tire-road friction using observer based identifiers," Vehicles Systems Dynamics, vol. 31, pp: 233–261, 1999.
86. C. Canudas de Wit and R. Horowitz, "Observers for tire/road contact friction using only wheel angular velocity information," Proceedings of IEEE Conference on Decision and Control, vol. 4, pp: 3932-3937, 1999.
87. J. Yi, L. Alvarez, and R. Horowitz, et. al, "Adaptive emergency braking control using a dynamic tire/road friction model," Proceedings of IEEE Conference on Decision and Control, vol. 1, pp: 456-461, 2000.
88. J. Yi, L. Alvarez, and X. Claeys, et. al, "Emergency braking control with an observer-based dynamic tire/road friction model and wheel angular velocity information," Proceedings of American Control Conference, vol. 1, pp: 19-24, 2001.
89. L. Li, F.-Y. Wang, and G. Shan, et. al, "Design of tire fault observer based on estimation of tire/road friction conditions, Automatica Sinica, vol.28, no. 5, pp: 689-694, 2003.
90. J. R. Zhang, S. J. Xu, and A. Rachid, "Robust sliding mode observer for automatic steering of vehicles," Proceedings of IEEE Intelligent Transportation Systems, pp: 89-94, 2000.
91. C. Lee, K. Hedrick, and K. Yi, "Real-time slip-based estimation of maximum tire-road friction coefficient," IEEE/ASME Transactions on Mechatronics, vol. 9, no. 2, pp: 454-458, 2004.
92. D. M. Bevly, J. C. Gerdes, and C. Wilson, "The use of GPS based velocity measurements for measurement of sideslip and wheel slip," Vehicle System Dynamics, vol. 38, no. 2, pp. 127-147, 2002.
93. R. Daily and D. M. Bevly, "The use of GPS for vehicle stability control systems," IEEE Transactions on Industrial Electronics, vol. 51, no. 2, pp: 270-277, 2004.
94. W. Sienel, "Estimation of the tire cornering stiffness and its application to active car steering," Proceedings of IEEE Conference on Decision and Control, vol. 5, pp: 4744-4749, 1997.
95. S. Saraf and M. Tomizuka, "Slip angle estimation for vehicles on automated highways," Proceedings of American Control Conference, vol. 3, pp: 1588-1592, 1997.
96. L. R. Ray, "Nonlinear tire force estimation and road friction identification: simulation and experiments," Automatica, vol. 33, no. 10, pp: 1819-1833, 1997.
97. L. R. Ray, "Experimental determination of tire forces and road friction," Proceedings of American Control Conference, vol. 3, pp: 1843-1847, 1998.
98. B. Samadi, R. Kazemi, and K. Y. Nikravesh, et. al, "Real-time estimation of vehicle state and tire-road friction forces," Proceedings of American Control Conference, vol. 5, pp: 3318-3323, 2001.
99. K. Huh, J. Kim, and K. Yi, et. al, "Monitoring system design for estimating the lateral tire force," Proceedings of American Control Conference, vol. 2, pp: 875-880, 2002.

100. Kimbrough, S., "A brake control strategy for emergency stops that involve steering: Part 1 theory," Proceedings of the 2nd Symposium on Transportation Systems, 1990 ASME Winter Annual Meeting, 1990.
101. Kimbrough, S., "A brake control strategy for emergency stops that involve steering: Part 2 implementation issues and simulation results," Proceedings of the 2nd Symposium on transportation Systems, 1990 ASME Winter Annual Meeting, 1990.
102. R. Emig, H. Goebels, and H. J. Schramm, "Antilock braking systems (ABS) for commercial vehicles - status 1990 and future prospects," Vehicle Electronics in the 90's: Proceedings of the International Congress on Transportation Electronics, pp: 515-523, 1990.
103. R. Bosch, Automotive handbook, Robert Bentley, Publisher, 4th ed, 1997.
104. M. Schinkel and K. Hunt, "Anti-lock braking control using a sliding mode like approach," Proceedings of American Control Conference, vol. 3, pp: 2386-2391, 2002.
105. P. E. Wellstead and N. B. O. L. Pettit, "Analysis and redesign of an antilock brake system controller," IEE Proceedings - Control Theory and Applications, vol. 144, no. 5, pp: 413-426, 1997.
106. P. Tsiotras and C. Canudas-de-Wit, "On the optimal braking of wheeled vehicles," Proceedings of American Control Conference, vol. 1, no. 6, pp: 569-573, 2000.
107. D. Zhang, H. Zheng, and J. Sun, et. al, "Simulation study for anti-lock braking system of a light bus," Proceedings of the IEEE International Vehicle Electronics Conference, pp: 70-77, 1999.
108. S. Drakunov, U. Ozguner, and P. Dix, et. al, "ABS control using optimum search via sliding modes," IEEE Transactions on Control Systems Technology, vol. 3, no. 1, pp: 79-85, 1995.
109. M. Krstik and H.-H. Wang, "Design and stability analysis of extremum seeking feedback for general nonlinear systems," Proceedings of IEEE Conference on Decision and Control, pp. 1743-1748, 1997.
110. I. Petersen, T. A. Johansen, and J. Kalkkuhl, et. al, "Wheel slip control using gain-scheduled LQ - LPV/LMI analysis and experimental results," European Control Conference, 2003.
111. S. Taheri and E. H. Law, "Slip control braking of an automobile during combined braking and steering manoeuvres," Advanced Automotive Technologies, vol. 40, pp: 209–227, 1991.
112. K. Yi, K. Hedrick, and S. Lee "Estimation of tire-road friction using observer based identifiers," Vehicle System Dynamics, vol. 31, no. 4, pp: 233-261, 1999.
113. T. Shim and D. Margolis, "Model-based road friction estimation," Vehicle System Dynamics, vol. 41, no. 4, pp: 249-276, 2004.
114. W. R. Pasterkamp and H. B. Pacejka, "Application of neural networks in the estimation of tire/road friction using the tire as a sensor," SAE #971122, 1997.

115. M. Beato, V. Ciaravola, and M. Russo, et. al, "Lateral tyre force by a milliken test on a flat track roadway simulator," Vehicle System Dynamics, vol. 34, pp. 117-129, 2000.
116. J. A. Cabrera, A. Ortiz, and E. Carabias, et. al, "An alternative method to determine the magic tyre model parameters using genetic algorithms," Vehicle System Dynamics, vol. 41, no. 2, pp: 109-127, 2004.
117. H. S. Bae, J. Ryu, and J. C. Gerdes, "Road grade and vehicle parameter estimation for longitudinal control using GPS," Proceedings of IEEE Conference on Intelligent Transportation Systems, pp: 166-171, 2001.
118. W. R. Patserkamp and H. B. Pacejka, "The tire as a sensor to estimate friction," Vehicle System Dynamics, vol. 27, pp. 409-422, 1997.
119. A. Pohl, R. Steindl, and L. Reindl, "The "intelligent tire" utilizing passive SAW sensors measurement of tire friction," IEEE Transactions on Instrumentation and Measurement, vol. 48, no. 6, pp: 1041-1046, 1999.
120. A. Pohl, "A review of wireless SAW sensors," IEEE Transactions on Ultrasonics, Ferroelectrics and Frequency Control, vol. 47, no. 2, pp: 317-332, 2000.
121. M. Mizuno, H. Sakai, and K. Oyama, et. al, "The development of the tire side force model considering the dependence of surface temperature of tire," Vehicle System Dynamics, vol. 41, pp. 361-370, 2004.
122. L. Li, and F.-Y. Wang, "Research advances in vehicle lateral motion monitoring and control," International Journal of Intelligent Control and Systems, vol. 9, no. 3, 2004.

Advanced Vehicle Lateral Motion Control

3.1 Introduction

Generally, wheeled vehicles can be divided into two types: single track vehicles and track-trailers; see Fig.3.1. Single track vehicles usually refer to passenger cars or car-like vehicles/robots which can be viewed as a single steering tractor. Track-trailer systems consist of a steering tractor and one (sometimes more than one) passive trailer(s) linked with rigid free joints. From the viewpoint of steering, single track vehicles can be further classified into two types. front steering vehicles (2WS) in which only the two front tires can be steered, and full steering vehicles (4WS) of which the front and rear tires can be steered independently. Since discussion track-trailers steering control requires a dedicated book chapter due to its broad range, only single track vehicles are discussed here.

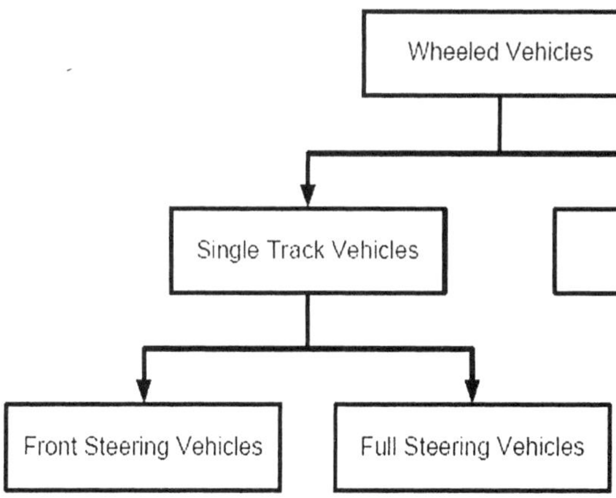

Fig.3.1 Classification of vehicles based on lateral motion model.

The prime task of vehicle lateral motion control is path/lane following, or more plainly, to keep the vehicle on the road/lane. Comparing to longitudinal dynamics, vehicle lateral dynamics are relatively easy to control, because they almost solely depend on steering subsystems control. However, to design a good steering controller still requires carefully handling several difficult problems. More specifically, researchers in this field address on the following topics [1]-[2].

(1) vehicle lateral motion modeling, which is tightly related to tire/road friction modeling that is discussed in *Chapter 2*;
(2) vehicle lateral motion related sensory including vehicle position sensory, lane detection rotation, and vehicle movement measurement etc.;
(3) vehicle steering actuator design and implementation. One hot topic in this area is steer-by-wire;
(4) vehicle lateral motion monitoring;
(5) autonomous vehicle lateral motion control;
(6) vehicle lateral motion control and driver assistance.

By scanning the quickly developing research within this area, this Chapter explains the initiatives and techniques for vehicle lateral motion (steering) control with an emphasis on lateral motion monitoring and steering controller design. Specially, robust steering controllers are carefully studied. This Chapter does not intend to make a thorough comparison among numerous different observers/controllers that had been proposed, since it is too difficult and almost impossible because of vehicle dynamic model variety and simulation/testing parameters difference.

3.2 Steer-By-Wire Systems

Quickly increasing requirements of safe and comfortable driving have led vehicle manufacturers and suppliers to actively pursue development programs in the so called "X-by-wire" subsystems. These computer-controlled subsystems include steer-by-wire, brake-by-wire, drive-by-wire and etc, which are connected through in-vehicle computer networks.

A steer-by-wire system replaces the traditional mechanical linkage between the steering wheel and the road wheel actuator (e.g., a rack and pinion steering system) with an electronic connection. Because it removes direct kinematical relationship between the steering and road wheels, it enables control algorithms to help enhance driver steering command [3]-[20]. For instance, Fig.3.2(b) shows a production model presented in [7], which was modified for full steer-by-wire capability by replacing the steering shaft with a brushless DC servomotor.

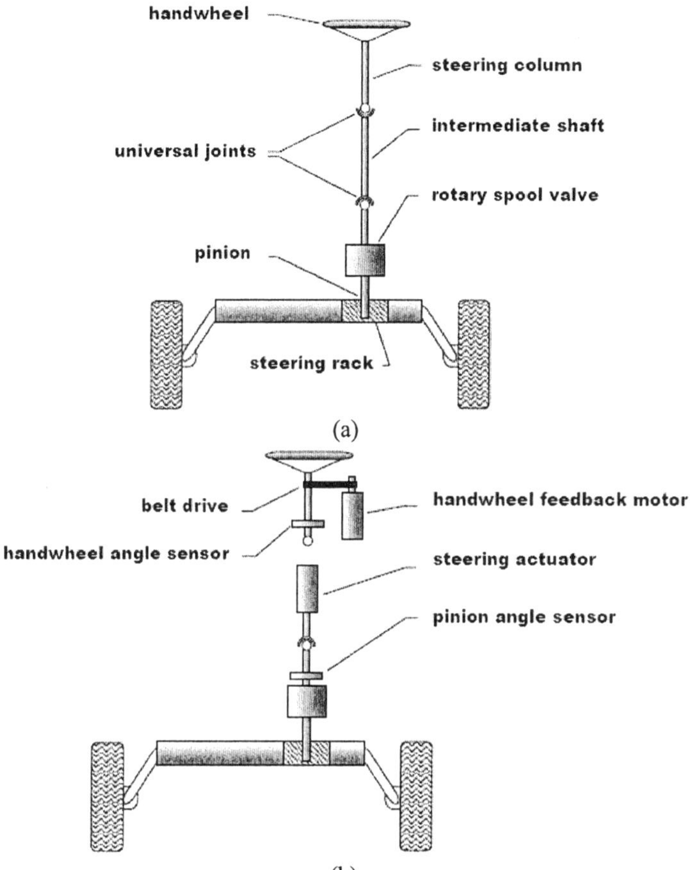

Fig.3.2 Diagram of steering systems. (a) conventional steering system, (b) steer-by-wire system, from [7]. (© [2003] IEEE)

In this system, a rotary position sensor measures the lower steering shaft angle, which is equal to the front wheel steer angle scaled by the steering ratio. An identical sensor attached to the upper steering shaft measures the hand wheel angle. The servomotor actuator specifications are chosen based on the maximum torque and speed necessary to steer the vehicle under typical driving conditions including moderate emergency maneuvers. The stock hydraulic power assist unit and rack-pinion mechanism is retained, since the incorporation of the power assist unit eliminates the need for extensive modifications to the existing.

The first direct benefit of steer-by-wire system is to help driver steer the vehicle more easily. In this proposed steer-by-wire system, on average, steering torque required at the hand-wheel during normal driving ranges from 0 to 2 Nm, while emergency maneuvers can demand up to 15 Nm of torque [7]. The actuator installed in the test vehicle provides a maximum steering torque of 17.1 Nm with a maximum steer rate of 700 degrees per second.

The second benefit of steer-by-wire system is to filter out the disturbance torque generated by road roughness, and parameter changes caused by tire pressure/temperature and loading variations. For instance, the driver control input in [19] was augmented by the adaptive component (force comp), which is based on the load experienced by the plant, to reject road feedback (disturbance).

The steering system dynamics for Fig.3.2(b) can be written as follows.

$$J\ddot{\theta} + b\dot{\theta} + F_c \operatorname{sgn} \dot{\theta} + k_a \tau_a = \tau \tag{3.1}$$

where θ is the pinion angle, J is the total moment of inertia of the system, b is viscous damping coefficient, F_c represents Coulomb friction force, k_a is a scale factor, τ_a is the tire self-aligning moment, and τ is the actuator torque.

The purpose of the steer-by-wire controller is to track commanded steer angle with minimal error; the control effort consists of three components.

$$\tau = \tau_{feedback} + \tau_{feedforwar\ d} + \tau_{aligning} \tag{3.2}$$

A proportional derivative (PD) feedback component is also designed in [7] as

$$\tau_{feedback} = K_p(\theta_d - \theta) + K_d(\dot{\theta}_d - \dot{\theta}) \tag{3.3}$$

where θ_d is the desired steer angle, K_p is the proportional feedback constant, and K_d is the derivative feedback constant. The feedback gains K_p and K_d are selected to give a fast closed loop system response without oscillatory behavior.

The purpose of steer-by-wire controller is to track driver commanded steer angle θ_d with minimal error. Because the system is second order, PD control alone results in some steady state error when tracking some types of commands. To obtain these measurements, the front wheels are raised off the ground so as to isolate the influence of J, b and F_c from static friction at the tire-ground interface.

The addition of feed-forward compensation is given as

$$\tau_{feedforwar\ d} = J\ddot{\theta}_d + b\dot{\theta}_d + F_c \operatorname{sgn} \dot{\theta}_d \tag{3.4}$$

This PD controller will cancel any tracking errors associated with the system dynamics and internal friction. Here, J, b and F_c are determined through closed-loop identification of the steering system.

When driving a vehicle over the road, however, an additional disturbance acts on the system causing a steering error that is directly attributable to tire self-aligning moment. The total aligning moment is given by

$$\tau_a = (t_p + t_m)F_{yf}\varpi_f \tag{3.5}$$

where t_p and t_m are the tire pneumatic and mechanical trails, respectively. Front tire slip angle, ϖ_f can be calculated from the following relationship involving estimated sideslip and other measurable parameters.

$$\varpi_f = \beta + \frac{l_f r}{v_x} - \delta_f \tag{3.6}$$

As what has been introduced in *Chapter 2*, β denotes the angle from the longitudinal axis of vehicle body to the direction of the vehicle velocity. δ_f is the front tires steering angle. l_f is distance from front tire and CG. Yaw rate is denoted as r. v_x is the longitudinal vehicle speed. But, it should be pointed out that the front tire slip angle here has different meaning from what we introduced in *Chapter 2*, see Fig.2.8. And their relationship can be expressed as

$$\alpha_f = \delta_f + \varpi_f \tag{3.7}$$

Aligning moment may also be directly approximated as an empirical function of slip angle. The approximation of aligning moment is added to the feedback and feed-forward control as

$$\tau_{aligning} = k_a \hat{\tau}_a \varpi_f \tag{3.8}$$

where k_a is a scale factor to account for torque reduction by the steering gear.

The addition of $\tau_{aligning}$ to the actuator effort effectively eliminates most of the steering disturbances that arise when turning at speed. Besides the approaches demonstrated above, many other tests also showed that the response in steer-by-wire system is more quickly and accurately than that in conventional steering system. This improves vehicle stability and provides a basis for further steering systems fault detection.

Moreover, steer-by-wire technique simplifies assembly and reduces vehicle's mass. For example, in the steer-by-wire system proposed in [7], only the stock hydraulic power assist unit and rack-pinion mechanism is retained. It simplifies assembly and reduces vehicle's mass, which simultaneously allows flexibility in packaging and transportation.

As summarized in [21], three things are still on the steer-by-wire wish list: reliability, smartness, and driver-friendly design. Currently, four problems have to be solved before the steer-by-wire system is applied into practical vehicles:

(1) how to choose fast and stable tracking algorithms or hardware for steer-by-wire systems;

(2) how to deal with disturbance torque;

(3) how to improve the driver usability and maneuverability;

(4) how to implement robust steer-by-wire systems with fault-tolerant hardware and software.

With the aid of steer-by-wire systems, the same vehicle can be made to handle differently by effectively changing front cornering stiffness. In other words, it is possible to maintain consistent handling characteristics under variable operating conditions. Therefore, driver-oriented steering ratio control becomes a hot topic in this area recently [22]-[28].

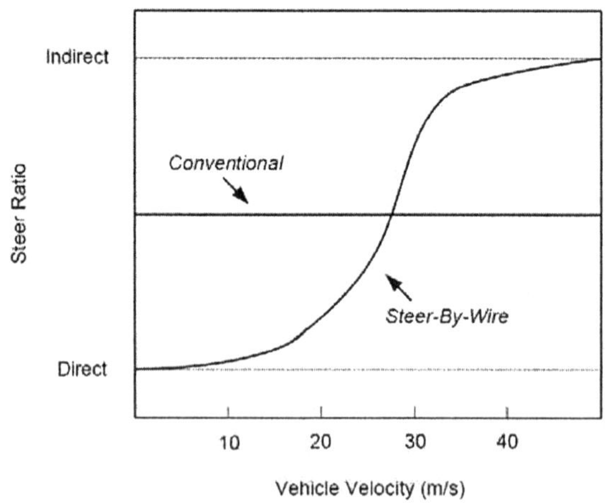

Fig.3.3 Variable steering ratio for different vehicle velocities.

In [26], the steer-by-wire system is composed of two motors controlled by Electronic Control Unit (ECU) instead of mechanical linkage. One motor in the steering wheel is to improve the driver's steering feel and the other motor in the steering linkage is to improve the vehicle stability under under-steer and over-steer situations concerning vehicle's velocity.

Some current products provide more flexible steering ratio for the driver. As mentioned in [27], at low vehicle speeds, the driver prefers a fixed steering ratio, after which the steering ratio increases (becomes increasingly indirect) as the speed increase. A steady-state constant Human-Machine Interaction yaw gain (yaw-rate divided by HMI angle) is derived from vehicle data or measurements at intermediate and high speeds; see Fig.3.3.

Fault detection and fault tolerant control are also essential parts in steer-by-wire systems. Designers must reduce the chance of faults in safety-critical steer-by-wire systems, and choose an acceptable compromising point between costs and level of risks. Many topics were presented and studied from the surrounding field in both hardware and software reliability [177]-[192], including

(1) self fault detection and self repair;

(2) online/offline diagnosis of actuators and sensors;

(3) online/offline diagnosis of inner-vehicle field bus and vehicle-infrastructure communication networks;

(4) online/offline diagnosis testing of control algorithms and modules.

It is also expected that smart steer-by-wire systems can reduce the driver's workload by compensating driver's phase lead maneuvering. As a preliminary examination in [28], the effect to the reactive torque was investigated when vehicle behavior parameters feedbacks were considered in that control logic. It showed that the relationship between the driver's characteristics and reactive torque including mechanical characteristics of the input device can be identified by supervised learning, and the proper relationship between vehicle behavior and reactive torque will be more investigated as well.

In general, steer-by-wire system is expected to provide a better operation platform of lateral motion controllers. It is estimated to take ten or twenty years to widely employ steer-by-wire subsystems into the real world to provide a better operation platform of lateral motion controllers.

3.3 Vehicle Lateral Motion Modeling and Control Strategies

Lateral vehicle dynamics has been studied since the 1950s [30]-[32]. In 1956, Segel presented a vehicle model with three degrees of freedom in order to describe lateral movements including roll and yaw. If roll movement is neglected, a simple model known as "bicycle model" is obtained. This model is widely used for studies of lateral vehicle dynamics (yaw and sideslip) now. A modern deduction procedure from "magic formula" to "bicycle model" can be found in *Section 2.3.2* and also [33]-[39]. Because "bicycle model" grasps the prime characteristics of vehicle steering movement and yields a relatively simple linear model for analyzing, it is now widely used in steering controller design. The following discussions will be primly carried out based on this model.

It should be pointed out that full steering vehicles significantly outperform front steering vehicles in handling and stability [40]-[45]. Usually, when the vehicle enters the curved path, the rear wheel will first steer in the opposite direction to the front wheel in order to generate sufficient yaw motion to follow the desired yaw rate. After that, the rear wheel steers will synchronize with the front wheel to keep the yaw rate with desired value and also control the lateral motion for path tracking. In [46]-[47] a generic control law for robust decoupling of lateral and yaw motion by yaw-rate feedback to front-wheel steering was derived. It showed that ideal steering dynamics were able to be achieved by velocity scheduled lateral acceleration feedback to front-wheel steering. For robust yaw stabilization a velocity-scheduled yaw-rate feedback to rear-wheel steering is given, by which the linearized system gets velocity-independent yaw eigenvalues. Since most steering controller design methods can be applied to both front and full steering vehicles without tedious modifications, the difference between front steering vehicles and full steering vehicles will not be emphasized in the rest of this chapter.

In "bicycle model", the design specifications of steering controller are often given in terms of maximal displacement from the guideline, maximal steering angles and steering angle rates. For instance, the benchmark problem mentioned in [22] mainly requires

(1) the steering angle is limited as

$$\left\|\delta_f\right\| \le 40 \text{ deg} \tag{3.9}$$

(2) the steering angle rate is limited as

$$\left\|\dot{\delta}_f\right\| \le 28 \text{ deg}/\ s \tag{3.10}$$

(3) the displacement from the guideline must not exceed 0.15m in transient state and 0.02m in steady state;

(4) the lateral acceleration must not exceed 2m/s for passengers comfort. The ultimate limit is 4m/s.

Besides "bicycle model", there were numerous other vehicle lateral motion models of different levels of complexities that had been proposed and analyzed [48]-[52]. These models usually build integrated dynamic models which consider vehicle longitudinal and lateral speed variations simultaneously in terms of steering angle. For instance, the vehicle dynamics proposed in [48] tried to incorporate longitudinal tracking force model into lateral motion model as well vehicle aerodynamics. It was represented by the following set of nonlinear equation sets

$$\begin{cases} \dot{v}_x = \dfrac{1}{m}[T + mv_y r - mfg + c_f \dfrac{v_y + l_f r}{v_x}\delta_f + v_x^2(fk_1 - k_2)] \\ \dot{v}_y = \dfrac{1}{m}[(c_f + T)\delta_f - (c_f + c_r)\dfrac{v_y}{v_x} - (l_f c_f - l_r c_r + mv_x^2)\dfrac{r}{v_x}] \\ \dot{r} = \dfrac{1}{I_z}[(l_f c_f + T)\delta_f - (l_f c_f - l_r c_r)\dfrac{v_y}{v_x} - (l_r^2 c_r + l_f^2 c_f)\dfrac{r}{v_x}] \end{cases} \tag{3.11}$$

where v_x and v_y are vehicle's longitudinal velocity and lateral velocity in vehicle coordinate respectively. T is the traction or braking force. f is the rotation coefficients. k_1 and k_2 are lift and drag parameters from aerodynamics respectively.

Usually, the following mapping relationship (3.12) is employed to describe the transformation function between vehicle velocities denoted in orthogonal coordinates and that denoted in world coordinates

$$\begin{cases} \dot{x} = v_x \cos(\psi) - v_y \sin(\psi) \\ \dot{y} = v_x \sin(\psi) + v_y \cos(\psi) \\ \dot{\psi} = r \end{cases} \tag{3.12}$$

where (x, y) denote vehicle center gravity's position in world coordinates. And the safe driving requirements were brought in terms of (x, y) straightforward.

Since it is impossible to discuss all these models, they are not enumerated in this chapter.

In general, vehicle lateral control strategies can be grouped into two kinds based on the employed sensors: look-ahead and look-down reference systems. Look-ahead reference systems resemble human driving by measuring the heading angle and lateral displacement from the lane reference ahead of the vehicle using vision sensors, i.e. camera [53]-[54]. Look-down reference systems calculate the heading angle and lateral displacement from a reference point, for example a magnetic marker, very close to the vehicle [55]-[61].

Look-down reference systems have the advantage of being reliable, yielding accuracy and good performance under any weather or light conditions. And they can work well when other vehicles or pedestrians occlude lane lines. Therefore, it is very useful in urban areas navigation. However, it is also reported that Look-down reference systems only yielding good experimental results only with speeds of less than 25m/s (55mi/h) which are impractical for highway automation. Look-ahead reference systems is suitable for such kind of driving scenarios, since current vehicle vision systems can effectively detect the upcoming road curvature variations and lateral displacement when driving at highway level 40 m/s (90 mi/h).

Some other critical vehicle state variables can also be measured via using appropriate sensors. For instance, vehicle steering angle can be directly measured from actuator [62]; vehicle inertial can be indirectly calculated from its driving behaviors [63]-[64]. In the last two decades, great efforts were also put into tire/ road friction coefficients estimation; see *Section 2.5* and literals [65]-[68]. Since this chapter addresses on vehicle lateral motion monitoring and control instead of vehicle sensors design, it is simply assumed in the rest of this chapter that all the information needed can be accurately measured. Detailed discussions on vehicle motion related sensing are presented in *Chapter 8-10*.

3.4 Vehicle Lateral Motion Monitors

Since a vehicle is a highly complex system that constitutes of varied mechanical, electronic and electromechanical elements, numerous sensors were designed and employed to measure vehicle movement information. Concentrated on vehicle lateral motion control, position and movement information of a vehicle need to be precisely measured.

In many recent approaches, not all the vehicle characteristics are directly measured due to high cost or certain other reasons. Instead, several special observers are employed to reconstruct the needed information. In literals, these observers were also called virtual sensors.

For example, knowledge of sideslip angle, yaw rate and lateral velocity is essential in vehicle control, but is difficult to obtain directly. In 1997 and 1999, Kiencke etc. proposed a linear observer and a nonlinear observer using reduced order bicycle model in [69] and [70]. Soon after that, Kalman filter in [71]-[74] were used for linearized vehicle models. In [72], Huh et. al. constructed the monitoring system based on KFMEC (Scaled Kalman Filter with Model Error Compensator) technique to improve the robustness of ordinary Kalman filters. In [74], Zhang, Xu and Rachid showed the feasibility of a special sliding mode observer for vehicle lateral motion. Similar conclusions were reached by Perruquetti and Barbot in [75].

To filter out the unexpected effect of disturbances from the observer output, different robust design methods had been introduced in Luenberger observer construction. In [76]-[78], H_∞ filer theory was employed to design the optimal observers to resist disturbance for reduced order bicycle model. In [79], H_∞ loop shaping was used for observer design of the linearized lateral motion model of a single-unit HDV (tractor-semitrailer type vehicle). In [80], the equivalent H_∞ LMI design method was applied in Luenberger observer and fault detection filter design.

The above Bicycle models (2.23) can be written into canonical form as

$$\dot{x} = Ax + Bu + Ew \tag{3.13}$$

where $x = [\beta \quad r \quad \psi \quad y_f \quad y_r]^T$ is the state variable, $u = [\delta_f \quad \delta_r]^T$ is the control input, and $w = [\rho_{ref} \quad f_w]^T$ is the disturbance. A, B and E are the corresponding system matrices.

The measurement output y is usually formulated as

$$y = Cx + Du \tag{3.14}$$

It can be easily proven that the system is observable if either displacements y_f or y_r can be measured. But the system is unobservable if only ψ is measured. These conclusions take valid for both front steering vehicle and full steering vehicle [81], [87].

Simulation also shows that to measure both y_f and y_r may increase the robustness of the observer. But many vehicles only equip front sensor because of cost consideration.

Usually, the linear Luenberger observer is formulated as

$$\begin{cases} \dot{\hat{x}} = A\hat{x} + Bu - L'(y - \hat{y}) \\ \hat{y} = C\hat{x} + Du \end{cases} \tag{3.15}$$

where $\hat{x}$ denotes observer state, $\hat{y}$ denotes observer output. L is the observer matrix.

Its performance index can be specified as

Minimize the H_∞ norm of the transfer function matrix from w to e. Or equivalently Choose the smallest $\gamma > 0$ such that

$$\int_0^{+\infty} e^T e dt \le \gamma^2 \int_0^{+\infty} w^T w dt \tag{3.16}$$

This naturally leads to the well-known H_∞ design problem, which was solved using a linear matrix inequality (LMI) problem by Boyd, Ghaoui, and E. Feron et. al in [82], or equivalently solved using a set of Algebraic Riccati Equations (ARE) shown by Nagpal and Khargonekar in [83]. Finally, it leads to the following LMI design problem

Design Task 3.I:
Min γ
with $\gamma > 0$, $\mu > 0$ and moderate $\tau > 0$ such that

$$\begin{bmatrix} P[A+L'C]+[A+L'C]^T P+I & PE \\ E^T P & -\gamma^2 \end{bmatrix} < 0 \tag{3.17}$$

or equivalently

$$\begin{bmatrix} PA+A^T P+X'C+C^T X'^T+I & PE \\ E^T P & -\gamma^2 \end{bmatrix} < 0 \tag{3.18}$$

with the observer matrix

$$L' = P^{-1} X' \tag{3.19}$$

Most above observers utilized the accurate dynamic model using nominal values including tire concerning stiffness, vehicle mass and moment of inertia and distances between center of mass and tires. Thus, these observers depend on an accurate knowledge of these parameters, and are affected by variations in them. For instance, Stephant et. al pointed out in [39] that the speed of center of gravity is not an indispensable variable.

One method to solve this problem is to choose the estimation method without utilizing the vehicle dynamic model, i.e. the observers proposed in [84]-[86]. But non model-based observers are hard to apply along with steering control system. Another method is to introducing robust observer that is not sensitive to system parameter changes or adaptive observer that can change itself according to parameter change. In [87], a robust observer was developed by including an extra term and adopting the Lyapunov stability theorem. Since dynamic models (2.23) highly depend on system velocity, the system dynamics should be modified by considering parameter variances and nonlinear properties

$$\dot{x} = (A + \Delta A)x + Bu + Ew \tag{3.20}$$

where ΔA denotes the variance matrix that is determined by variance of mass, velocity, tire-road friction coefficients and nonlinear characteristics.

The norm of ΔA can be assumed to be bounded as

$$\|\Delta A\|_2 \le \varepsilon \tag{3.21}$$

Simulation reveals that ε mainly depends on estimation error of velocity v and road adhesion factor μ. In [85], the observer for (3.20) was

$$\begin{cases} \dot{\hat{x}} = A\hat{x} + Bu - L(y - \hat{y}) + \tau\alpha \\ \hat{y} = C\hat{x} + Du \\ \alpha = \dfrac{\varepsilon^2 \hat{x}^T \hat{x}}{(y-\hat{y})^T (y-\hat{y})} P^{-1} C^T (y - \hat{y}) \end{cases} \tag{3.22}$$

where $\hat{x}$ denotes observer state, $\hat{y}$ denotes observer output. L is the observer matrix. τ is a positive real scalar that needs to be determined.

Based on Eq.(3.20)-(3.22), we have the dynamics of observer error $e = x - \hat{x}$ as

$$\dot{e} = (A + LC)e + \Delta A x + Ew - \alpha = (A + \Delta A + LC)e + \Delta A\hat{x} + Ew - \tau\alpha \tag{3.23}$$

This naturally leads to the following design task:

Design Task 3.II:
Min γ
with $\gamma > 0$, $\mu > 0$ and moderate $\tau > 0$ such that

$$\left[\begin{array}{c|c} PA + A^T P + XC + C^T X^T + (1 + \lambda\varepsilon^2)I + (1/\lambda + 1/\tau)2P^2 & PE \\ \hline E^T P & -\gamma^2 \end{array}\right] < 0 \tag{3.24}$$

with the observer matrix

$$L = P^{-1} X \tag{3.25}$$

And the main result in [87] is stated as follows.

Theorem 3.1. H_∞ constraints (3.16) will be held if there exist a symmetry positive matrix P and two positive real numbers τ and λ that satisfy (3.24).

Proof: Define a Lyapunov like function $V(t) = e^T Pe$, where P is a symmetry positive matrix. Thus, the time derivative of $V(t)$ along the trajectories of (3.23) should be

$$\begin{aligned} \dot{V} &= e^T P[A + \Delta A + LC]e + e^T Ew + e^T \Delta A\hat{x} \\ &+ e^T [A + \Delta A + LC]^T Pe + w^T E^T e + \hat{x}^T \Delta A^T e \\ &- \tau e^T P\alpha - \tau\alpha^T Pe \end{aligned} \tag{3.26}$$

Thus, the design objective can be transformed as finding the smallest γ^2 that satisfies.

$$T = \dot{V} + e^T e - \gamma^2 w^T w \le 0 \tag{3.27}$$

Substituting (3.26) into (3.27) yields

$$\begin{aligned} T &= \begin{bmatrix} e \\ w \end{bmatrix}^T \left[\begin{array}{c|c} P[A + LC] + I + [A + LC]^T P & PE \\ \hline E^T P & -\gamma^2 \end{array}\right] \begin{bmatrix} e \\ w \end{bmatrix} \\ &+ 2e^T P\Delta Ae + 2e^T P\Delta A\hat{x} - 2\tau e^T P\alpha \le 0 \end{aligned} \tag{3.28}$$

Note that

$$e^T P\Delta Ae \le \frac{1}{\lambda} e^T P^2 e + \lambda e^T \varepsilon^2 e \tag{3.29}$$

and

$$e^T P\Delta A\hat{x} \le \frac{1}{\tau} e^T P^2 e + \tau\hat{x}^T \varepsilon^2 \hat{x} \tag{3.30}$$

together with

$$\hat{x}^T \varepsilon^2 \hat{x} \le e^T P\alpha \tag{3.31}$$

(3.24) is a sufficient condition for (3.28). □

From the discussions in [88]-[90], the observer (3.22) should be a forced, nonautonomous, dynamic system, and contains an unknown, perturbation term. Fortunately, we find that this condition can be naturally met here. Besides, τ should not be too large due to the constraints on implementation of the observer. In the practice, τ should be determined by compromising all the requirements together. But simulations prove that constraints on τ will directly limit the range of γ at the same time, too. It is proven in [87] that this nonlinear observer (3.22) maintains the good disturbance rejection property that derived from [76]-[80], while also provides tolerance to model variance as the observer too. To demonstrate the feasibility of the proposed observer, it was compared with a linear Luenberger observer that is only optimized using H_∞ method to reject disturbance.

Suppose the vehicle is front steering and only the front sensor is applied. Thus

$$C = \begin{bmatrix} 0 & 0 & 0 & 1 \end{bmatrix}, \; D = \begin{bmatrix} 0 & 0 & 0 & 0 \end{bmatrix} \tag{3.32}$$

The parameters are assigned as.

$\tilde{m} =$ 1500 kg, $\tilde{I}_z$ =2500 kg.m.m, $l_f =$ 1.2 m, l_r =1.5 m, $l_{fs} =$ 5 m, $l_w =$ 0.5 m, $c_f = c_r =$ 80000kN/rad.

Here τ is chosen as $\tau = 10$. The variance bound of velocity is $\Delta v =$ 1 m/s, $\Delta \tilde{m} =$ 100 kg, $\Delta \mu = 0.1$. Thus it chooses $\varepsilon = \sqrt{5}$.

Suppose v is incorrectly estimated as 16m/s when it equals to 15m/s and μ is misestimated as 0.7 instead of 0.8 during the simulation.

The Luenberger observer matrix is

$$L' = [0.2869 \quad -3.6024 \quad -0.1984 \quad -0.0019\,]^T \times 10^4$$

The robust observer matrix is

$$L = [0.3687 \quad -5.4516 \quad -0.2345 \quad -0.0025\,]^T \times 10^4$$

and symmetry matrix P is chosen as

$$P = \begin{bmatrix} 3.1771 & 0.3082 & -2.1728 & 0.4347 \\ 0.3082 & 0.0443 & -0.5471 & 0.1093 \\ -2.1728 & -0.5471 & 9.3211 & -1.8620 \\ 0.4347 & 0.1093 & -1.8620 & 0.3823 \end{bmatrix} \times 10^{10}$$

Fig.3.4 shows the tracking result of the side-slip angle β using the linear Luenberger observer optimized by solving Design Task 3.I. Fig.3.5 shows the direct tracking output of β using the proposed observer optimized by solving Design Task 3.II. Fig.3.6 shows the smoothed tracking result of β using a 10 points average filter. The maximum estimation error of the linear Luenberger observer is around 12%, while the estimation error of the proposed robust observer is around 2%. It is apparent that the proposed observer will yield better tracking performance. Simulation also reveals that there is not a global optimal observer matrix that can deal with all the velocity settings. Thus, the observer matrix L and nonlinear term $\tau\alpha$ should change with velocity v to maintain good tracking result. Fortunately, drivers will not change the vehicle's speed too much when he/she is steering. Thus the change of feedback matrix will not violate the observer system's stability.

Recently, Stephant, Charara and Meizel carefully compared four observers including linear Luenberger observer and three nonlinear observers: extended Luenberger observer, extended Kalman filter and sliding-mode observer in [39]. Based on simulation results and practical experiments, they showed that all four observers can yield acceptable estimation results if the observer's parameters are appropriately assigned.

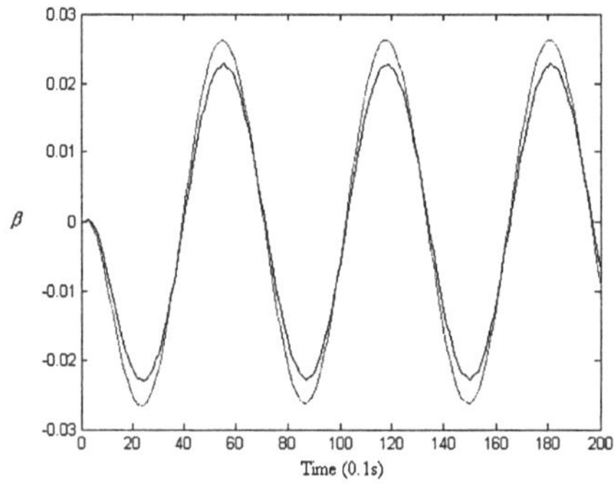

Fig.3.4 Tracking result using the linear Luenberger observer, from [87]. (© [2005] IEEE)

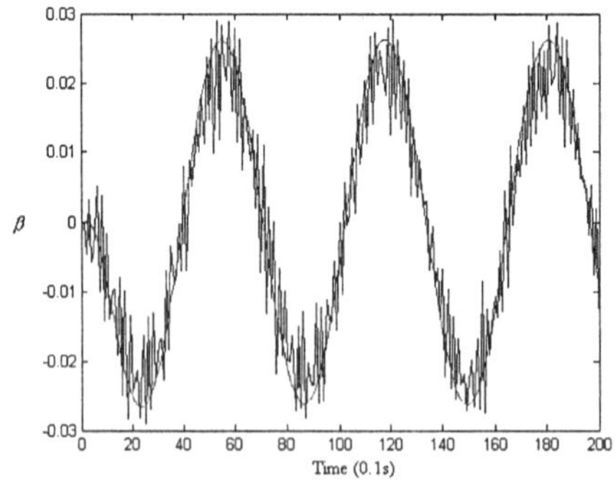

Fig.3.5 Tracking result using the proposed robust observer, from [87]. (© [2005] IEEE)

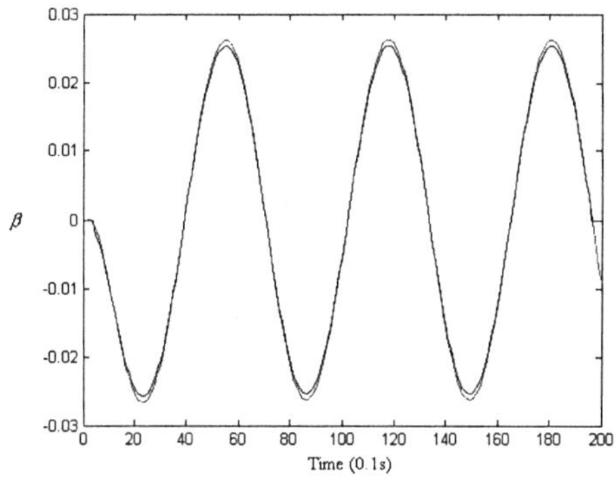

Fig.3.6 Smoothed tracking result using the proposed robust observer, from [87]. (© [2005] IEEE)

3.5 Vehicle Steering Controller Design

3.5.1 Vehicle Lateral Motion Control Objectives

As early as 1969, Kasselmann and Keranen [31] had developed an active steering system based on feedback from a yaw rate sensor. With accumulating efforts, people gradually realized that the difficulties of steering control mainly lie in the following five aspects.

(1) how to avoid skidding, which is one frequently encountered hazardous situation for green hand drivers;

(2) how to reject the disturbance caused by wind or some other reasons;

(3) how to deal with vehicle dynamics uncertainty and variation;

(4) how to handle actuator rate limits during steering;

(5) how to handle time delay exists in feedback block during steering.

To answer the five questions, numerous different designed methods had been proposed in the last two decades. Among them, varied linear robust controller design methods attract continuous interest. And most of them had been proposed especially for Bicycle steering models. Considering actuator saturation, system (2.23) can be rewritten as

$$\begin{bmatrix} \dot{\beta} \\ \dot{r} \\ \dot{\psi} \\ \dot{y}_f \\ \dot{\delta}_f \end{bmatrix} = \begin{bmatrix} a_{11} & a_{12} & 0 & 0 & b_{11} \\ a_{21} & a_{22} & 0 & 0 & b_{21} \\ 0 & 1 & 0 & 0 & 0 \\ v & l_{fs} & v & 0 & 0 \\ 0 & 0 & 0 & 0 & 0 \end{bmatrix} \begin{bmatrix} \beta \\ r \\ \psi \\ y_f \\ \delta_f \end{bmatrix} + \begin{bmatrix} 0 \\ 0 \\ 0 \\ 0 \\ 1 \end{bmatrix} u_f + \begin{bmatrix} 0 & \tilde{d}_1 \\ 0 & \tilde{d}_2 \\ -v & 0 \\ -vl_{fs} & 0 \\ 0 & 0 \end{bmatrix} \begin{bmatrix} \rho_{ref} \\ \tilde{f}_w \end{bmatrix} \tag{3.33}$$

whose canonical form is written as

$$\dot{x} = Ax + Bu_f + Ew \tag{3.34}$$

where $x = \begin{bmatrix} \beta & r & \psi & y_f & \delta_f \end{bmatrix}^T$ is state variable, $u = u_f$ is control input, and $w = \begin{bmatrix} \rho_{ref} & \tilde{f}_w \end{bmatrix}^T$ is taken as disturbance. A, B and E denote the corresponding system matrices.

To simplify induction process below, the wind disturbance is normalized to the same amplitude level as road curvature with

$$\tilde{d}_1 = \sigma d_1, \ \tilde{d}_2 = \sigma d_2 \tag{3.35}$$

where $\sigma > 0$ is a scale number determined by

$$\sigma = \|\rho_{ref}\|_{\max} / \|f_w\|_{\max} \tag{3.36}$$

It is easy to check that is controllable pair for most vehicle parameter setting. In the rest of this chapter, it is assumed that the system is controllable even under model uncertainty.

These linear robust controller design studies provide some basic solutions of skidding, disturbance, parameter uncertainties and steer angle saturation. Based on these results, different sliding model controller, fuzzy controller and adaptive controller were developed as time goes on. These nonlinear controllers usually outperform linear ones in several different aspects. But it should be reemphasized that it does not intend to make a thorough comparison among numerous different observers/controllers that had been studied in this book.

3.5.2 Robust Vehicle Steering Controllers

1) Frequency Domain Linear Robust Steering Controllers

Originated in the later 80s, frequency domain robust design techniques soon became and remained as one of the most important techniques in field of vehicle lateral motion control. It had been proven to be a practical and efficient approach by lots of literature [33]-[36], [91]-[95].

One direct idea to avoid skidding is to remove the influence of r on the lateral acceleration. The lateral and yaw motions of a car with active steering is decoupled by Ackermann in [46]. It was proved that for an ideal longitudinal mass distribution, the decoupling by yaw rate feedback is robust with respect to uncertain nonlinear tire side force characteristic, velocity and vehicle mass. But Ackermann later showed that this was not a simple and cheap control system since it requires measuring longitudinal velocity of vehicle $v_x \approx v \sin\beta$, yaw rate r and its derivative, slip angle β simultaneously. Thus, a practical controller was proposed in [33], in which only v_x needs to be measured. This simplified controller was proved to have similar steady-state behavior to a car.

In [34], it was further shown that additional feedback of the yaw rate r leads to a significant reduction of the deviation from the guideline in nearly all driving maneuvers compared to earlier controllers which used solely feedback of the deviation y_f. Therefore, a feedback controller with respect to both r and y_f were proposed. It is written as

$$\dot{\delta}_f = u_f - k_r \cdot r \tag{3.37}$$

As revealed in [92], using careful poles and zeros assignment, the system can be well stabilized. Besides it, the redundancy of the design parameters can be used to count off the variance of vehicle dynamics. For instance, the Γ-stability boundaries was analyzed for parameters (k_D, k_{DD}) in [93]. In [47], it was further proven that decoupling by yaw rate feedback is robust with respect to uncertain nonlinear tire side force characteristic, velocity and mass, if we assumed an ideal longitudinal mass distribution.

More precisely, the corresponding root locus of the transfer function from u_f to r for increasing gain k_r is shown in Fig.3.7, where the operating point with maximal virtual mass and maximal velocity was selected, which turned out to be the most critical operating condition in an eigenvalue analysis. Based on Fig.3.7, k_r is chose as 0.89 so as to implement an approximately cancellation of a pole/ zero-pair in the transfer function.

After applying feedback controller (3.37), one pole has been shifted from $s = 0$ to $s = -0.8934$ and the other two remain as $s = 0$. As shown in Fig.3.8, in order to attract the two braches in the revised root locus plot, a compensator transfer function is introduced as well as the above feedback controller

$$\frac{u_f(s)}{y_f(s)} = \omega_c^2 \frac{k_{DD}s^2 + k_D s + k_P}{s^2 + 2D\omega_c s + \omega_c^2} \tag{3.38}$$

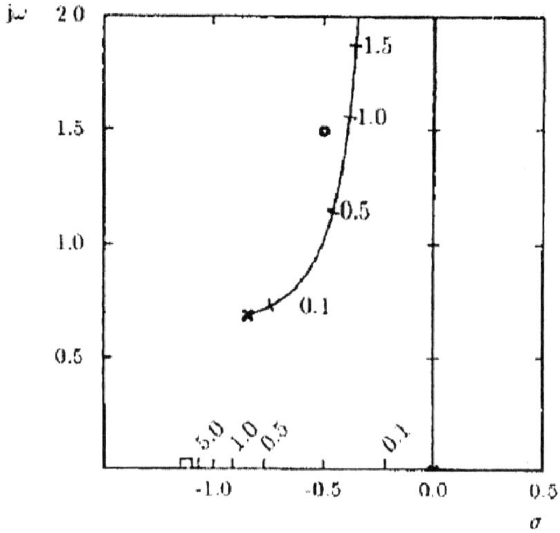

Fig.3.7 Root locus of the transfer function from u_f to r for increasing gain k_r, The zeros of the transfer function $y(s)/u_f(s)$ are marked with circles, and the zeros of $r(s)/u_f(s)$ with a cross [92]. (© [1995] IEEE)

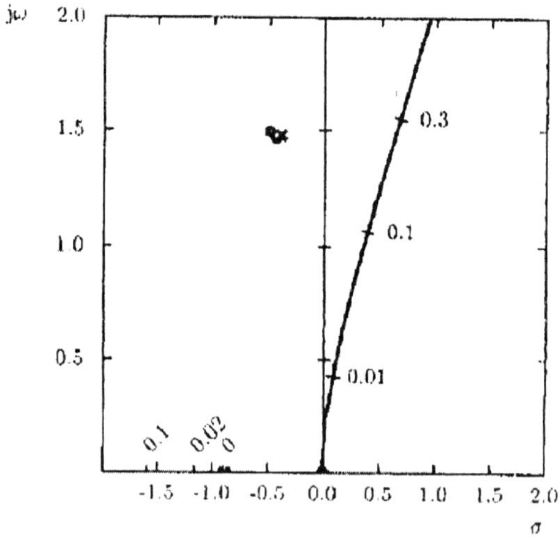

Fig.3.8 Root locus of the transfer function from u_f to r with yaw rate feedback k_r [92]. (© [1995] IEEE)

where inside the bandwidth ω_c, k_P denotes a proportional part, k_D denotes a differential part and k_{DD} denotes the double differential part.

Combining (3.37) and (3.38), we can get the final controller structure shown in Fig.3.9.

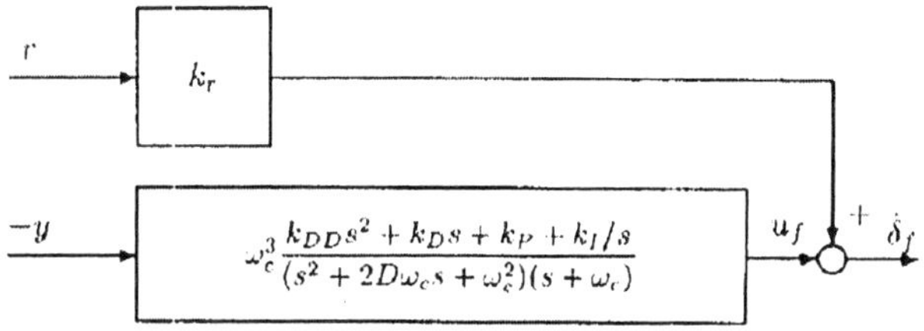

Fig.3.9 Controller structure [92]. (© [1995] IEEE)

To explore the effects of parameter variations of the system performance, Ackermann et. al 1995 analyzed the above controller with the possible operating domain Q shown in Fig.3.10.

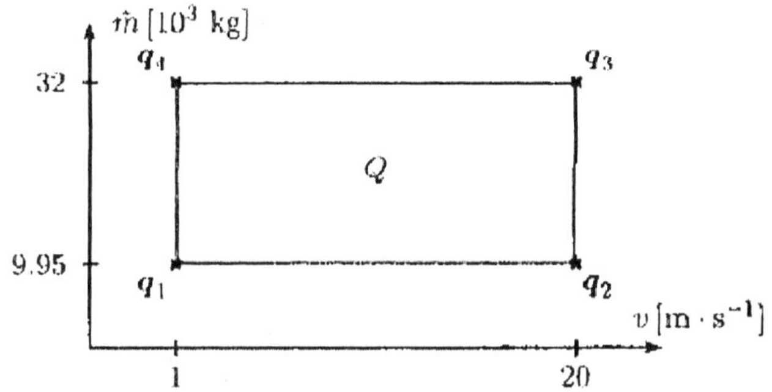

Fig.3.10 Operating domain (possible parameter variation range) Q of the city bus model [92]. (© [1995] IEEE)

Suppose the parameter vector is denoted by h, which consist of uncertain plant parameters q, like v and m in current system. And all the eigenvalues of the original system are located in the admissible region Γ, which is called Γ-stable in several literals.

Generally, the procedure for determining the stability regions in h space for Hurwitz-stability can be summarized as follows. First, the real and imaginary parts of the characteristic polynomial at $s = j\omega$ are

$$p(j\omega,h) = \text{Re } p(\omega,h) + j \text{ Im } p(\omega,h) \tag{3.39}$$

where frequency ω is girded in the interval $\omega \in [0,\infty)$ and for each grid point $\omega = \omega^*$, the simultaneous solution of the following two equations are calculated.

$$\text{Re } p(\omega^*,h) = 0 \tag{3.40}$$

and

$$\text{Im } p(\omega^*,h) = 0 \tag{3.41}$$

For each of the four extremal plants, the boundary $\partial\Gamma$ of the region Γ in the complex eigenvalue plane will be mapped into controller coefficient space resulting as four Γ-stabilizing regions. Their intersection is the set of Γ-stabilizers for the four extremal plants.

The Γ-stability boundaries divide the parameter plane into a finite number of separated regions. By selecting one arbitrary point from each region, the set of Γ-stabilizing controllers is determined. if one point of a region turns out to be, then the entire region is, and vice versa.

Based on (3.38), the initial compensator transfer function is chosen as

$$f_c(s) = 40^3 \frac{0.27s^2 + 1.3s + 1.9 + 0.75/s}{\left(s^2 + 2 \cdot 0.6 \cdot 40 \cdot s + 40^2\right)\left(s + 40\right)} \tag{3.42}$$

Assume $k_p = 1.9$, $k_I = 0.75$, $D = 0.6$ and $\omega_c = 40$ kept unchanged in (3.38), and vary k_D and k_{DD} only. The associated -stability boundaries for operating point q_3 is shown in Fig.3.11. The compensator (3.42) that stabilizes the given operating domain Q can be determined. For instance, the point (1.3,0.27) marked with a cross in Fig.3.11 is Γ-stable and, hence, all controller parameters from the entire region indicated by the cross Γ-stabilize the plant.

But the compensator (3.42) does not meet all specifications mentioned above. Therefore, Ackermann et. al investigated in [92] how to further improve it by small modification of the remaining parameters k_p, k_I and D. The desired direction in this three-dimensional subspace is to let the set of Γ-stabilizing controllers grows. Once the largest set is determined, the controller parameter ω_c is increased stepwise and in each step the free parameters k_p, k_I and D are again turned such that the set of simultaneously Γ-stabilizing controllers gets as large as possible.

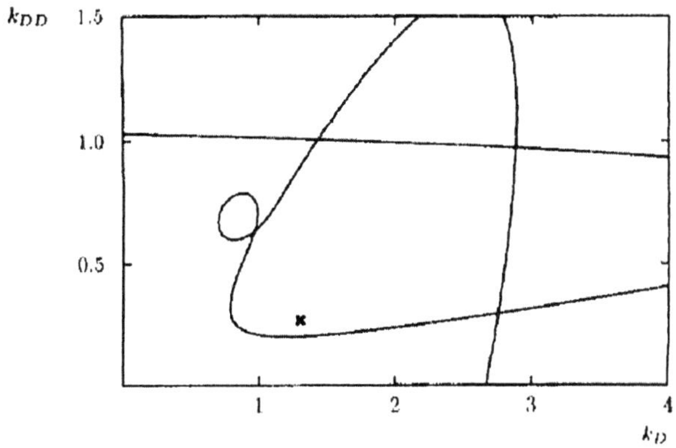

Fig.3.11 Set of Γ -stability controller for operation point q_3 [92]. (© [1995] IEEE)

After each iteration, several selected controllers from the simultaneously Γ -stabilizing set are tested in simulations and the specifications are checked. If there is no significant rise in the simulation, we stop searching the parameters.

In the simulation, when switch transition from manual to automatic operation, the bus is assumed to drive in a distance of $y = 0.15m$ parallel to the guideline. The side wind force is simulated as a gust which attacks the bus from the lateral direction. It is assumed that the wind velocity increases with $T = 0.5s$, $v_w = (1 - e^{-t/T})20m/s$. And the reference input ρ_{ref} is shown in Fig.3.12. The admissible velocity should be $2.5mg^{-1}$ or higher.

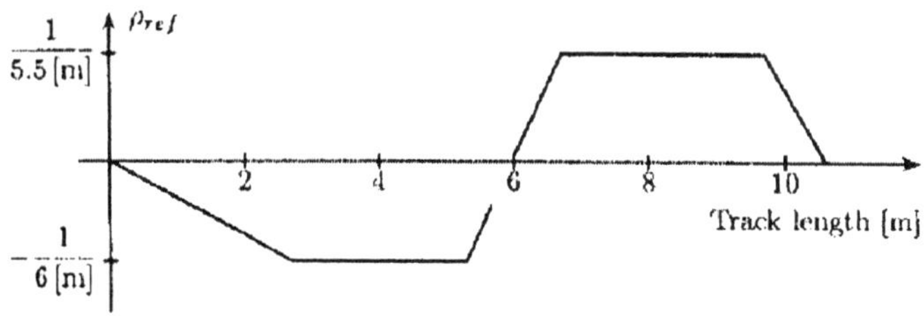

Fig.3.12 Simulated reference input ρ_{ref} for entering a bus stop bay [92]. (© [1995] IEEE)

Fig.3.13 shows a wide range in the parameters k_D and k_{DD} that had been obtained in searching. Finally, the controller was chosen as

$$f_c'(s) = 100^3 \frac{0.6s^2 + 13s + 10 + 3/s}{\left(s^2 + 2 \cdot 100 \cdot s + 100^2\right)\left(s + 100\right)} \tag{3.43}$$

Simulation shows that this new compensator can satisfy all the requirements and yield better performance than the previous one.

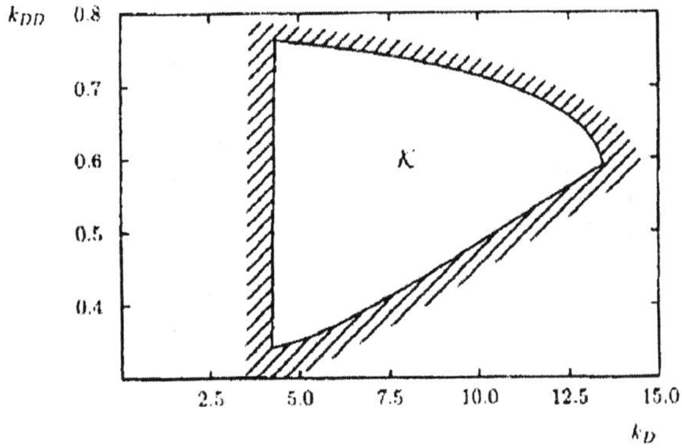

Fig.3.13 Set of Γ -stability of the controller (3.44) at $\omega_c = 100$, $D = 0.5$, $k_I = 3$, and $k_p = 10$ [92]. (© [1995] IEEE)

In the last 90's, H_∞ robust analysis method was introduced into steering controller design field to reduce the unexpected effect of wind disturbance. H_∞ theory is constructed to handle the deterministic disturbance model consisting of bounded energy (square-integrable) L_2 signals and allows controller design for narrow-band disturbance rejection, see Francis 1987 [96] and Zames 1981 [97]. In [98], Guvenc, Bunte and Odenthal etc. designed a disturbance observer, whose model regulation capability allows the specification and achievement of desired yaw dynamics. The proposed integrated control model is shown in Fig.3.14, where G_{ref} was the transfer function from disturbance torque ρ_{ref} to yaw rate r . $\tilde{G}_u$ is the nominal system model and G_u is the un-modeled dynamics.

It was shown that the model regulation and disturbance rejection property of this proposed observer can be considered as a special H_∞ loop shaping for path following. The filtering effect is chosen to satisfy the classical loop shaping constraint

$$|W_s S| + |W_T T| < 1 \text{, for } \forall \omega \tag{3.44}$$

where W_s and W_T are the sensitivity function weight and the complementary sensitivity function weight respectively. S and T denote the sensitivity function and transfer function, respectively.

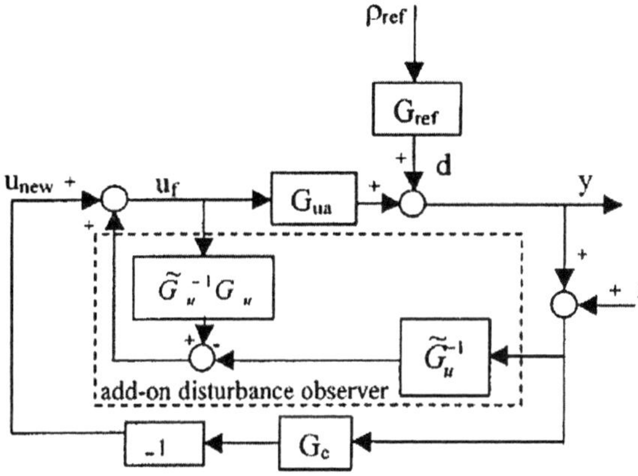

Fig.3.14 Diagram of system architecture with add-on disturbance observer [98]. (© [2004] IEEE)

In Fig.3.14, $G = G_n(1 + \Delta_m)$ is the single track yaw dynamics model with multiplicative uncertainty Δ_m, G_n is the nominal model or a desired vehicle yaw dynamics model. G_a is the auxiliary steering actuator model. r is the vehicle yaw rate and δ_s and δ_c are the steering commands coming from the driver via the steering wheel and the auxiliary steering angle coming from the steering controller, respectively. G_d is the transfer function from yaw disturbance torque M_d to yaw rate. Disturbance observer design requirements are specified in terms of the unity gain low pass filter Q which should be small at high frequencies for sensor noise attenuation and robustness of stability in the presence of high frequency un-modeled dynamics. At low frequencies, Q is usually chosen as unity for good steady state accuracy, disturbance rejection and model regulation. Then, due to the specific controller structure, the input-output behavior of the controlled system including its steady state behavior will be the same as that of the nominal (or desired) model G_n up to the bandwidth of the low pass filter Q. The low frequency design requirements are, however, remarkably different in the vehicle steering control application considered here where the driving task should be left to the drier at low frequencies (i.e. the fading effect). The result is that, contrary to standard disturbance observer design practice, a band-pass Q filter has to be designed as it illustrated in Fig.3.15.

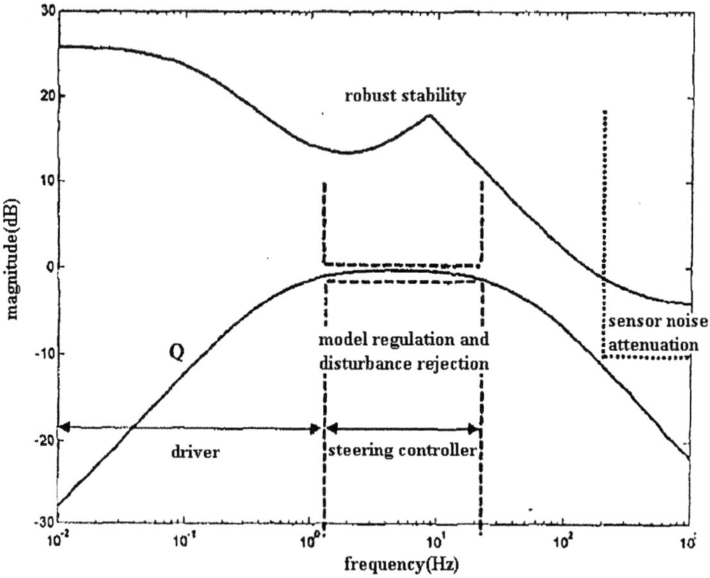

Fig.3.15 Diagram of Q filter design specifications.

The Q filter is chosen as a simple band pass filter as

$$Q(s) = \frac{\tau_{bp} s}{(\tau_{bp} s + 1)(\tau_Q s + 1)} \tag{3.45}$$

where the band of $Q(s)$ in (5.30) is between the frequencies $1/\tau_{bp}$ and $1/\tau_Q$ in Hz. More precisely τ_{bp} is chosen as 0.25sec.

The desired yaw dynamics model was set as a first order system here given by

$$G_n(s) = \frac{K_n(v)}{\tau_n s + 1} < 1 \tag{3.46}$$

where $K_n(v) = \lim_{s \to 0} G(s)\big|_{\mu=1}$.

Thus, the sensitivity function weight W_s and the complementary sensitivity function weight W_T used in (3.44) were assigned as

$$W_s(s) = \frac{0.3333\,s + 4.2}{1.8s + 1.26} \tag{3.47}$$

and

$$W_T(s) = \max\left\{\left|\frac{1.667\,s + 6.2833}{s + 188.5}\right|, \left|\frac{0.04268\,s^2 + 1.8977\,s + 0.90719}{s^2 + 9.006\,s + 17.6494}\right|\right\} \tag{3.48}$$

where the two free controller parameter τ_u and τ_Q are tuned in the design effort to meet the mixed sensitivity requirement is designed to penalize parametric uncertainty at low frequencies and un-modeled uncertainty at high frequencies.

Equivalently, Mammar etc. developed several two degree of freedom (2DOF) steering controller in [99]-[101] using the H_∞ loop shaping technique, Fig.3.16. The design task is directly assigned as finding a robust feedback control $\tilde{K}$ to guarantee the stabilizing of system and minimize energy bound of the transfer function from ρ_{ref} to the pre-selected measurement output z_1, z_2 and z_3. Based on the works of Kuzuya and Shin that was reported in [47], these robust 2DOF steering controller can be easily digital implemented.

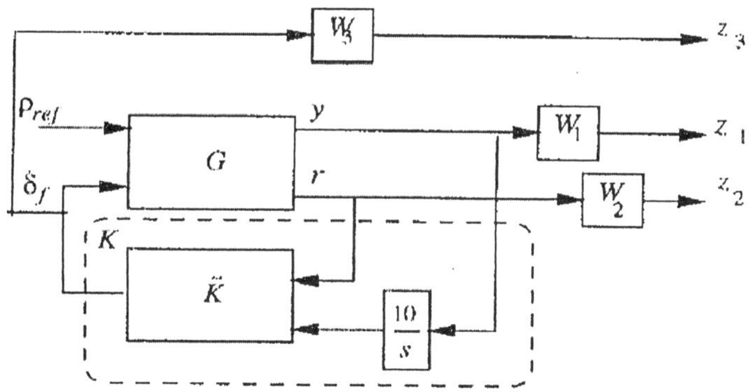

Fig.3.16 Diagram of two degree of freedom (2DOF) steering controller [99]. (© [1997] IEEE)

In [98] and [101], the saturation properties of steering actuator were studied. The actual feedback control architecture considering steering actuator rate limits. Simulations and experiments pointed out that the steering angle rate actuator saturation forms a major limitation of performance. In [101], the undesirable limit cycles caused by saturations were analyzed by a describing function approach in combination with the representation of limit-cycle-free regions in a parameter plane of velocity and road/tire friction coefficient.

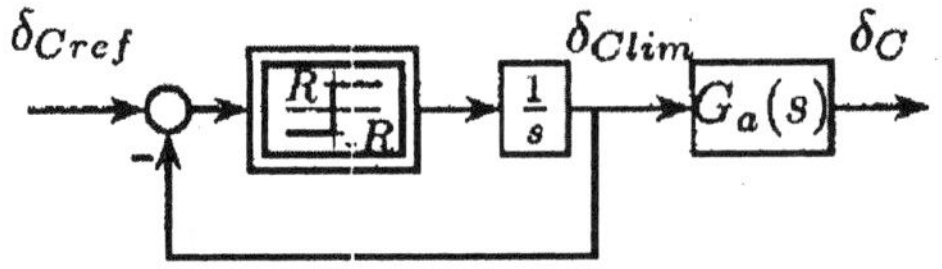

Fig.3.17 Feedback control architecture considering actuator saturation, where the reference input is denoted by δ_{cref} [101]. (© [1997] IEEE)

In their suggestion, a feedback signal δ_f in an active steering system should be entered as

$$\delta_f = \delta_L + \delta_c \tag{3.46}$$

The additional steering angle δ_c is assumed to be ideal in the first derivation of the ideal decoupling controller. And a more realistic actuator transfer function $G_a(s)$ is analyzed, i.e. we have

$$\delta_c = \frac{1}{s}(r_{ref}(s) - r(s))G_a(s) \tag{3.50}$$

where $r_{ref}(s) = F(s)\delta_L$ is generated from the hand wheel input δ_L via a pre-filter $F(s)$.

Further suppose the steering actuator is modeled as a linear dynamic part as the following (3.51), we can have system model

$$G_a(s) = \frac{\omega_a^2}{s^2 + \sqrt{2}\omega_a s + \omega_a^2} \tag{3.51}$$

The output of the rate limiter δ_{clim} is analyzed by describing function as

$$\left|\dot{\delta}_{\text{clim}}\right| \le R \tag{3.52}$$

Using the above actuator limit model, Ackermann and Bunte 1997 analyzed the uncertainty of the parameters v and μ, where v denotes the velocity of the vehicle and μ is the parameter used to normalized the vehicle mass m .

Based on (3.33), the modified actuator transfer function is written as

$$G_{\delta_f - r}(s) = \frac{r(s)}{\delta_f(s)} = \frac{\mu v(f_1 vs + f_0 \mu)}{e_2 v^2 s^2 + e_1 \mu vs + (e_{01} \mu^2 + e_{02} \mu v^2)} \quad (3.53)$$

where

$$f_0 = c_f c_r (l_f + l_r), \; f_1 = c_f m l_f, \; e_{01} = f_0 (l_f + l_r),$$

$$e_{02} = (c_r l_r - c_f l_f) m, \; e_1 = m(c_r l_r + c_f l_f)(l_f + l_r), \; e_2 = m^2 l_r l_f$$

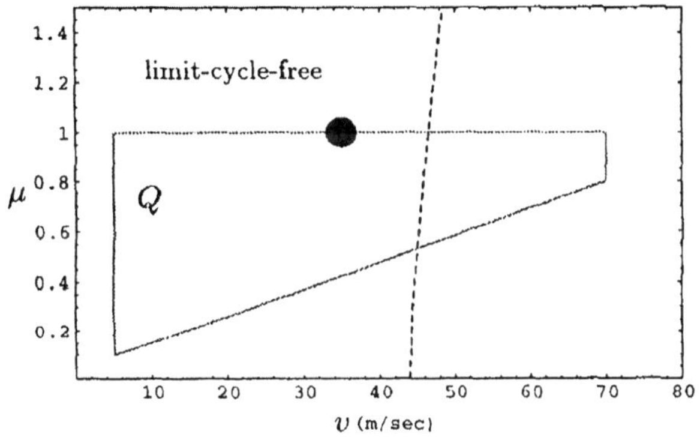

Fig.3.18 Stability boundaries as limit cycles, the black dot indicates the limit-cycle-free operating point, from [101]. (© [1997] IEEE)

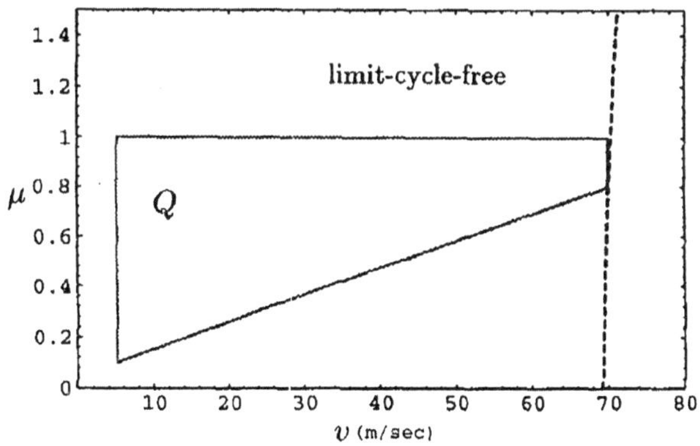

Fig.3.19 Stability boundaries as limit cycles, the black dot indicates the limit-cycle-free operating point, from [101]. (© [1997] IEEE)

Fig.3.18 shows the result for such an actuator with the bandwidth $\omega_a = 4\pi$ Hz. The gain condition boundary (dashed line style) intersects Q, i.e. not the entire operating range is limit-cycle-free. But if we increase the actuator bandwidth to $\omega_a = 6.3\pi$ Hz, Fig.3.19 shows that this bandwidth suffices to get Q limit-cycle-free.

However, in the steady state the feedback signal δ_c should vanish, such that with $\delta_f = \delta_L$ the steering angle authority is fully transferred back to the driver. Thus, an additional fade integrator is introduced to the system in [101], whose transfer function is written as

$$G_{FI}(s) = \frac{s}{s^2 + 3s + 1} \tag{3.54}$$

The actuator system is shown in Fig.3.20. With this implementation of the fading integrator the rate limiter never gets activated, see Fig.3.21.

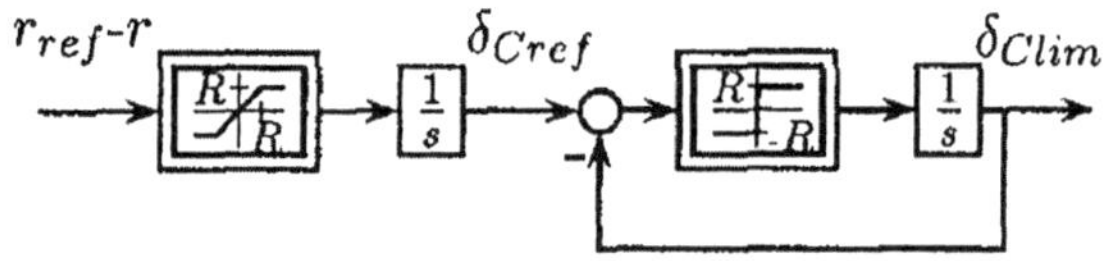

Fig.3.20 The new system dynamic model, form [101]. (© [1997] IEEE)

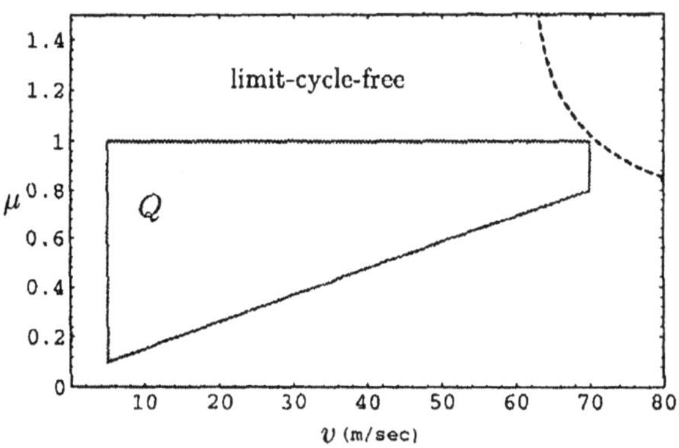

Fig.3.21 The reshaped operating range, from [101]. (© [1997] IEEE)

The results were formulated in terms of required actuator bandwidth that achieves robustness in the entire operating range. It turned out that the use of a fading integrator can reduce the required actuator bandwidth. Based on similar ideas, a compensator with high order was investigated in [98] to achieve better performance.

Some other approaches based on H_∞ theory can be found in [102]-[114]. Most of them were mainly devoted to find an optimal compromise point between steering performance and wind disturbance rejection. Specially, the effect of time delay that exists in steering actuator system is studied in [115]-[116]. Constrained by the length, no detailed discussions are given here.

2) Time Domain Linear Robust Steering Controllers

Recently, time domain robust design technique was used in vehicle lateral control too [117]-[119]. Although time domain robust design is intrinsically equivalent to frequency domain robust design, the obtained controllers differ in many aspects. A systematical discussion in this direction was presented in [120].

In general, the system (3.33) with disturbance $w \in C^{n_w}$, saturated control input and bounded state can be extended as

$$\begin{cases} \dot{x} = Ax + Bu' + Ew \\ z_\infty = C_\infty x \\ z_1 = C_1 x \\ z_u = u' = sat(Kx) \end{cases} \tag{3.55}$$

where $z_\infty \in C^{n_{z\infty}}$ is the H_∞-performance output, $z_1 \in C^{n_{z1}}$ is the L_1-performance output, $z_u \in C^{n_u}$ is the auxiliary performance output for bounded control input. C_∞ and C_1 are rid performance measurement matrices.

Notice that in Ineq.(3.9) and (3.10), the safe ride requirement on y_f and δ_f can be measured by choosing

$$C_1 = \begin{bmatrix} 0 & 0 & 0 & 1 & 0 \\ 0 & 0 & 0 & 0 & 1 \end{bmatrix} \text{ and } C_\infty = I \tag{3.56}$$

The actuator saturation limit on $\dot{\delta}_f$ can be hold by keeping

$$u' = sat(Kx) = \begin{cases} Kx, & |Kx| \le \dot{\delta}_{\max} \\ sign(Kx) * \dot{\delta}_{\max}, & |Kx| > \dot{\delta}_{\max} \end{cases} \tag{3.57}$$

Thus, the multi-objective design is to find a feedback control to achieve the following three objectives.

Design Task 3.III

1) guarantee the quadratic stability of system (9);

2) minimize the norm of the transfer function matrix from w to z_{∞} so as to reject disturbance;

3) keep bounded to satisfy the offset constraints.

In [121], Abedor, Nagpal and Poolla used the star-norm approach proposed in [122] to measure the actuator saturation. In [123]-[124], Nguyen and Jabbari further showed that the saturation limits could be explicitly taken into account by constraining the linear feedback within an ellipsoid reachable set determined by the upper bound of disturbance. This new method has the following main properties: controllers are designed so that actuators are used at or near capacities, and the guaranteed performance bound is a function of the actuator capacity. This method is similar to the H_{∞} loop shaping proposed in [98], since both of them tried to keep the linear feedback within saturation limits. However, L_1-norm is more suitable than H_{∞}-norm to describe the point-wise- in-time bound of actuator saturation and peek offset.

Extending the conclusions in [121]-[124], a mixed L_1/H_{∞} robust controller is formulated as below

$$\begin{cases} \dot{x} = Ax + Bu + Ew \\ z_{\infty} = C_{\infty}x \\ z_1 = C_1 x \\ z_u' = u = Kx \end{cases} \tag{3.58}$$

The design objective is reformulated to find a control $u = Kx$ to achieve the following four objectives

Design Task 3.IV

1)-3) same as 3.III.1)-I.3);

4) keep bounded to guarantee the inputs to remain less than or equal to the saturation limits.

Before entering the main part of this chapter, let's introduce two lemmas in which only one disturbance input is considered.

Lemma 3.1 [121]: Consider system $\begin{cases} \dot{\chi} = A_{\chi}\chi + E_{\chi}\omega \\ z_{\chi} = C_{\chi}\chi \end{cases}$ with A_{χ} stable. If there exists a constant symmetry matrix $Q_{\chi} > 0$, for some scalar $\alpha > 0$, such that

$$A_\chi Q_\chi + Q_\chi^T A_\chi + \alpha Q_\chi + \alpha^{-1} E_\chi E_\chi^T \le 0 \tag{3.59}$$

the reachable set is contained inside the ellipsoid $\{\chi : \chi^T Q_\chi^{-1} \chi \le \omega_{\max}^2\}$, and $\max_{t\ge 0} |z_\chi| \le \|C_\chi^T Q_\chi^{-1} C_\chi\|^{1/2}$.

Lemma 3.2 [120]: Consider system (3.58) with $z_{ui} = u_i$ for $i = 1,\ldots, n_u$. Given some desired level of performance $\gamma_\infty > 0$ and $\gamma_{u,i} > 0$ associated with each input, a observer-based state feedback controller

$$\begin{cases} \dot{\hat{x}} = A\hat{x} + Bu + LC(y - C\hat{x}) \\ u_i = -\dfrac{\lambda}{2} B_i^T Q^{-1} \hat{x} \end{cases} \tag{3.60}$$

guarantees the quadratic stability of (3.55) with L_2-gain γ_∞ from w to z_∞ and bounded control input if there exist a symmetry constant matrix $Q > 0$ and a scalar $\lambda > 0$ for some $\alpha > 0$ such that the following LMIs are feasible

$$\begin{bmatrix} AQ + QA^T - \lambda BB^T & E & QC_\infty^T \\ E^T & -\gamma_\infty^2 & 0 \\ C_\infty Q & 0 & -I \end{bmatrix} < 0 \tag{3.61}$$

and

$$\begin{bmatrix} AQ + QA^T + \alpha Q - \lambda BB^T & E \\ E^T & -\alpha \end{bmatrix} \le 0 \tag{3.62}$$

together with

$$\begin{bmatrix} 4Q & \lambda B_i \\ \lambda B_i^T & \gamma_{u,i}^2 \end{bmatrix} > 0 \tag{3.63}$$

where B_i is the ith column of the input matrix B.

Moreover, the control input is in the ellipsoid $\xi_F = \{x : x^T Q^{-1} x \le w_{\max}^2\}$, or equivalently $\|u_i\|_\infty \le \gamma_{u,i} w_{\max}$. Here L is an appropriately chosen observer matrix, whose design methods can be found in *Section 3.4*.

Since the L_1-norm of w is defined to be the supremum over all time of the two-norm of w at each time instant, the multiple disturbance inputs case can then be considered by simply over bounding w in Lemma 2 with $w \le \sqrt{n_w} w_{\max}$. These lemmas lead to our main results naturally. □

Theorem 3.2. Consider system (3.58) with given some desired performance level $\gamma_\infty > 0$, $\gamma_{u,i} > 0$ and $\gamma_{1,j} > 0$ associated with each input, the state feedback controller (3.60) guarantees the quadratic stability of (3.55) with L_2-gain γ_∞ from w to z_∞ and bounded control input if there exist a symmetry constant matrix $Q > 0$ and a scalar $\lambda > 0$ for some $\alpha > 0$ such that the LMIs (3.61)-(3.64) are feasible

$$\begin{bmatrix} Q & QC_{1,j}^T \\ C_{1,j}Q & \gamma_{1,j}^2 \end{bmatrix} > 0 \tag{3.64}$$

where $C_{1,j}$ is the jth row of matrix C_1.

Moreover, it guarantees

$$\left\|u_f\right\|_{\max} \leq \sqrt{2}\gamma_{uf} w_{\max} \tag{3.65}$$

and

$$\left\|y_f\right\|_{\max} = \left\|z_{1,1}\right\|_{\max} \leq \sqrt{2}\gamma_{1,1} w_{\max} \tag{3.66}$$

together with

$$\left\|\delta_f\right\|_{\max} = \left\|z_{1,2}\right\|_{\max} \leq \sqrt{2}\gamma_{1,2} w_{\max} \tag{3.67}$$

where $w_{\max} = \left\|\rho_{ref}\right\|_{\max}$.

Brief proof: LMI (3.61) is necessary and sufficient for the existence of a state feedback controller for system (3.58) which guarantees quadratic stability with

$$\int_0^{+\infty} z^T z dt \leq \gamma_\infty^2 \int_0^{+\infty} w^T w dt$$

Inequalities in (3.62) and (3.63) are the necessary and sufficient conditions for $\left\|u_i\right\|_\infty \leq \sqrt{n_w}\gamma_{u,i} w_{\max}$. Similarly to the proof for Lemma 2, the sufficiency of (3.64) for $\left\|z_j\right\|_\infty \leq \sqrt{n_w}\gamma_{1,j} w_{\max}$ can be directly obtained from (3.63) and Lemma 1.

The controller given in (3.60) is the simplest controller that can be obtained for system (3.55). Considering generic control feedback matrix will lead to a bilinear matrix inequality design problem that has not been thoroughly solved yet.

Moreover, the system uncertainties are mainly caused by vehicle mass and tire-road-friction variation. The steering model with these two uncertainties can be taken as a linear parameter-varying plant, where the robust control techniques proposed in [125]-[126] can be applied here. □

Considering system uncertainties, model (3.49) is rewritten as

$$\begin{cases} \dot{x} = A(\theta_t)x + Bu + E(\theta_t)w \\ z_\infty = C_\infty x \\ z_1 = C_1 x \\ z'_u = u = Kx \end{cases} \tag{3.68}$$

where θ_t is a time-varying vector of parameters $\mu(t)$ and $1/m(t)$ such that

$$\mu(t) \in [\mu_{\min}, \mu_{\max}],\ 1/m(t) \in [1/m_{\max}, 1/m_{\min}] \tag{3.69}$$

It is apparent that each time-varying matrix in Eq.(24) locate in a certain matrix polytope, which is defined as the convex hull of a finite number of matrices with the same dimensions. For example,

$$A(\theta_t) \in Co\{A_1, A_2, \ldots, A_4\} = \left\{\sum_{k=1}^{4} \pi_k A_k, \pi_k \geq 0, \sum_{k=1}^{4} \pi_k = 1\right\} \tag{3.70}$$

where

$$A_1 = A(\mu_{\min}, 1/m_{\max}),\ \ldots,\ A_4 = A(\mu_{\max}, 1/m_{\min}) \tag{3.71}$$

Notice the convex property of matrix polytope, it leads to

Theorem 3.3. Consider system (3.68) with given some desired performance level $\gamma_\infty > 0$, $\gamma_{u,i} > 0$ and $\gamma_{1,j} > 0$ associated with each input, the state feedback controller (3.60) guarantees the quadratic stability of (3.68) with L_2-gain γ_∞ from w to z_∞ and bounded control input, if there exist a symmetry constant matrix $Q > 0$ and a scalar $\lambda > 0$ for some $\alpha > 0$ such that the following LMIs are all feasible

$$\begin{bmatrix} A_k Q + Q A_k^T - \lambda B B^T & E_k & Q C_\infty^T \\ E_k^T & -\gamma_\infty^2 & 0 \\ C_\infty Q & 0 & -I \end{bmatrix} < 0 \tag{3.72}$$

$$\begin{bmatrix} A_k Q + Q A_k^T - \lambda B B^T & E_k & Q C_\infty^T \\ E_k^T & -\gamma_\infty^2 & 0 \\ C_\infty Q & 0 & -I \end{bmatrix} < 0 \tag{3.73}$$

Moreover, the safety constraints (3.65)-(3.67) still holds.

Brief proof: Notice that

$$\begin{bmatrix} A(\theta_t)Q + QA^T(\theta_t) - \lambda BB^T & E(\theta_t) & QC_\infty^T \\ E^T(\theta_t) & -\gamma_\infty^2 & 0 \\ C_\infty Q & 0 & -I \end{bmatrix}$$

$$= \sum_{k=1}^{4} \pi_k(t) \begin{bmatrix} A_k Q + QA_k^T - \lambda BB^T & E_k & QC_\infty^T \\ E_k^T & -\gamma_\infty^2 & 0 \\ C_\infty Q & 0 & -I \end{bmatrix} < 0$$

and

$$\begin{bmatrix} A(\theta_t)Q + QA^T(\theta_t) + \alpha Q - \lambda BB^T & E(\theta_t) \\ E^T(\theta_t) & -\alpha \end{bmatrix}$$

$$= \sum_{k=1}^{4} \pi_k(t) \begin{bmatrix} A_k Q + QA_k^T + \alpha Q - \lambda BB^T & E_k \\ E_k^T & -\alpha \end{bmatrix} \le 0$$

The collusion follows Theorem 1 directly. □

To demonstrate the feasibility of the proposed controller, simulation tests were carried out in [120]. Suppose the vehicle is front steering and only the front vision sensor is applied. By solving Design Task I, the feedback matrix is obtained by choosing $\gamma_{u,1} = 0.4$, $\gamma_{1,1} = 0.2$ as

$$K = \begin{bmatrix} -1.2797 & -0.5277 & -1.4316 & -1.3468 \end{bmatrix}^T$$

Calculation shows that the maximum displacement can be kept less than $\pm 0.15\text{m}$ and the control input is kept within $\pm 0.3\text{rad}$. A typical simulation example is shown below, where Fig.3.22 shows wind disturbance, road curvature variation and control feedback, respectively; and Fig.3.23 displays the variation of β, r, ψ and y_f, respectively. Here v=15m/s.

Different from [120], Mobus and Zomotor considered the actuator saturation as a multi-parametric H_∞ Quadratic Program with linear constraints in [127]. Using the iterative solving methods proposed in [128]-[129], they obtained an optimal switch controller. There were also some other approaches using H_∞ design techniques. For instance, nonlinear H_∞ controller were proposed in [130]. However, they are neglected here, since to examine such approaches requires too much preliminary discussions.

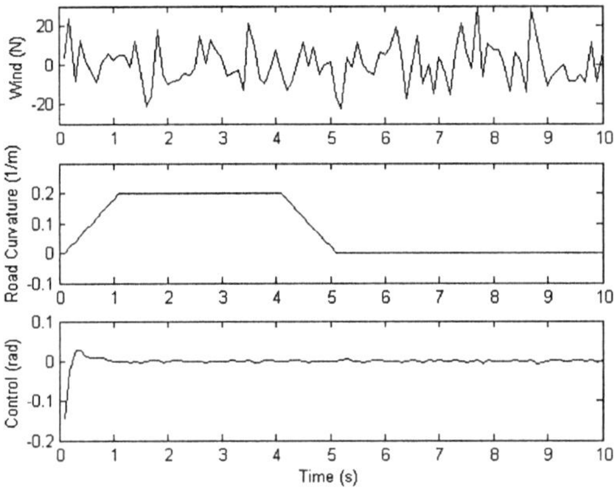

Fig.3.22 Wind disturbance, road curvature profile and control used in Fig.3.23 below, from [120]. (© [2005] IEEE)

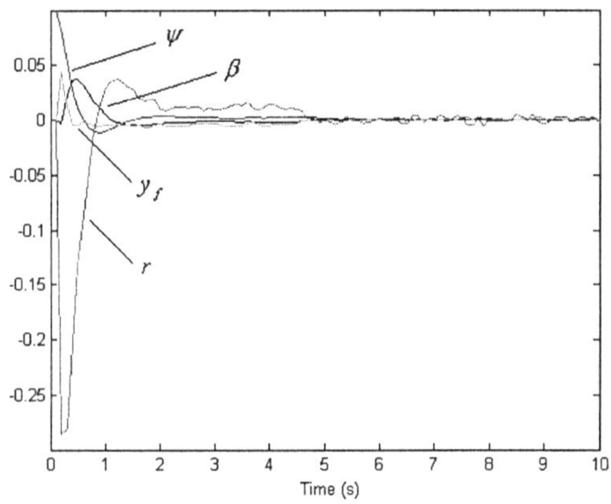

Fig.3.23 System state output under disturbances, from [120]. (© [2005] IEEE)

3.5.3 Sliding Mode Steering Controllers

Sliding mode steering controller is another frequently used approach in vehicle lateral motion control. Generally speaking, the basic idea of sliding mode control is to restrict the state space trajectories of the dynamic system to a manifold called "sliding manifold" which is usually denoted by

$S = 0$. This is achieved by directing the system trajectories towards this manifold "from both sides". In [92], the sliding manifold was chosen as

$$S = c\Delta r + \Delta \dot{r} \tag{3.74}$$

where $c > 0$ is a constant gain that determines system behavior once the motion of system (3.74) has been restricted to the neighborhood of the manifold $S = 0$.

The structure of the proposed controller was shown in Fig.3.24, in which the system ideal feedback control strategy is written as

$$r_d = -\frac{1}{l_s}\left[v(\beta + \psi) + Ky\right] \tag{3.75}$$

where $K > 0$ determines the desired rate of decay of y.

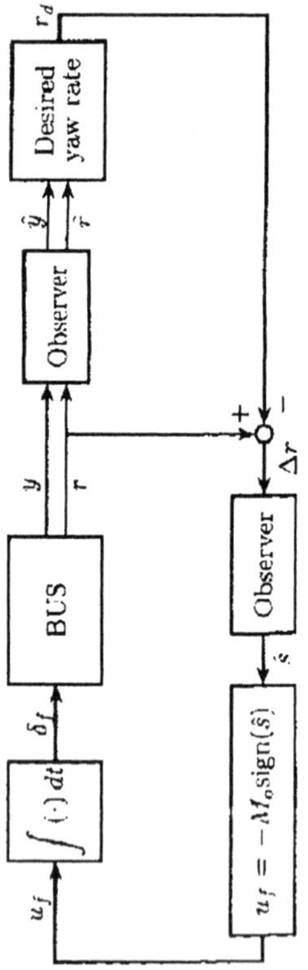

Fig.3.24 Diagram of a typical sliding mode steering controller [92]. (© [1995] IEEE)

Notice both the states β and ψ are unknown, the following observer (3.76)- (3.77) was introduced for estimating these two state variables

$$\dot{\hat{y}} = \hat{q} + l_s r + l_1 \bar{y}, \; l_1 > 0 \tag{3.76}$$

and

$$\dot{\hat{q}} = l_2 \bar{y}, \; l_2 > 0 \tag{3.77}$$

Thus the actual feedback control was obtained as

$$r_d = -\frac{1}{l_s}[\hat{q} + K\hat{y}] \tag{3.78}$$

and the feedback control was chosen as

$$u_f = -M_u \text{sign}(S) \tag{3.79}$$

where M_u is the available steering angle rate.

To further improve passenger comfort, it was proved to be advantageous to replace linear term $K\hat{y}$ in (3.75) by a saturation function

$$r_d = -\frac{1}{l_s}\left[\hat{q} + \lambda \frac{\hat{y}}{\sqrt{\hat{y}^2 + \varepsilon}}\right], \; \lambda > 0, \; \varepsilon > 0 \tag{3.80}$$

and the feedback control can be substituted by a continuous approximation as

$$u_f = -M_u \frac{S}{\sqrt{S^2 + 0.0001}} \tag{3.81}$$

Comparing to the above linear controllers, it was proven in [92] that this sliding mode controller yields smaller deviations from the guideline and it has a more oscillatory behavior that shows up particularly in the lateral acceleration and in the steering angle rate, but not in the derivations from the guideline. Regarding settling times, there are no significant differences between the two controllers.

There were several other sliding mode steering controllers which chose different sliding manifold [131]-[135]. [131] studied the lateral and longitudinal control of vehicle using a PID typed sliding surfaces, whose stability was proven using Lyapunov theory. In [132], a velocity related sliding mode controller was proposed to deal the input couple problem. Another special sliding mode integral action controller and the corresponding sliding mode observer are used to enhance vehicle stability in a split-μ maneuver in [51]. In [133], it was shown that the sliding mode controller can also be used to deal with the nonlinear front steering model considering track force control. Moreover, sliding mode steering controller is also an important approach in tractor-trailer vehicles lateral motion control [134].

3.5.3 Adaptive Steering Controllers

Because of its capability of handling model uncertainty and parameter variation, adaptive steering controllers achieve continuous interest in the last twenty years. Generally, there are two different methods: model concentrated approaches [136]-[141] and non-model concentrated approaches [142]-[145].

In [115], Brennan and Alleyne proposed a steering controller based on model reference control (MRC) with a modification based on rejection of known disturbance dynamics. Model reference control was initialed by Astrom and Wittenmark in 1997 [137]. This method was shown in [136] to be effective for steering control systems consist of dynamic uncertainty and disturbance. In [138], a special adaptive observer was proposed to deal with system parameters variations. In [118], an adaptive rule was proposed to make the controller flexible with velocity change. However, how to keep the balance between steering control precision and cost of implementation facility still needs to further discussion.

In [142]-[144], the concepts of Selected Adaptive Critic (AC) and Dual Heuristic Programming (DHP) were used to design steering controller. Selected adaptive critic methods are known to be capable of designing (approximately) optimal control policies for non-linear plants (in the sense of approximating Bellman Dynamic Programming). The present research focuses on an AC method known as dual heuristic programming. There were lots of issues related to the pragmatics of successfully applying the AC methods.

In [142], a straight forward utility function to meet these requirements would take the following form:

$$\delta_f = -Ay_{f_{error}}^2 - B\dot{y}_{f_{error}}^2 - Cv_{y_{error}}^2 - D\dot{v}_{y_{error}}^2 \tag{3.82}$$

where A, B, C and D were determined from programming to indicate the human designer's judgment about the relative importance of each term, according to desired plant response characteristics (e.g., the derivative terms encourage more "damped" responses). In [79], it was further shown that DHP can be employed to optimize fuzzy steering controllers. In [146], reinforcement learning method was used to supervise vehicle lateral motion controller.

3.5.4 Fuzzy Steering Controllers

Fuzzy set theory and Fuzzy inference was first presented by Zadeh in 1965. Recently, some new approaches take advantage of Fuzzy inference to avoid addressing complex vehicle dynamics. Moreover, these approaches were proven to able to incorporate and utilize human steering skills to improve the automatic driving performance [147]-[157].

For example, a direct Fuzzy control strategy was proposed by Brown and Hung 1994 for the above 4WS car model in [148]. The corresponding Fuzzy control rule is something as

"IF

Yaw_Rate_Error is Negative_Large AND
Front_Slip is Negative_Large

THEN

Command_Front_Steering_Angke is Positive_Large"

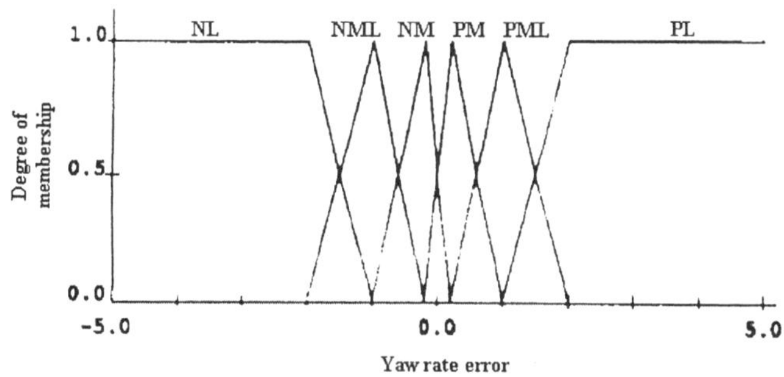

Fig.3.25 Yaw rate Fuzzy membership function, from [148]. (© [1994] IEEE)

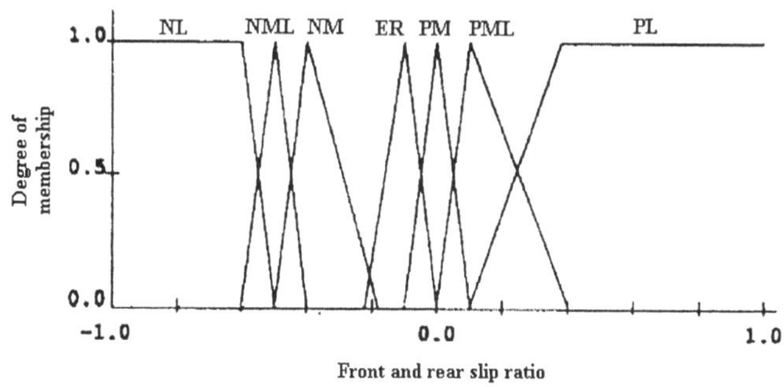

Fig.3.26 Front and rear slip ratio's Fuzzy membership function, from [148]. (© [1994] IEEE)

The front steering command and the rear steering command were designed to cooperate appropriately to avoid skipping. For example in [77], a direct Fuzzy control strategy was proposed by Brown and Hung. The proposed corresponding The complete Fuzzy control rules were shown in Fig.3.27.

		Fuzzy Input					
		NL	NML	NM	PM	PML	PL
Front Slip Fuzzy Input	NL	PL	PL	PM	NM	NML	NL
	NML	PML	PML	PM	NM	NML	NML
	NM	PM	PM	PM	NM	NM	NM
	ZR	NM	NM	NM	PM	PM	PM
	PM	NM	NM	NM	PM	PM	PM
	PML	NM	NM	NM	PM	PM	PM
	PL	NML	NML	NM	PM	PM	PML

		Fuzzy Input					
		NL	NML	NM	PM	PML	PL
Rear Slip Fuzzy Input	NL	NL	NL	NM	PM	PML	PL
	NML	NML	NML	NM	PM	PML	PML
	NM	NM	NM	NM	PM	PM	PM
	ZR	PM	PM	PM	NM	NM	NM
	PM	PM	PM	PM	NM	NM	NM
	PML	PM	PM	PM	NM	NM	NM
	PL	PML	PML	PM	NM	NM	NML

Fig.3.27 Rule table for front steering compensation and rear steering compensation, from [148]. (© [1994] IEEE)

From the rule table, we can see that the front steering command and the rear steering command were designed to cooperate appropriately in order to avoid skipping. Brown and Hung also claimed in [148] that the 4WS car using this Fuzzy controller was quite robust to wind and road perturbations.

There were several other directly fuzzy controllers reported in [150]-[154]. For instance, Cai, Rad and Chan et. al 2003 proposed two robust fuzzy PD steering controllers for those 2Ws car that equips both front and rear sensors. The input of the first controller is the average of y_f and y_r, and the input of the other fuzzy controller is the difference of y_f and y_r. The first controller is used to keep the center of vehicle following the desired path, and the second controller to make the angle between centerline of vehicle and path tangent to zero. It was reported that the Fuzzy PD controller exhibit significant robustness than conventional PD controller, especially when it was optimized by GA. But it should be remarked that, in spite of the well functioning of these fuzzy controllers, they were still heuristic controllers. Since no stability proof had was presented, and then they were less reliable.

In some recent literals, the proposed fuzzy controllers were constructed as follows. First, an optimal steering controller was constructed for each local model, which indeed constructed a part of the fuzzy model. Then, local controllers were combined using fuzzy rules to form a fuzzy logic controller. Therefore, the performance of the fuzzy controller can be analyzed using linear matrix inequalities (LMI) or algebra Riccati equation (ARE).

In [155]-[157], these methods were shown to be effective by both theoretical analysis and simulations. In [157], the fuzzy describing function analysis and traditional frequency domain method are applied to determine the stability condition when steering system has perturbation or adjustable parameters. And the stable boundary of the fuzzy controller can be finally determined with the obtained operating range similar to what had been discussed in report [92].

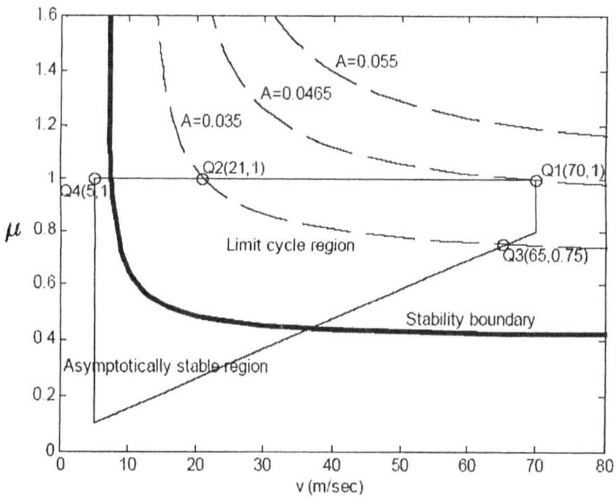

Fig.3.28 The designed operating range Q for μ [157]. (© [2005] IEEE)

There were some other special nonlinear steering controllers, i.e. in [158]-[161]. For example, preview control method was used in [158] to build steering controller. However, most such kind of approaches did not consider robustness and actuator saturation.

In [162], a simple proportional controller was compared with H_∞ robust controllers, fuzzy controller and adaptive controller. This proportional feedback was found to yield the largest offset with respect to other controller, although it was only slightly affected by the wind force. On the other hand, the self-tuning regulator presents the smallest errors. The responses of H_∞ and fuzzy controllers are comparable in most tests regarding to self-tuning regulator. Although these conclusions were made for special controllers, they were widely expected to be true in general situations.

3.6 Summary

The recent trends of research on development of vehicle steering control are reviewed in this chapter. The focus was on vehicle controller design techniques. Advances in steer-by-wire, vehicle dynamic modeling and observer design were also briefly investigated since they are tightly related to our main topic.

However, some important contents leave unmentioned in this chapter due to length limits and their premature. For instance,

(1) the relationship between vehicle longitudinal and lateral motion will be further analyzed in *Chapter 4* and *Chapter 6*; Especially how to guarantee ride safety when both steering and braking occurs is an important problem that will be carefully studied;

(2) the effect of steering sensor installment and measurement error needs further discussions [163];

(3) moreover, fault tolerant steering control device/algorithm and fault detection algorithms are archiving increasing considerations recently [164]-[176]. With rapidly increasing demands on driving safety, a boom in this research field is expected in the near future;

(4) Although great efforts had been put in steer-by-wire research, how to make intelligent steering controller cooperate appropriately with driver command still remains as a difficult problem to be fully conquered. Some previous works [177]-[192] revealed that this problem is quite complex, in which both drivers' characteristics, feelings, habits and driving status should be further analyzed.

For example, as pointed out in [186], guidance information is significant to vehicle steering, when the available information to the driver is restricted by the external situations. An example situation is the snowplow operation under white-out condition, in which it is extremely difficult for the driver to safely operate the vehicle. In such situations, a supplemental human-machine interface (HMI) that displays guidance information to assist the driver for safe vehicle handling may become an important tool to. However, how to appropriately notify driver about the desired steer action and the actual steer action needs to be solved. In [192], an attempt was presented as follows: the blue curves display the road boundaries on the screen. The green arrow in the right top corner indicates the optimal front steering angle, while the red arrow shows the actual front steering angle. Thus, the driver will try to adapt the right steering policy by making the red arrow overlap the green arrow. It is quite straightforward and drivers can easily get accustomed with it.

3.7 Reference

1. L. Li and F.-Y. Wang, "Research advances in vehicle lateral motion monitoring and control," International Journal of Intelligent Control and Systems, vol. 10, no. 1, pp. 60-76, 2005.
2. L. Li, F.-Y. Wang, and Q. Zhou, "Integrated longitudinal and lateral tire/road friction modeling and monitoring for vehicle motion control," IEEE Transactions on Intelligent Vehicle Systems, 2005.
3. K. Huh, C. Seo, and J. Kim, et. al, "Active steering control based on the estimated tire forces," Proceedings of American Control Conference, vol. 1, pp. 729-733, 1999.
4. V. Claesson, S. Poledna, and J. Soderberg, "The XBW model for dependable real-time systems," Proceedings of International Conference on Parallel and Distributed Systems, pp. 130-138, 1998.
5. W. Harter, W. Pfeiffer, and P. Dominke, et. al, "Future electrical steering systems. realizations with safety requirements," SAE#2oW-01-0822, 2000.
6. R. Hayama, K. Nishizaki, and S. Nakano, et. al, "The vehicle stability control responsibility improvement using steer-by-wire," Proceedings of IEEE Intelligent Vehicles Symposium, pp. 596-601, 2000.
7. P. Yih, J. Ryu, and J. C. Gerdes, "Modification of vehicle handling characteristics via steer-by-wire," Proceedings of American Control Conference, vol. 3, pp. 2578-2583, 2003.
8. M. Segawa, S. Kimura, and T. Kada, et. al, "A study on the relationship between vehicle behavior and steering wheel torque on steer by wire vehicles," Vehicle System Dynamics Supplement, vol. 41 pp. 202-211, 2004.
9. M. Nagai, H. Mouri, and P. Paksincharoensak, "Vehicle lane-tracking control with steering torque input," Vehicle System Dynamics Supplement, vol. 37, pp. 267-278, 2002.
10. R. Cortesao and N. Bajcinca, "Model-matching control for steer-by-wire vehicles with under-actuated structure," Proceedings of IEEE/RSJ International Conference on Intelligent Robots and Systems, vol. 2, pp. 1148-1153, 2004.
11. L. Dacko, R. F. Darlington, and D. Shindman, "Landing gear steer-by-wire control system; digital vs. analog study," Proceedings of Annual Reliability and Maintainability Symposium, pp. 215-220, 1990.

12. K. Inoue, J. Yoshitsugu, and S. Shirogane, et. al, "DC brush-less servo motor drive systems using automatic learning control-based auto gain parameter tuning scheme," Proceedings of International Conference on Industrial Electronics, Control and Instrumentation, vol. 3, pp. 1006-1011, 1997.
13. I. Camuffo, G. Caviasso, and L. Pascali, et. al, "Simulation tools and evaluation criteria for steering wheel feel improvement of an electric power steering system," SAE #2002-01-1593, 2002.
14. F. Rodriguez, E. Uy, and A. Emadi, "Brushless DC motor drive for steer-by-wire and electric power steering applications," Proceedings of Electrical Insulation Conference and Electrical Manufacturing & Coil Winding Technology Conference, pp. 535-541, 2003.
15. P. Setlur, J. Wagner, and D. Dawson, et. al, "A hardware-in-the-loop and virtual reality test environment for steer-by-wire system evaluations," Proceedings of American Control Conference, vol. 3, pp. 2584-2589, 2003.
16. C. Wilwert, Y. Song, and F. Simonot-Lion, et. al, "Evaluating quality of service and behavioral reliability of steer-by-wire systems," Proceedings of IEEE Conference on Emerging Technologies and Factory Automation, vol. 1, pp. 193-200, 2003.
17. J. Jarlmark, "Modeling and validation of steering system response to road and driver induced forces," Vehicle System Dynamics Supplement, vol. 41, pp. 371-380, 2004.
18. S. Amberkar, F. Bolourchi, and J. Demerly, et. al, "A control system methodology for steer by wire systems," SAE #2004-01-1106, 2004.
19. M. Tai, P. Hingwe, and M. Tomizuka, "Modeling and control of steering system of heavy vehicles for automated highway systems," IEEE/ASME Transactions on Mechatronics, vol. 9, no. 4, pp. 609-618, 2004.
20. J. Langenwalter, "Embedded automotive system development process - steer-by-wire system," Proceedings of Design, Automation and Test in Europe, vol. 1, pp. 538-539, 2005.
21. J. Ogando, "A steer-by-wire wish list, reliable hardware, better software tools must come first," Design News, 2003.
22. J. Post, and E. Law, "Modeling, characterization and simulation of automobile power steering systems for the prediction of on-center handling," SAE#960178, 1996.
23. S. Mammar, V. B. Baghdassarian, and L. Nouveliere, "Speed scheduled vehicle lateral control," Proceedings of IEEE/IEEJ/JSAI International Conference on Intelligent Transportation Systems, pp. 80-85, 1999.
24. T. Wong, "Hydraulic power steering system design and optimization simulation," SAE#2001-01-0479, 2001.
25. W. Harter, W. Pfeiffer, and P. Dominke, et. al, "Future electrical steering systems. realizations with safety requirements," SAE#2000-01-0822, 2000.
26. S.-H. Jang, T.-J. Park, and C.-S. Han, "A control of vehicle using steer-by-wire system with hardware-in-the-loop-simulation system," Proceed-

ings of IEEE/ASME International Conference on Advanced Intelligent Mechatronics, vol.1, pp. 389-394, 2003.

27. P.-S. Huang, H. Smakman, and J. Guldner, "Control concepts for lateral vehicle guidance including HMI properties," Proceedings of IEEE International Conference on Systems, Man and Cybernetics, vol. 1, pp. 1-6, 2004.
28. F.-Y. Wang, L. Li and P. Li, "Adaptive vehicle driving control and driver assistance using Neuro-Fuzzy networks," International Journal of Intelligent Control and Systems, 2004.
29. B. Tongue, "Two brains, one car - actively controlled steering," IEEE Control Systems Magazine, vol. 25, no. 5, pp. 14-16, 2005.
30. M. Segel, "Theoretical prediction and experimental substantiation of the response of the automobile to steering control," Proceedings of Automobile division of the institute of mechanical engineers, vol. 7, pp. 310–330, 1956.
31. J. Kasselmann and T. Keranen, "Adaptive steering," Bendix Technical Journal, vol. 2, pp. 26-35, 1969.
32. R. E. Fenton, G. C. Melocik, and K. W. Olson, "On the steering of automated vehicles: theory and experiment," IEEE Transactions on Automatic Control, vol. 21, pp. 306–315, 1976.
33. J. Ackermann, "Robust car steering by yaw rate control," Proceedings of the 29th IEEE Conference on Decision and Control, pp. 2033-2034, 1990.
34. J. Ackermann and W. Dareberg, "Automatic track control of a city bus," IFAC Theory Report on Benchmark Problems for Control Systems Design, 1990.
35. J. Ackermann and W. Sienel, "robust automatic steering of a bus," Proceedings of European Control Conference, pp. 1534-1539, 1993.
36. J. Ackermann and W. Sienel, "Robust yaw damping of cars with front and rear wheel steering," IEEE Transactions on Control Systems Technology, vol. 1, no. 1, pp. 15-20, 1993.
37. E. Ono, S. Hosoe, and D. Tuan, et. al, "Bifurcation in vehicle dynamics and robust front wheel steering control," IEEE Transactions on Control Systems Technology, vol. 6, no. 3, pp. 412-420, 1998.
38. D.-C. Liaw, H.-H. Chiang, and T.-T. Lee, "A bifurcation study of vehicle's steering dynamics," Proceedings of IEEE Intelligent Vehicles Symposium, pp. 388-393, 2005.
39. J. Stephant, A. Charara, and D. Meizel, "Virtual sensor: application to vehicle sideslip angle and transversal forces, " IEEE Transactions on Industrial Electronics, vol. 51, no. 2, pp. 278-289, 2004.
40. Y. Furukawa, N. Yuhara, and S. Sano, et. al, "A review of four-wheel steering studies from the viewpoint of vehicle dynamics and control," Vehicle System Dynamics, vol. 18, pp. 151-186, 1989.
41. C. B. Winkler, "Simplified analysis of the steady-state turning of complex vehicles," Vehicle System Dynamics, vol. 29, pp. 141-180, 1998.

42. S. Takezono, H. Minamoto and K. Tao, "Two-dimensional motion of four-wheel vehicles," Vehicle System Dynamics, vol. 32, pp. 441-458, 1999.
43. H. Kuzuya and S. Shin, "Development of robust motor servo control for rear steering actuator based on two-degree-of-freedom control system," Mechatronics, vol. 10, no. 1-2, pp. 53-66, 2000.
44. P. Raksincharoensak, M. Nagai, and H. Mouri, "Investigation of automatic path tracking control using four-wheel steering vehicle," Proceedings of the IEEE International Vehicle Electronics Conference, pp. 73-77, 2001.
45. S. Mammar and D. Koenig, "Vehicle handling improvement by active steering," Vehicle System Dynamics, vol. 38, no. 3, pp. 211-242, 2002.
46. J. Ackermann, "Robust decoupling, ideal steering dynamics, and yaw stabilization of 4WS cars," Automatica, vol. 30, pp. 1761-1768, 1994.
47. J. Ackermann, "Robust decoupling of car steering dynamics with arbitrary mass distribution," Proceedings of American Control Conference, vol. 2, pp. 1964-1968, 1994.
48. A. Alloum, "Modelisation et commande dynamique d'un vehicule pour la securite de conduite," Ph.D. Thesis, U.T.C., Compiegne, France, 1994.
49. R. Outbib and A. Rachid, "Control of vehicle speed. a nonlinear approach," Proceedings of IEEE Conference on Decision and Control, vol. 1, pp. 462-463, 2000.
50. F. Tahami, S. Farhangi, and R. Kazemi, "A Fuzzy logic direct yaw-moment control system for all-wheel-drive electric vehicles," Vehicle System Dynamics, vol. 41, no. 3, pp. 203-221, 2004.
51. R. G. Hebden, C. Edwards, and S. K. Spurgeon, "Automotive steering control in a split-μ maneuver using an observer-based sliding mode controller," Vehicle System Dynamics, vol. 41, no. 3, pp. 181-202, 2004.
52. H. Lee and M. Tomizuka, "Coordinated longitudinal and lateral motion control of vehicles for IVHS," ASME Journal of Dynamic Systems, Measurement, and Control, vol. 123, pp. 535-543, 2001.
53. J.-C. Hsu and M. Tomizuka, "Analyses of vision-based lateral control for automated highway system," Vehicle System Dynamics, vol. 30, pp. 345-373, 1998.
54. Y. Narita, S. Katahara,and M. Aoki, "Lateral position detection using side looking line sensor cameras," Proceedings of IEEE Intelligent Vehicles Symposium, pp. 271-275, 2003.
55. J. Guldner, H.-S. Tan and S. Patwardhan, "Analysis of automatic steering control for highway vehicle with look-down lateral reference systems," Vehicle System Dynamics, vol. 26, no. 4, pp. 243-269, 1996.
56. H.-S. Tan, J. Guldner, and S. Patwardhan, et. al, "Development of an automated steering vehicle based on roadway magnets -a case study of mechatronic system design," IEEE/ASME Transaction on Mechatronics, vol. 4, no. 3, pp. 258–272, 1999.

57. S. L. Miller, B. Youngberg, and A. Millie, et. al, "Calculating longitudinal wheel slip and tire parameters using GPS velocity," Proceedings of American Control Conference, vol. 3, pp. 1800-1805, 2001.
58. D. M. Bevly, R. Sheridan, and J. C. Gerdes, "Integrating INS sensors with GPS velocity measurements for continuous estimation of vehicle sideslip and tire cornering stiffness," Proceedings of American Control Conference, vol. 1, pp. 25-30, 2001.
59. C.-Y. Chan, "Magnetic sensing as a position reference system for ground vehicle control," IEEE Transactions on Instrumentation and Measurement, vol. 51, no. 1, pp. 43-52, 2002.
60. S. M. Donecker, T. A. Lasky, and B. Ravani, "A mechatronic sensing system for vehicle guidance and control," IEEE/ASME Transactions on Mechatronics, vol. 8, no. 4, pp. 500-510, 2003.
61. J. I. Hernandez and C.-Y. Kuo, "Steering control of automated vehicles using absolute positioning GPS and magnetic markers," IEEE Transactions on Vehicular Technology, vol. 52, no. 1, pp. 150-161, 2003.
62. E. Dilge, M. Gulbins, and T. Ohnesorge, et. al, "On a redundant diversified steering angle sensor," Proceedings of IEEE On-Line Testing Symposium, pp. 191-196, 2003.
63. C. Doniselli, M. Gobbi, and G. Mastinu, "Measuring the inertia tensor of vehicles," Vehicle System Dynamics Supplement, vol. 37, pp. 301-313, 2002.
64. J. Ryu and J. C. Gerdes, "Integrating inertial sensors with Global position system (GPS) for vehicle dynamics control," ASME Journal of Dynamic Systems, Measurement, and Control, vol. 126, pp. 243-254, 2004.
65. R. McCann, L. R. Pujara, and J. Lieh, "Influence of motor drive parameters on the robust stability of electric power steering systems," Power Electronics in Transportation, pp. 103-108, 1998.
66. S. L. Miller, B. Youngberg, and A. Millie, et. al, "Calculating longitudinal wheel slip and tire parameters using GPS velocity," Proceedings of American Control Conference, vol. 3, pp. 1800-1805, 2001.
67. D. M. Bevly, R. Sheridan, and J. C. Gerdes, "Integrating INS sensors with GPS velocity measurements for continuous estimation of vehicle sideslip and tire cornering stiffness," Proceedings of American Control Conference, vol. 1, pp. 25-30, 2001.
68. S. Brennan, A. Alleyne, "Real-time identification of vehicle chassis dynamics using a novel reparameterization based on sensitivity invariance," International Journal of Adaptive Control and Signal Processing, vol. 18, no. 2, pp. 103-123, 2004.
69. U. Kiencke and A. Daiß, "Observation of lateral vehicle dynamics," Control Engineering Practice, vol. 5, no. 8, pp. 1145–1150, 1997.
70. U. Kiencke and L. Nielsen, Automotive Control System for Engines, Driveline and Vehicle, Berlin, Germany. Springer-Verlag, 2000.
71. P. J. T. Venhovens and K. Naab, "Vehicle dynamics estimation using Kalman filters," Vehicle Systems Dynamics, vol. 32, pp. 171–184, 1999.

72. C.-F. Lin, A. G. Ulsoy, and D. J. LeBlanc, "Vehicle dynamics and external disturbance estimation for vehicle path prediction," IEEE Transactions on Control Systems Technology, vol. 8, no. 3, pp. 508-518, 2000.
73. K. Huh, J. Kim, and K. Yi, et. al, "Monitoring system design for estimating the lateral tire force," Proceedings of American Control Conference, vol. 2, pp. 875-880, 2002.
74. J. R. Zhang, S. J. Xu, and A. Rachid, "Robust sliding mode observer for automatic steering of vehicles," Proceedings of IEEE Intelligent Transportation Systems, pp. 89-94, 2000.
75. W. Perruquetti and J.-P. Barbot, Sliding Mode Control in Engineering, New York. Marcel Dekker, 2002.
76. R. E. Fenton and I. Selim, "On the optimal design of an automotive lateral controller," IEEE Transactions on Vehicular Technology, vol. 37, no. 2, pp. 108-113, 1988.
77. J. Farrelly and P. Wellstead, "Estimation of vehicle lateral velocity," Proceedings of the 1996 IEEE International Conference on Control Applications, pp. 552-557, 1996.
78. J.-Y. Wang and M. Tomizuka, "Gain-scheduled H_∞ loop-shaping controller for automated guidance of tractor-semitrailer combination vehicles," Proceedings of American Control Conference, vol. 3, pp. 2033-2037, 2000.
79. S. Ibaraki and M. Tomizuka, "Taming internal dynamics by mismatched and H_∞-optimized state observer," Proceedings of American Control Conference, vol. 1, no. 6, pp. 715-719, 2000.
80. S. Ibaraki, S. Suryanarayanan, and M. Tomizuka, "H_∞ optimization of Luenberger state observers and its application to fault detection filter design," Proceedings of the 40th IEEE Conference on Decision and Control, vol. 2, pp. 1011-1016, 2001.
81. G. Lu and M. Tomizuka, "Vehicle lateral control with combined use of a laser scanning radar sensor and rear magnetometers," Proceedings of American Control Conference, vol. 5, pp. 3702-3707, 2002.
82. S. Boyd, L. El Ghaoui, and E. Feron, et. al, Linear Matrix Inequalities in System and Control Theory, SIAM Studies in Applied Mathematics, Philadelphia, 1994.
83. K. M. Nagpal and P. P. Khargonekar, "Filtering and smoothing in an H_∞ setting," IEEE Transaction on Automatic Control, 1992, vol. 36, no. 2, pp: 152-166.
84. Y. Aoki, T. Inoue, and Y. Hori, "Robust design of gain matrix of body slip angle observer for electric vehicles and its experimental demonstration," The 8th IEEE International Workshop on Advanced Motion Control, pp. 41-45, 2004.
85. S. Saraf and M. Tomizuka, "Slip angle estimation for vehicles on automated highways," Proceedings of the American Control Conference, vol. 3, pp. 1588-1592, 1997.

86. Y. U. Yim and S.-Y. Oh, "Modeling of vehicle dynamics from real vehicle measurements using a neural network with two-stage hybrid learning for accurate long-term prediction," IEEE Transactions on Vehicular Technology, vol. 53, no. 4, pp. 1076-1084, 2004.
87. L. Li, F.-Y. Wang and Q. Zhou, "A robust observer designed for vehicle lateral motion estimation," Proceedings of IEEE Intelligent Vehicles Symposium, pp. 417-422, 2005.
88. D.-W. Gu and F. W. Poon, "A robust state observer scheme," IEEE Transactions on Automatic Control, 2001, vol. 46, no. 12, pp: 1958-1963.
89. M. Boutayeb and M. Darouach, "Comments on 'A robust state observer scheme'," IEEE Transactions on Automatic Control, 2003, vol. 48, no. 7, pp: 1292-1293.
90. D.-W. Gu and F. W. Poon, "Authors' reply [Comment on 'A robust state observer scheme']," IEEE Transactions on Automatic Control, 2003, vol. 48, no. 7, pp: 1293-1294.
91. H. Peng and M. Tomizuka, "Lateral control of front-wheel-steering rubber-tire vehicles," Publication of PATH Project, UCB-ITS-PRR-90-5, 1990.
92. J. Ackermann, J. Guldner, and W. Sienel, et. al, "Linear and nonlinear controller design for robust automatic steering," IEEE Transactions on Control Systems Technology, vol. 3, no. 1, pp. 132-143, 1995.
93. J. Ackermann, "Robust control prevents car skidding," IEEE Control Systems Magazine, vol. 17, no. 3, pp. 23-31, 1997.
94. J. Guldner, W. Sienel, and H.-S. Tan, et. al, "Robust automatic steering control for look-down reference systems with front and rear sensors," IEEE Transactions on Control Systems Technology, vol. 7, no. 1, pp. 2-11, 1999.
95. F. N. Koumboulis and M. G. Skarpetis, "Robust control of cars with front and rear wheel steering," IEE Proceedings of Control Theory and Applications, vol. 149, no. 5, pp. 394-404, 2002.
96. B. A. Francis, A course in H_∞ control theory, New York. Springer-Verlag, 1987.
97. G. Zames, "Feedback and optimal sensitivity. Model reference transformations, multiplicative seminorms, and approximate inverses," IEEE Transactions on Automatic Control, vol. 26, pp. 301-320, 1981.
98. B. A. Guvenc, T. Bunte, and D. Odenthal, et. al, "Robust two degree-of-freedom vehicle steering controller design," IEEE Transactions on Control Systems Technology, vol. 12, no. 4, pp. 627-636, 2004.
99. S. Mammar, "Lateral vehicle control using gain scheduled H_∞ controllers," Proceedings of IEEE Conference on Intelligent Transportation System, pp. 248-253, 1997.
100. S. Mammar, "Two-degree-of-freedom H_∞ optimization and scheduling for robust vehicle lateral control," Vehicle System Dynamics, vol. 34, pp. 401-422, 2000.

101. J. Ackermann and T. Bunte, "Actuator rate limits in robust car steering control," Proceedings of the 36th IEEE Conference on Decision and Control, vol. 5, pp. 4726-4731, 1997.
102. S. E. Shladover, C. A. Desoer, and J. K. Hedrick, et. al, "Automated vehicle control developments in the PATH program," IEEE Transactions on Vehicular Technology, vol. 40, no. 1, pp. 114-130, 1991.
103. R. T. O'Brien, P. A. Iglesias, and T. J. Urban, "Lane change maneuver via H_∞ steering control methods," Proceedings of the 4th IEEE Conference on Control Applications, pp. 131-136, 1995.
104. N. Sugitani, Y. Fujuwara, and K. Uchida, "Electric power steering with H-infinity control designed to obtain road information," Proceedings of American Control Conference, vol. 5, pp. 2935-2939, 1997.
105. K.-T. Feng, H.-S. Tan and M. Tomizuka, "Automatic steering control of vehicle lateral motion with the effect of roll dynamics," Proceedings of American Control Conference, pp. 2248-2252, 1998.
106. J.-Y. Wang and M. Tomizuka, "Robust H_∞ lateral control of heavy-duty vehicles in automated highway system," Proceedings of American Control Conference, pp. 3671-3675, 1999.
107. H.-S. Tan and C.-Y. Chan, "Design of steering controller and analysis of vehicle lateral dynamics under impulsive disturbances," Proceedings of American Control Conference, vol. 3, pp. 2023-2027, 2000.
108. H.-S Tan, B. Bougler, and A. Steinfeld, "Snowplow steering guidance with gain stabilization," Vehicle System Dynamics, vol. 36, no. 4-5, pp. 279-305, 2001.
109. S. Mammar, D. Koenig, and L. Nouveliere, "Combination of feedforward and feedback H_∞ control for speed scheduled vehicle automatic steering," Proceedings of American Control Conference, vol. 2, pp. 684-689, 2001.
110. Hatipoglu, U. Ozguner, and K. A. Redmill, "Automated lane change controller design," IEEE Transactions on Intelligent Transportation Systems, vol. 4, no. 1, pp. 13-22, 2003.
111. G. Lu and M. Tomizuka, "Vehicle following as backup control schemes for magnet-magnetometer-based lateral guidance," IEEE Transactions on Control Systems Technology, vol. 13, no. 2, pp. 274-285, 2005.
112. C. S. Kim, J. Y. Choi, and S. P. Hong, et. al, "H_∞ steering control for the unmanned vehicle system," Proceedings of Annual Conference of the IEEE Industrial Electronics Society, vol. 3, pp. 2139-2143, 2001.
113. J. H. Park and W. S. Ahn, "H_∞ yaw-moment control with brakes for improving driving performance and stability," Proceedings of IEEE/ASME International Conference on Advanced Intelligent Mechatronics, pp. 747-752, 1999.
114. M. T. Raharijaona, M. G. Duc, and M. S. Mammar, "Linear parameter-varying control and H-infinity synthesis dedicated to lateral driving as-

sistance," Proceedings of IEEE Intelligent Vehicles Symposium, pp. 407-412, 2004.

115. C. Hatipoglu, H. Ozbay, and U. Ozguner, "H_∞ controller design for automatic steering of vehicles with modeled time delays," Proceedings of IEEE Conference on Intelligent Transportation System, pp. 260-265, 1997.

116. S. Brennan and A. Alleyne, "H-infinity vehicle control using nondimensional perturbation measures," Proceedings of American Control Conference, vol. 3, pp. 2534-2539, 2002.

117. C. S. Kim, J. Y. Choi, and S. P. Hong, et. al, "H_∞ steering control for the unmanned vehicle system," Proceedings of Annual Conference of the IEEE Industrial Electronics Society, vol. 3, pp. 2139-2143, 2001.

118. Y. Fujiwara and S. Adachi, "Control design of steering assistance system for driver characteristics," Proceedings of SICE Annual Conference, vol. 5, pp. 2879-2882, 2002.

119. Y. Fujiwara and S. Adachi, "Steering assistance system for driver characteristics using gain scheduling control," European Control Conference, 2003.

120. L. Li, F.-Y. Wang, G. Lai, "An LMI approach to robust vehicle steering controller design," Proceedings of IEEE Intelligent Transportation Systems, 2005.

121. J. Abedor, K. Nagpal, and K. Poolla, "A linear matrix inequality approach to peak-to-peak gain minimization," International Journal of Robust and Nonlinear Control, vol. 6, pp: 899–927, 1996.

122. S. R. Venkatesh and M. A. Dahleh, "Does star norm capture l1 norm?" Proceedings of American Control Conference, vol. 1, pp. 944-945, 1995.

123. T. Nguyen and F. Jabbari, "Output feedback controllers for disturbance attenuation with bounded control," Proceedings of the 36th IEEE Conference on Decision and Control, vol. 1, pp. 177-182, 1997.

124. T. Nguyen and F. Jabbari, "Disturbance attenuation for systems with input saturation: An LMI approach," IEEE Transactions on Automatic Control, vol. 44, no. 4, pp. 852-857, 1999.

125. P. Apkarian and P. Gahinet, "A convex characterization of gain-scheduled H_∞ controllers," IEEE Transactions on Automatic Control, vol. 40, no. 5, pp. 853-864, 1995.

126. P. Gahinet, P. Apkarian and M. Chilali, "Affine parameter-dependent Lyapunov functions and real parameter uncertainty," IEEE Transactions on Automatic Control, vol. 41, no. 3, pp. 436-442, 1996.

127. R. Mobus and Z. Zomotor, "Constrained optimal control for lateral vehicle guidance," Proceedings of IEEE Intelligent Vehicles Symposium, pp. 429-434, 2005.

128. A. Bemporad, M. Morari, and V. Dua, et. al, "The explicit linear quadratic regulator for constrained systems," Automatica, vol. 38, no. 1, pp. 3-20, 2002.

129. P. Grieder, F. Borrelli, and F. Torrisi, et. al, "Computation of the constrained infinite time linear quadratic regulator," Proceedings of American Control Conference, vol. 6, pp. 4711-4716, 2003.
130. K. Moriwaki, "Autonomous steering control for electric vehicles using nonlinear state feedback H_∞ control," Nonlinear Analysis, 2005.
131. S. Drakunov and R. DeCarlo, "Sliding mode control design for automated steering via Lyapunov approach," Proceedings of IEEE Conference on Decision and Control, vol. 4, pp. 4086-4088, 1995.
132. J. R. Zhang, A. Rachid, and S. J. Xu, "Velocity controller design for automatic steering of vehicles," Proceedings of American Control Conference, vol. 2, pp. 696-697, 2001.
133. J. R. Zhang, S. J. Xu, and A. Rachid, "Sliding mode controller for automatic path tracking of vehicles," Proceedings of American Control Conference, pp. 3974-3979, 2002.
134. M. Tai, "Experimental study of lateral control of heavy vehicles for automated highway systems (AHS)," Proceedings of American Control Conference, vol. 2, pp. 851-856, 2002.
135. O. Mokhiamar and M. Abe, "Combined lateral force and yaw moment control to maximize stability as well as vehicle responsiveness during evasive maneuvering for active vehicle handling safety," Vehicle System Dynamics Supplement, vol. 37, pp. 246-256, 2002.
136. Q. Zhou, F.-Y. Wang, and L. Li, "Robust Sliding Mode Control of 4WS Vehicles for Automatic Path Tracking," Proceedings of IEEE Intelligent Vehicles Symposium, pp. 819-826, 2005.
137. S. Brennan and A. Alleyne, "Driver assisted yaw rate control," Proceedings of American Control Conference, vol. 3, pp. 1697-1701, 1999.
138. K. J. Astrom and B. Wittenmark, Computer Controlled Systems. Theory and Design, Prentice-Hall, 1997.
139. S. B. Choi, "The design of a look-down feedback adaptive controller for the lateral control of front-wheel-steering autonomous highway vehicles," IEEE Transactions on Vehicular Technology, vol. 49, 6, pp. 2257-2269, 2000.
140. A. Halanay, A. Ionita, and V. Rasvan, "Stability and manoeuvrability analysis of vehicle with four wheel steering system," Proceedings of the Third IEEE Conference on Control Applications, 1994, pp. 385-390.
141. E. Freund and R. Mayr, "Nonlinear path control in automated vehicle guidance," IEEE Transactions on Robotics and Automation, vol. 13, no. 1, pp. 49-60, 1997.
142. T. Fukao, S. Miyasaka, and K. Mori, "Active steering systems based on model reference adaptive nonlinear control," Proceedings of IEEE Conference on Intelligent Transportation Systems, pp. 502-507, 2001.
143. G. G. Lendaris, L. Schultz, and T. Shannon, "Adaptive critic design for intelligent steering and speed control of a 2-axle vehicle," Proceedings of IEEE-INNS-ENNS International Joint Conference on Neural Networks, vol. 3, pp. 73-78, 2000.

144. L. J. Schultz, T. Shannon, and G. G. Lendaris, "Using DHP adaptive critic methods to tune a fuzzy automobile steering controller," IFSA World Congress and 20th NAFIPS International Conference, vol. 1, pp. 557-562, 2001.
145. L. Beji and Y. Bestaoui, "An adaptive control method of automated vehicles with integrated longitudinal and lateral dynamics in road following," Proceedings of the Second International Workshop on Robot Motion and Control, pp. 201-206, 2001.
146. S.-Y. Oh, J.-H. Lee, and D.-H. Choi, "A new reinforcement learning vehicle control architecture for vision-based road following," IEEE Transactions on Vehicular Technology, vol. 49, no. 3, pp. 997-1005, 2000.
147. T. Hessburg and M. Tomizuka, "Fuzzy logic control for lateral vehicle guidance," IEEE Control Systems Magazine, vol. 14, no. 4, pp. 55-63, 1994.
148. G. L. Brown and J.C. Hung, "Vehicle steering assistance in extreme situations using a fuzzy logic controller," Proceedings of the IEEE International Conference on Industrial Technology, pp. 225-229, 1994.
149. Z. Zalila, F. Bonnay, and F. Coffin, "Lateral guidance of an autonomous vehicle by a fuzzy logic controller," IEEE International Conference on Systems, Man, and Cybernetics, 1998, vol. 2, pp. 1996-2001.
150. N. E. Hodge and M. B. Trabia, "Steering fuzzy logic controller for an autonomous vehicle," Proceedings of 1999 IEEE International Conference on Robotics and Automation, vol. 3, pp. 2482-2488, 1999.
151. A. El Kamel, J.-Y. Dieulot, and P. Borne, "Fuzzy controller for lateral guidance of busses," Proceedings of the 2002 IEEE International Symposium on Intelligent Control, pp. 110-115, 2002.
152. K. R. S. Kodagoda, W. S. Wijesoma, and E. K. Teoh, "Fuzzy speed and steering control of an AGV," IEEE Transactions on Control Systems Technology, vol. 10, no. 1, pp. 112-120, 2002.
153. L. Cai, A. B. Rad, and W. L. Chan, et. al, "A robust fuzzy pd controller for automatic steering control of autonomous vehicles," Proceedings of The 12th IEEE International Conference on Fuzzy Systems, vol. 1, pp. 549-554, 2003.
154. A. B. Will, M. C. M. Teixeira, and S. H. Zak, "Four wheel steering control system design using fuzzy models," Proceedings of IEEE International Conference on Control Applications, pp. 73-78, 1997.
155. O. Pages, A. El Hajjaji, and R. Ordonez, "H_∞ tracking for fuzzy systems with an application to four wheel steering of vehicles," Proceedings of IEEE International Conference on Systems, Man and Cybernetics, vol. 5, pp. 4364-4369, 2003.
156. A. El Hajjaji and S. Bentalba, "Fuzzy path tracking control for automatic steering of vehicles," Robotics and Autonomous Systems, vol. 43, pp. 203-213, 2003.
157. J.-W. Perng, H.-I. Chin, B.-F. Wu, T.-T. Lee, "Stability analysis of a robust fuzzy vehicle steering control system," Proceedings of the IEEE Intelligent Vehicles Symposium, pp. 382-387, 2005.

158. H. Peng and M. Tomizuka, "Preview control for vehicle lateral guidance in highway automation," ASME Journal of Dynamic Systems, Measurement and Control, vol. 115, no. 4, pp. 679-686, 1993.
159. M. Fu, J. Ruan, and Y. Li, "A new kind of robust design method of intelligent vehicle lateral control," Proceedings of Fifth World Congress on Intelligent Control and Automation, vol. 3, pp. 2438-2442, 2004.
160. J.-O. Hahn, J.-W. Hur, and K. Yi, et. al "Nonlinear vehicle stability control using disturbance observer," Proceedings of International Conference on Control Applications, vol. 1, 2002, pp. 441-446, 2002.
161. S. J. Xu and J. R. Zhang, "Nonlinear automatic steering control of vehicles based on Lyapunov approach," Proceedings of IEEE International Conference on Intelligent Transportation Systems, pp. 183-187, 2002.
162. S. Chaib, M. S. Netto, and S. Mammar, "H_∞, adaptive, PID and fuzzy control. a comparison of controllers for vehicle lane keeping," Proceedings of IEEE Intelligent Vehicles Symposium, pp. 139-144, 2004.
163. A. Alleyne and M. DePoorter, "Lateral displacement sensor placement and forward velocity effects on stability of lateral control of vehicles," Proceedings of American Control Conference, vol. 3, pp. 1593-1597, 1997.
164. J. T. Spooner and K. M. Passino, "Fault tolerant longitudinal and lateral control for automated highway systems," Proceedings of American Control Conference, vol. 1, pp. 663-667, 1995.
165. V. Krishnaswami and G. Rizzoni, "Model based health monitoring of vehicle steering system using sliding mode observers," Proceedings of American Control Conference, vol. 3, pp. 1652-1656, 1995.
166. V. Garg, A. E. Lindsey, "Fault detection for combined lateral and longitudinal control of vehicles for AHS," Proceedings of IEEE Decision and Control, vol. 2, pp. 2301-2302, 1996.
167. R. Rajamani, A. S. Howell, and C. Chen, et. al, "A complete fault diagnostic system for automated vehicles operating in a platoon," IEEE Transactions on Control Systems Technology, vol. 9, no. 4, pp. 553-564, 2001.
168. J. Huang and M. Tomizuka, "H_∞ controller for vehicle lateral control under fault in front or rear sensors," Proceedings of American Control Conference, vol. 2, pp. 690-695, 2001.
169. A. Shrivastava and R. Rajamani, "Fault diagnostics for GPS-based lateral vehicle control," Proceedings of American Control Conference, vol. 1, pp. 31-36, 2001.
170. R. Isermann. R. Schwarz. and S. Stolzl, "Fault-tolerant drive-by-wire systems," IEEE Control Systems Magazine, vol. 22, no. 5, 2002.
171. C. Scherrer and A. Steininger, "Dealing with dormant faults in an embedded fault-tolerant computer system," IEEE Transactions on Reliability, vol. 52, no. 4, pp. 512-522, 2003.
172. S. Suryanarayanan, M. Tomizuka, and T. Suzuki, "Design of simultaneously stabilizing controllers and its application to fault-tolerant lane-

keeping controller design for automated vehicles," IEEE Transactions on Control Systems Technology, vol. 12, no. 3, pp. 329-339, 2004.
173. S. You and L. Jalics, "Hierarchical component-based fault diagnostics for by-wire systems," SAE#2004-01-0285, 2004.
174. S. Ibaraki, S. Suryanarayanan, and M. Tomizuka, "Design of Luenberger state observers using fixed-structure H_∞ optimization and its application to fault detection in lane-keeping control of automated vehicles," IEEE/ASME Transactions on Mechatronics, vol. 10, no. 1, pp. 34-42, 2005.
175. T. Hsiao and M. Tomizuka, "Sensor fault detection in vehicle lateral control systems via switching kalman filtering," Proceedings of American Control Conference, pp. 5009-5014, 2005.
176. J. Huang and M. Tomizuka, "LTV controller design for vehicle lateral control under fault in rear sensors," IEEE/ASME Transactions on Mechanics, vol. 10, no. 1, pp. 1-7, 2005.
177. D. McRuer, and D.H. Weir, "Theory of manual vehicular control," Proceedings of International Symposium on Manmachine Systems, vol. MMS-10, no. 4, 1969.
178. D. L. Kleinman, S. Baron, and W. H. Levison, "A control theoretic approach to manned-vehicle systems analysis," IEEE Transactions on Automatic Control, vol. 16, no. 6, pp. 824-832, 1971.
179. J. R. McLean, and E. R. Hoffmann, "The effects of restricted preview on driver steering control and performance," Human Factors, vol. 15, no. 4, pp. 1228-1232, 1973.
180. R. A. Hess and A. Modjtahedzadeh, "A control theoretic model of driver steering behavior," IEEE Control Systems Magazine, vol. 10, no. 5, pp. 3-8, 1990.
181. A. Modjtahedzadeh and R. A. Hess, "A model of driver steering control behavior for use in assessing vehicle handling qualities," ASME Journal of Dynamic Systems, Measurement and Control, vol. 115, pp.456-464, 1993.
182. M. Plochl and P. Lugner, "Theoretical investigations of the interaction driver-feedback-controlled automobile," Proceedings of International Symposium on Advanced Vehicle Control, pp. 42-48, 1994.
183. A. Liu and S. Chang, "Force feedback in a stationary driving simulator," Proceedings of the IEEE International Conference on Systems, Man, and Cybernetics, vol. 2, pp. 1711-1716, 1995.
184. K.-T Feng, H.-S. Tan and M. Tomizuka, "Future predictor for vehicle steering guidance - sensitivity analysis and experimental results," Proceedings of IEEE Conference on Decision and Control, pp. 3722-3727, 1999.
185. T. Fujioka, Y. Shirano, and A. Matsushita, "Driver's behavior under steering assist control system," Proceedings of 1999 IEEE/IEEJ/JSAI International Conference on Intelligent Transportation Systems, pp. 246-251, 1999.

186. K.-T. Feng, H.-S. Tan, and M. Tomizuka, "Design of vehicle lateral guidance system for driver assistance," Proceedings of American Control Conference, vol. 4, pp. 2553-2557, 2000.
187. Z. Gu, G. Fang, and J. Wu, et. al, "Investigation of the transient state steering stability in side winds for a high-speed vehicle," International Journal of Vehicle Design, vol. 26, no. 5, pp. 562 - 572, 2001.
188. T. Miyazaki, T. Kodama, and T. Furuhashi, et. al, "Modeling of human behaviors in real driving situations," Proceedings of IEEE Intelligent Transportation Systems, pp. 643-646, 2001.
189. D. Lynch and A. Alleyne, "Velocity scheduled driver assisted control," International Journal of Vehicle Design, vol. 29, no. 1/2, pp. 1-22, 2002.
190. P. Seflur, D. Dawson, and J. Chen, et. al, "A nonlinear tracking controller for a haptic interface steer-by-wire systems," Proceedings of IEEE Conference on Decision and Control, vol. 3, pp. 3112-3117, 2002.
191. N. Bajcinca, R. Cortesao, and M. Hauschild, et. al, "Haptic control for steer-by-wire systems," Proceedings of IEEE/RSJ International Conference on Intelligent Robots and Systems, vol. 2, pp. 2004-2009, 2003.
192. L. Li, G. Lai, and F.-Y. Wang, "Safe steering speed estimation and optimal trajectory planning for intelligent vehicles," Proceedings of IEEE International Conference on Networking, Sensing and Control, pp. 722-727, 2005.

Advanced Vehicle Longitudinal Motion Control

4.1 Introduction

Generally, a vehicle's longitudinal dynamics is mainly determined by its internal kinematical components:

(1) engine, which is the source of power and convert chemical energy of the fuel to kinetic energy of a flywheel;

(2) driveline/transmission, which carries power from the engine to the wheels;

(3) The braking system, which can slow down or stop the car.

Engine is the power source of a vehicle. Almost all the ground vehicles use internal combustion engines, which converts the fuel into motion. However, great efforts had been put into the searching process for their alternatives, due to the severe energy crisis that every country needs to face now.

Sometimes, the term powertrain only refers to the combination of engine and transmission. However, in many other literals, powertrain consists of all the kinematical components of a vehicle's longitudinal drive system, which includes engine, transmission, differential(s), hubs, any interconnecting or separate shafts and clutch, etc., see Fig.4.1, Fig.4.2. Usually, drivetrain has the same meaning of powertrain in most situations.

In most literals, a transmission subsystem simply refers to the gear sets and/or hydraulic system that transmit mechanical power from the engine to some form of propeller shaft. Usually, gears in the transmission alter torque and speed of power and pass it onto the propeller shaft. A transmission subsystem needs to be carefully designed to generate sufficient torque; because the engine provides its highest torque outputs approximately in the middle of its range, while often the greatest torque is required when the vehicle is moving from rest or traveling slowly.

Driveline indicates everything in the powertrain, let alone the engine and the transmission subsystems. Modern drivcline often provides a five and six speed transmission conversion specialist. How to further improve power transmission efficiency and torque adjustment performance is still under intensive research.

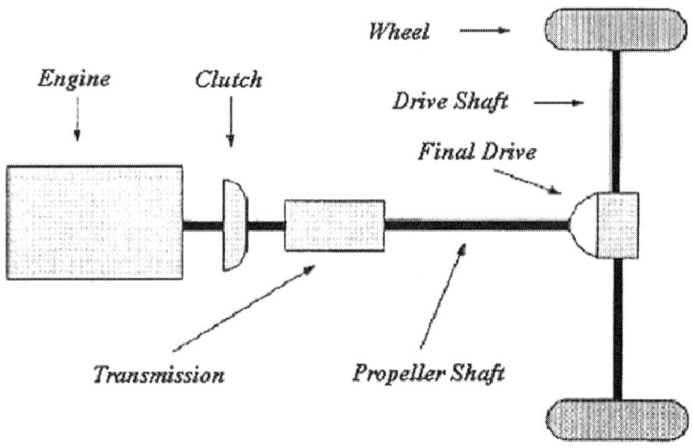

Fig.4.1 Bird-view diagram of a rear-driven vehicular powertrain.

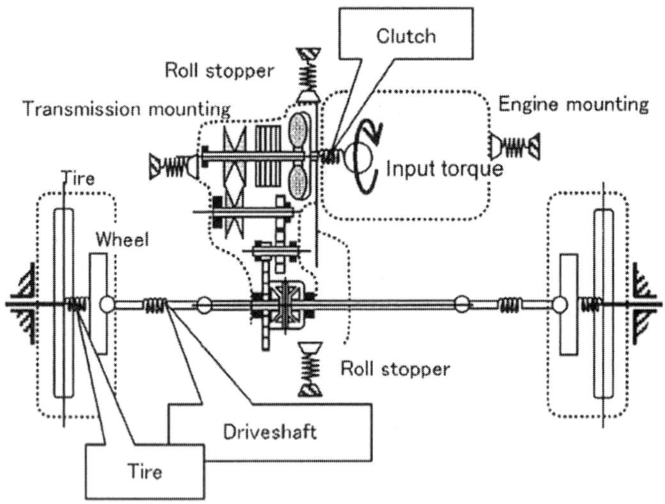

Fig.4.2 Front-view diagram of a rear-driven vehicular powertrain, from [2].

The braking subsystem is another essential component to a vehicle. It is well known that to stop a car quickly on slippery road surface is challenging for conventional braking subsystems. Thus, most nowadays vehicles are equipped with Anti-lock Braking Systems (ABS). The fundamental theory behind ABS is that a skidding wheel (where the tire contact patch is sliding relative to the road) has less traction than a non-skidding wheel. So the anti-lock braking system needs to know when a wheel is about to lock up and apply appropriate control strategy to avoid it. This naturally leads to great developments of related sensory and actuator design.

Besides, vehicle aerodynamics also heavily affects a vehicle's longitudinal dynamics, especially when its navigation speed is high. Thus, a brief discussion is given in this Chapter to show the research trend of vehicle shape design.

Upon the developments of vehicle internal kinematical components, vehicle external longitudinal driving behavior attracts great attentions in the last three decades. The objective of vehicle longitudinal motion control is to keep proper navigation speed and spacing from the leading vehicle. Particularly, it is expected that intelligent or adaptive cruise controllers (ICC or ACC) using sensing devices such as radar can be used to determine and maintain safe headway/gap among vehicles at given velocities and vehicle conditions (brake, throttle, vehicle wind drag, tire traction, vehicle weight/weight distribution) while maximizing road capacity and driving efficiency. Moreover, ICC or ACC are required to minimize the switching between acceleration and brake to obtain a smooth and fuel efficient ride. Since the emphasis is given to the longitudinal motion control strategy, it is assumed that all the driving information needed can be correctly measured in the rest of this Chapter unless intentionally pointed out.

4.2 Advanced Vehicle Powertrain Control

4.2.1 Advanced Engine Control

Under growing performance requirements for today's vehicles, sophisticated sensory and control schemes were proposed to regulate the engine's operations. Now, controlling the working condition of engine becomes a processor-intensive job, which is based on additional electric devices installed in the engine including engine control unit (ECU), various sensors [3] and actuators [4].

Since different automotive company use varied engines, there were countless apparently different engine models had been proposed in the literature. It is thus difficult to choose a general formulation which is fitful to a particular design task [5]-[17].

Generally, engine designers need to consider several interlaced problems: fuel burning efficiency, energy transfer performance, output torque equilibrium and calibration, engine crankshaft rotational kinetics. For example, a practical full-scale mathematical model for advanced SI engine control systems design can be formulated to count the following six parts:

(1) Dynamics of the fuel injector;

(2) Dynamics of intake manifold which include airflow and fuel flow dynamics;

(3) Torque generation during combustion;

(4) Engine crankshaft rotational kinetics;

(5) Process delay;

(6) Dynamics of the Universal Exhaust Gas Oxygen sensor (UEGO).

Apparently, it is impossible to discuss some many problems in this Chapter; therefore only two interesting special problems: engine air/fuel ratio control and driveline vibration control, are studied as follows, since to discuss them does not require complex preliminary introductions.

1) Engine Air/Fuel Ratio Control

The past 20 years have revealed all the limits of the spark-ignition engine system in terms of pollution and fuel consumption. Therefore extensive research has been carried out to reduce waste gas emissions and improve fuel efficiency [18]-[27].

The catalytic converter is one of the most important parts of a car's exhaust system to reduce emissions. Typical characteristics of a catalytic converter are shown in Fig.4.3. It is clear that there is only a very narrow range of air-fuel ratios (A/F) around stoichiometry over which simultaneously high conversion efficiencies can be attained for the three main emission constituents HC , CO and NO_x.

Under normal conditions the stoichiometric ratio for gasoline is

$$\lambda_{st} = (A/F)_s = \frac{m_{a,th}}{m_{f,th}} = 14.66 \tag{4.1}$$

where $m_{a,th}$, $m_{f,th}$ are the ideal theoretical air and fuel mass respectively.

However, it is found that deviations of 1% A/F from the value λ_{st} cause the conversion efficiency to drastically decrease for at least one of the emission components.

Define the equivalent air-fuel ratio (AFR) as $\lambda = \lambda_{ax} / \lambda_{st}$, where λ_{ax} is the actual air-fuel ratio. Based on Fig.4.3, it leads to

(1) Lean operation ($\lambda > 1$): Less fuel is injected than needed for ideal combustion. This results in high emissions of NO_x;

(2) Rich operation ($\lambda < 1$): More fuel is injected than needed for stochiometric combustion. This results in high emissions of hydrocarbon HC and in a decreasing fuel economy.

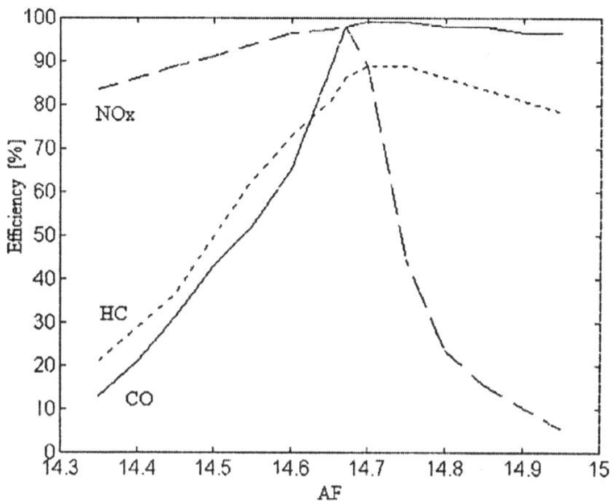

Fig.4.3 Efficiency of a catalytic converter as a function of AF, from [27]. (© [2003] IEEE)

Obviously, to minimization of emissions in automotive fuel-injection engines requires the design of accurate control systems to keep the A/F at the optimal stoichiometric value $(A/F)_s$ or keep AFR at 1.

In many Engine Management Systems (EMS) designs, the Fuel Pulse Width (FPW) of the fuel injector is read from simple look-up tables as a function of current engine conditions to match the air flow commanded by the driver. Classic feedback controllers for A/F also provide a simple control action, which using information coming from the universal exhaust gases oxygen sensor (UEGO) installed on the exhaust pipes as feedback signal to correct the fuel amount injected into intake manifold to match the air flow commanded by the driver. However, the duration of the process delay renders these simple approaches for effective AFR control. Thus, more sophisticated predictive control strategy is required based upon a model of the engine process.

Various model based AFR control approaches are proposed in the past two decades. Fig.4.4 shows the AFR related system dynamics. The key system inputs are: Throttle angle θ and Fuel injection rate via FPW. And the first input is considered a disturbance while the second is the manipulated variable of the control system.

The engine states of concern are: Intake manifold pressure P_m, In-Cylinder AFR $\lambda_c = \frac{1}{\lambda_{st}} \cdot \frac{\dot{m}_{ao}}{\dot{m}_{fo}}$, Engine speed N, and UEGO sensor output λ_s, in which only the last is measured.

One important AFR control approach is time-domain sliding mode AFR control method. Sliding mode control is a technique to achieve perfect robust performance with magnitude limited uncertainty that is in the range space of the control derivative of the plant by using state feedback and high-gain actuation. The method is appropriate for SI engine control applications due to its uncertain and nonlinear nature [11].

For example, Pieper and Mehrotra incorporated a sliding mode control based on a linearized engine model in [26]. The controller calculates the fuel to be injected so that the fuel entering the cylinder will match the air inducted into the cylinder, taking into account the process delays.

Considering the relative dynamic elements above, Pieper and Mehrotra gave the combined linearized state space model as

$$\dot{X} = AX + Bu + B_\theta \theta \tag{4.2}$$

where $X = \left[m_{ai}, m_{ao}, P_m, m_{ff}, m_{fo}, N, \lambda_c\right]^T$ and $u = \dot{m}_{fi}$. θ is the throttle angle commanded by the driver which is taken as a measurable disturbance.

Define $(\dot{m}_{fi})_s$ is the stochiometric input or the theoretical amount fuel flow rate calculated as

$$(\dot{m}_{fi})_s = \frac{\dot{m}_{ao}}{\lambda_{st}} = A_s X + B_{s\theta}\theta \tag{4.3}$$

then the true fuel injection rate can be written as

$$\dot{m}_{fi} = (\dot{m}_{fi})_s + (\dot{m}_{fi})_c \tag{4.4}$$

where $(\dot{m}_{fi})_c$ is the difference between the actual fuel injection rate and the stoichiometric due to load disturbances and throttle commands.

By Substituting Eq.(4.2)-(4.3) to Eq.(4.1), the system with the new control input $u_c = (\dot{m}_{fi})_c$ should be written as

$$\dot{X} = A_1 X + Bu_c + B_{1\theta}\theta \tag{4.5}$$

where $A_1 = A + B \cdot A_s$ and $B_{1\theta} = B_\theta + B \cdot B_{s\theta}$.

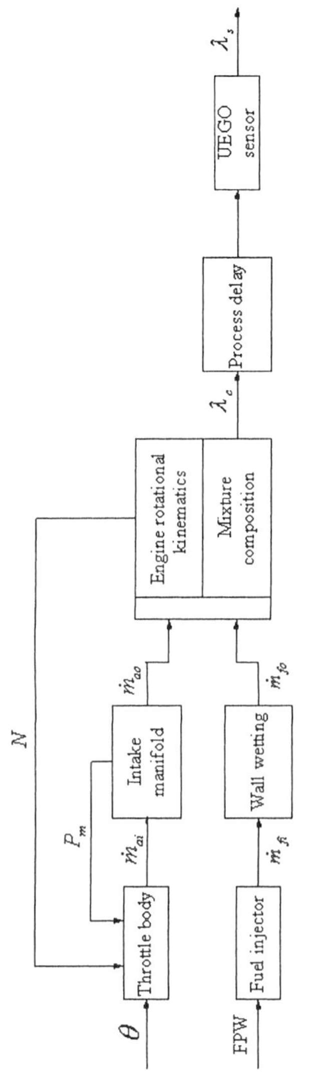

Fig.4.4 AFR related engine dynamics, from [24].

Construct a high pass filter as

$$\dot{q} = \tau_q \cdot ((m_{fi})_c - q) \tag{4.6}$$

It is clear that

$$\lim_{t\to\infty}\frac{1}{\tau_q}\dot{q}=\lim_{t\to\infty}((m_{fi})_c-q)=0 \tag{4.7}$$

and choose $\tau_q=1000$ for practical purposes yields

$$(\dot{m}_{fi})_c\approx\dot{q}=\tau_q\cdot((m_{fi})_c-q) \tag{4.8}$$

Define $Z=\begin{bmatrix}X\\ q\end{bmatrix}$, the system become

$$\dot{Z}=\begin{bmatrix}A_1 & -\tau_q B\\ 0 & -\tau_q\end{bmatrix}Z+\begin{bmatrix}\tau_q B\\ \tau_q\end{bmatrix}u_c+\begin{bmatrix}B_{1\theta}\\ 0\end{bmatrix}\theta=A_N Z+B_N u_c+A_N\theta \tag{4.9}$$

For the only state that is measured is the air fuel equivalence ratio coming from the UEGO as

$$y=CZ \tag{4.10}$$

where $C=[0\quad 0\quad 0\quad 0\quad 0\quad 0\quad 1\quad 0]^T$.

Therefore a state estimator

$$\dot{\hat{Z}}=A_N\hat{Z}+B_N u_c+L\left[y(t)-C\hat{Z}\right] \tag{4.11}$$

is proposed to eliminate the cost of sensors, where $\hat{Z}$ is the estimated state vector and L is a feedback gain computed using a Kalman filter.

According to [28], for the revised model (4.11), control feedback should be

$$u_c=-(C_{sl}B_N)^{-1}(C_{sl}A_N Z+\kappa\,\mathrm{sgn}(C_{sl}Z)) \tag{4.12}$$

will achieve a sliding mode behavior in that the hyperplane $S_{sl}=C_{sl}Z=0$ will be induced provided

$$\kappa>|B_1\theta|_{\max} \tag{4.13}$$

where $C_{sl}\in R^{1\times 8}$ and $|B_1\theta|_{\max}$ is the maximal magnitude disturbance.

The first order filtering and linear quadratic optimal methods are used to design C_{sl} so that sliding surface $S_{sl}=C_{sl}Z=0$ has the suitable dynamics to keep $\dot{m}_f$ near $\dot{m}_{ao}/\lambda_{st}$ in [26].

There were also some other approaches, i.e. the neural network based one proposed in [27]. Fig.4.5 demonstrates the validation results over experiments. By providing an irregular input for θ, it can be seen that under different engine speeds, the neural controller provides better performance than conventional TFC controllers proposed in [29] and [30].

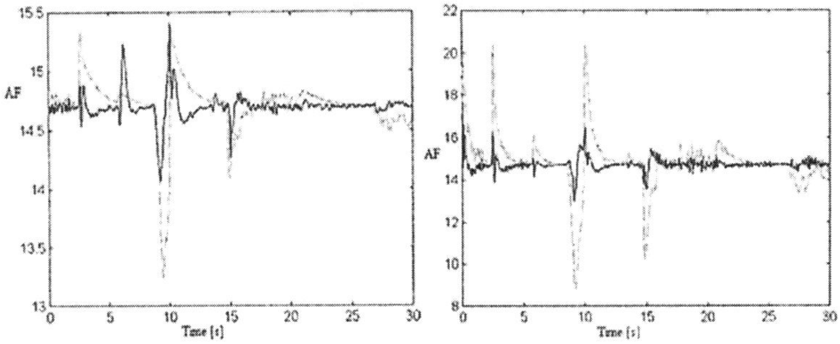

Fig.4.5 Controlled AFR when: (left) N=1000rpm (right) N=4000rpm. Continuous line: results of neural controller, dashed line: results of TFC controller, from [27]. (© [2003] IEEE)

2) Engine Idle Speed Control

Idle speed control (ISC) is another important operating mode for automotive engines [31]-[33]. This is because vehicles consume about 30 percent of their fuel in city driving and it is expected that with increased traffic loads, this percentage will further increase in the future. It is well-known that the engine idle speed should be as low as possible for improved fuel economy, since the fuel consumed while idling makes no contribution to work done by the engine. But, these low idle speeds often cause not only unpleasant vibrations of the vehicle under rapid external torque load changes but also deterioration of accessory performance and engine combustion quality. In the worst case, this would cause the engine to stall.

Thus the primary goal of the ISC is to remain a desired engine speed despite of both the system nonlinearities or uncertainties and external torque disturbances due to the following reasons: air conditioning, power steering, alternator torque, and engagement of automatic transmission. This desired engine speed N_0 is determined on the basis of operating conditions, minimum energy dissipation, etc.

For Idle speed control, the related engine dynamics include: Intake manifold dynamics, Torque generation, and Engine crankshaft rotational kinetics. The key system inputs are: Throttle angle θ and Spark advance δ. The engine states of concern are: Intake manifold pressure P_m, Engine speed N, and Engine torque $T = T_{ind}$.

There has been a significant amount of published materials that focus on idle speed control, which include the combined LQR idle speed control method proposed by Livshiz et. al [32], the sliding mode controller with a high gain observer by Bhatti et. al [33]. Besides, neural networks and fuzzy logic have also been proved as efficient methods to solve nonlinearity and uncertainty problems involved in engine dynamics. For instance, a attempt engine idle controller was proposed by Vachtsevanos et. al in [22], whose control loop is shown in Fig.4.6.

The engine model is formulated in the following form

$$\dot{X} = F(X,U) \tag{4.14}$$

where the state vector $X \in R^2$, and the control vector $U \in R^2$ are defined as

$$X = \begin{bmatrix} x_1 \\ x_2 \end{bmatrix} = \begin{bmatrix} P_m \\ M \end{bmatrix}, \quad U = \begin{bmatrix} u_1 \\ u_2 \end{bmatrix} = \begin{bmatrix} \theta \\ \delta \end{bmatrix} \tag{4.15}$$

In Fig.4.6, the nominal (equilibrium) operating point $X_0 = [P_0, N_0]^T$ is determined on the basis of operating conditions, minimum energy dissipation, etc., for example $P_0 = 34.25 kPa$, $N_0 = 750 r/\min$. The speed and pressure errors $e = [e_p, e_N]^T$ are then denoted as $e_P = P_0 - P$ and $e_N = N_0 - N$.

Most fuzzy control routines have relied upon heuristic information (based on operators' experience) to design the knowledge base. But the lack of formalism, in the control-theoretic sense, has limited severely the introduction of such conventional performance metrics as stability, robustness, and optimality, especially when the control problem is viewed as an expert solution paradigm for a mechanical system. So, a data-based method of automatic rule generation is introduced. The above engine model is used to generate required datasets. To infer the discrete fuzzy rules from the simulations, firstly, the description of engine performances should be quantized. A finite set of representative center points in the state space is chosen to anticipate the trajectories from one subspace to another when applying a certain control inputs. Each state variable or control variable ranges a permissible set of permissible values. The engine speed N, for example, can take values in [350, 1150] r/min. Now each interval associated with a variable is then divided into smaller subintervals. The number of the subintervals for each state variable is chosen based on: a) the specified ranges of each variable and the requirement of accuracy and resolution; b) the number of rules that are considered manageable in the sense of minimizing the search time and reducing computation complexity.

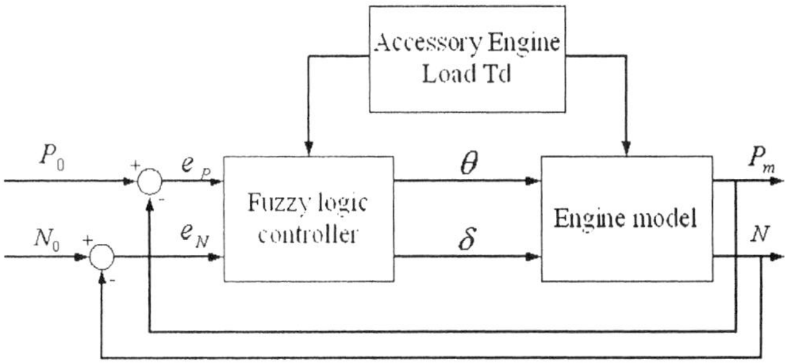

Fig.4.6 Scheme of fuzzy idle speed control, from [22]. (© [1993] IEEE)

Here, cell L_{mn} is defined as the finite region in the state space which is the m th and n th subintervals of the x_1 and x_2 respectively. Intervals of control variables are quantized into N_{u_1}, N_{u_2} equal subintervals, for u_1, u_2 respectively, and to derive (N_{u_1} +1) intermediate values for u_1, and (N_{u_2} +1) intermediate values for u_2. This scheme results in a set S_1 of intermediate values θ_1, ..., $\theta_{N_{u_1}}$ for the control variable u_1 and another set S_2 of values δ_1, ..., $\delta_{N_{u_2}}$ for u_2.

Furthermore, to associate these discrete regions with the fuzzy law, the borders of the cells or subintervals are carefully fuzzified and the fuzzy rules are set as the following form:

If (P is P_m) and (N is N_n) THEN (θ is θ_i) and (δ is δ_j)

which is correspondent to

If States P and N are in cell L_{mn}, THEN Control θ and δ are in U_{ij}

To find the "best " control at each cell in the state space, two steps are taken below. Firstly is to gather the data reflecting the control effects on engine states. The idea now is the following: for each cell-group, consider the center point. This point is the initial state for the simulation. Apply to the system the constant input (θ_i, δ_j), for all (i, j) in the set $S = S_1 \times S_2$, and perform a simulation run until system enters into another cell in the state space, we perform (N_{u_1} +1)(N_{u_2} +1) simulations, which result in (N_{u_1} +1)(N_{u_2} +1) transitional relations of the type

$$(\theta_1, \delta_1): L_{1,1} \to L_{1,2}, \cdots\cdots, (\theta_1, \delta_{N_{u2}}): L_{1,N_{u2}+1} \to L_{2,N_{u2}+1}$$

$$(\theta_{N_{u1}}, \delta_1): L_{N_{u1}+1,1} \to L_{N_{u1}+1,2}, \cdots\cdots, (\theta_{N_{u1}}, \delta_{N_{u2}}): L_{N_{u2}+1,N_{u2}+1} \to L_{N_{u2},N_{u2}+1}$$

The time required for the transition is stored, as well as the energy $U^T U$ of the control law and the Euclidean distance of the state from the equilibrium point, at the end of each simulation. After all pairs of control values in S have been used, continue with the next cell. Repeat until all cells in the state space have been simulated.

At the end of this phase, the following information is stored: if the system starts at cell L_{mn} and the constant control values θ_i, δ_j are applied, then the system will enter cell L_{kl} in some time costing an amount of energy and being at some distance from the equilibrium. This is done for all cells and for all control values in S.

The next step is to search this data file and come up with the following results: if the system starts in cell L_{mn}, then in order to drive it eventually to the equilibrium point, the "best" control to be applied is $\lfloor \theta_i, \delta_j \rfloor$, and such a rule has to be derived for all cells in the state space. "Best" can be in the sense of minimum time or minimum energy or minimum error (distance from the equilibrium). The general form of the cost of each transition is given as

$$J = [\alpha(e^T e) + \beta(u^T u) + \tau]t \tag{4.16}$$

where α, β and τ are binary coefficients for selecting a criterion as squared error, energy or time, respectively.

More precisely, the search procedure is stated by Vachtsevanos et. al as:

Step1: Enter the node corresponding to the equilibrium cell into the main list.

Step2: Expand the last node entered into the main list, i.e., find all arcs that end up in the node that was entered last into the main list. Then enter the starting node of each one of these arcs provided that it is not already in the main list, into the non-basic list. If a node is already in the non-basic list, then there already exists a path from this node to the equilibrium node. If the new path has less total cost (from this node to the equilibrium node) than the existing one, then update the total cost associated with that node.

Step3: If the non-basic list is not empty, then find the node in the non basic list with the smallest cost, enter it into the main list, and remove it from the non-basic list, else stop.

Step4: Go to Step 2.

At the end, the graph is a tree and there exists unique path from every node to the equilibrium node and each is associated with a certain control values. The information suggests the following strategy: if the system starts in cell L_{mn}, apply the control corresponding to the first arc of the path connecting L_{mn} with the equilibrium cell, until the system enters another cell.

Once the search procedure generates the optimal tree, every single transition that is part of this tree, will become a rule in the fuzzy controller. It is obvious that using different cost function as searching criteria will lead different fuzzy rule base and eventually different control performance. In [22], Vachtsevanos et. al compared the phase trajectories for minimum time, minimum energy and minimum squared error control for an instance.

4.2.2 Advanced Vehicle Transmission Control

Besides the engine, transmission subsystem is also an important component in the powertrain systems. Every engine has a so called red-line - a maximum rpm value and narrow rpm ranges where horsepower and torque are at their maximum; see Fig.4.7. Transmission's primary job is to allow the engine to operate in its narrow range of speeds while providing a wide range of output speeds by shifting the gear ratios between the engine and the drive wheels [34]-[43].

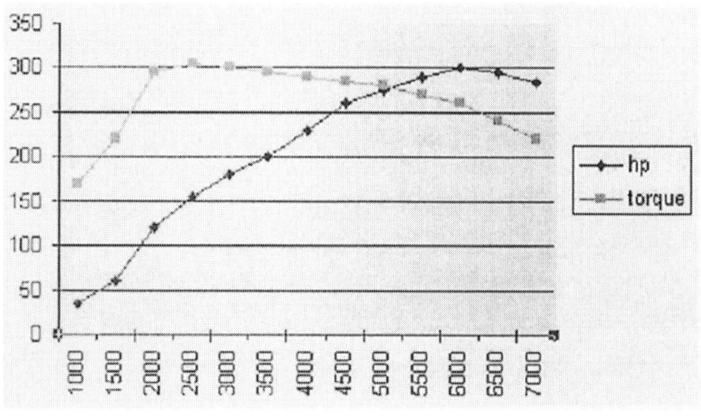

Fig.4.7 Typical horsepower curve for an engine.

Modern passenger cars usually have four or five gear sets. Fig.4.8 shows the engine speed–torque characteristics after the transmission, where each line represents the engine-speed–torque curve at each gear ratio. The transmission allows a vehicle to be able to run at any speed even though the engine operating region is small. A recapitulative shift rule is that at low vehicle speed or high acceleration the transmission is placed at lower gear and on the other hand, at high vehicle speed or low acceleration the transmission is placed at higher gear. Besides vehicle speed and throttle input, many other factors especially driving conditions and driving style may greatly influence the proper gear shifting.

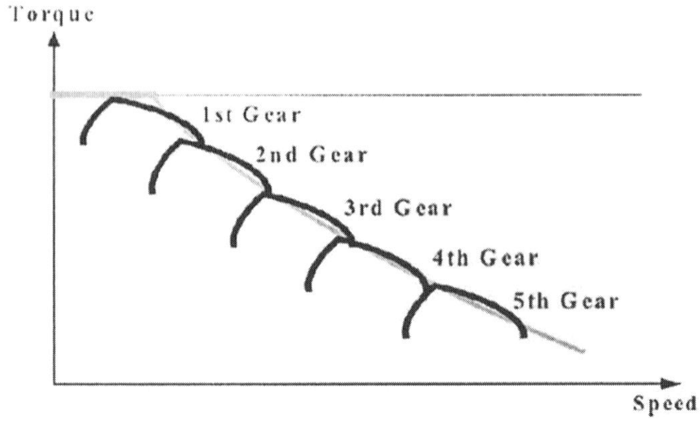

Fig.4.8 Engine speed–torque characteristics after the transmission, from [56]. (© [2004] IEEE)

To further reduce fuel consumption, related research is now extended from independent engine control into the cooperative control of an engine and the so called continuously variable transmission (CVT). The application of dynamic fuel consumption characteristics enables the optimization of fuel consumption in the transient condition as well as in the static condition [44]-[59].

On the other hand, as pointed out in [55], "a vehicle's driveability can be improved at the expense of increased fuel consumption by increasing the power reserve, i.e., by generating the required engine power in high-speed low-torque operating points (far) below the E-line. The driveability can also be improved by incorporating a second power source in the powertrain. Modern hybrid electric vehicles combine a combustion engine with a powerful electric motor and a moderate capacity battery. Unlike purely electric vehicles with their inherent drawbacks of large weight, small driving range and large recharging time, the hybrid electric vehicle is a very attractive concept." Fig.4.9 shows the diagram of engine map that contains some curves of constant fuel mass flow per unit engine power [brake specific fuel consumption (bsfc)] and the E-line [the set of stationary operating points the delivered power is generated with minimum bsfc].

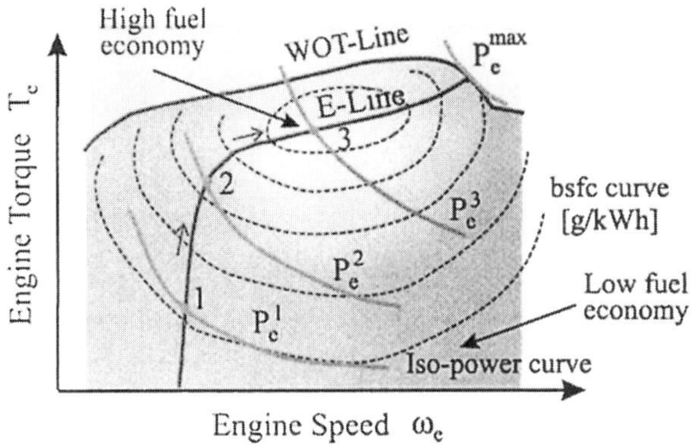

Fig.4.9 Typical engine map, from [57]. (© [2004] IEEE)

The implementation of a CVT requires harmonious control of engine control unit (ECU) and transmission control unit (TCU) under various driving conditions: i.e. to maintain engine speed while changing the pulleys' radio to achieve torque multiplication with acceptable efficiency; or to allow the engine vary within a specified speed range and simultaneously adjust the engine throttle and CVT gear ratio to adjust output torque/speed. To clearly discuss how a CVT works needs to a dedicate book. Besides, different engines and transmission systems requires specific control strategies. Therefore, no detailed discussions on nonlinear CVT control strategy are provided here.

Another interesting problem that is relatively easy to understand and solve is transmission vibration control. It is widely found that vehicles equipped with continuously variable transmission (CVT) usually have relatively low resonance frequencies and thus are prone to generate annoying vibrations [2], [60]-[63]. In [63], various sources that can generate vibrations were listed as follows:

(1) rapid engine torque change;

(2) capacity characteristics of torque converter, Fig.4.10. The characteristics of torque converters can lead to self-induced vibration;

(3) inappropriate engine speed and/or CVT gear ratio control;

(4) unavoidable gear shift.

One way to solve the problem is apply optimal mounting method, i.e. [64]. More approaches applied vibration-suppressing controls [63]:

(1) canceling out vibration by generating anti-phase force;

(2) compensating filters with throttle control;

(3) apply appropriate engine speed and CVT control;

(4) apply damped engine torque control.

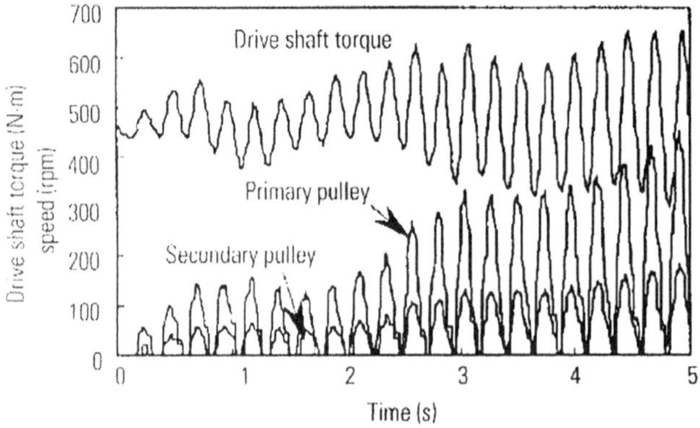

Fig.4.10 Self-induced vibration immediately after start-off, from [63].

For instance, a close-loop H_∞ based active damping system was proposed for vehicle oscillation suppressing as shown in Fig.4.11. The only measured signal is the engine speed ω_{eng}. The controller modifies the torque required by the driver in order to limit the rise of oscillations. In other words, the damping ratio ξ_p of the vibration transfer path $G_{acc}(s)$ is modified by the feedback action.

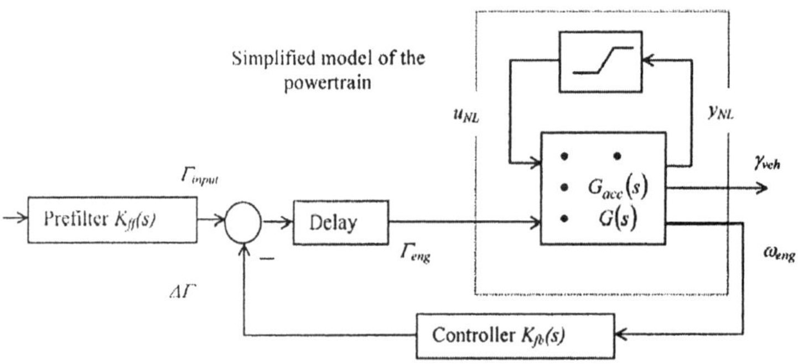

Fig.4.11 Self-induced vibration immediately after start-off, from [62]. (© [2003] IEEE)

More precisely, the feedback control is chosen as active when the input torque Γ_{output} is modified by more than 10% by the corrective torque $\Delta\Gamma$. This method had been validated with real experiments on a French car 406 of PSA Peugeot-Citroen. Fig.4.12 depicts the simulation result of the vehicle acceleration during transients with and without the controller. It is clear that the first oscillation is weaken by using the active compensator when the H_∞ filtering weight functions were carefully selected.

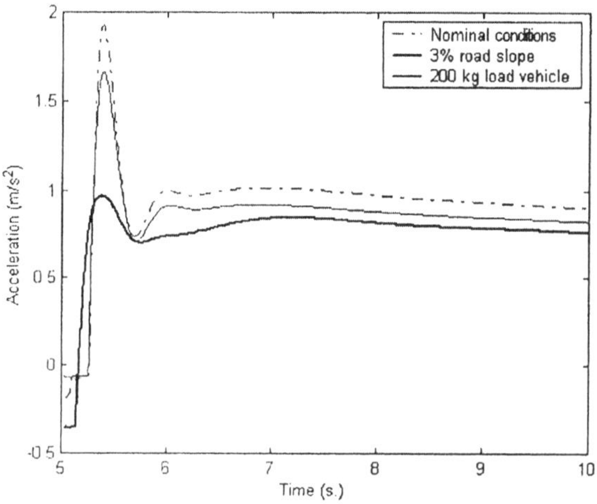

Fig.4.12 Step response of the vehicle acceleration, from [62]. (© [2003] IEEE)

4.3 Vehicle Aerodynamics

Body shape is the main factor imposing on the vehicle aerodynamics. Although not being tightly related to our discussion on the controller design and decision algorithms within this book, vehicle aerodynamics should not be neglected for intelligent vehicle development. Common conclusions have been reached that appropriate vehicle body structure may significantly reduce the aerodynamic resistance and improve vehicle engine combustion process [65]-[75].

Kelly and Holcombe pointed out in [65] that the power required to overcome the aerodynamic resistance is greater than that required to overcome the rolling resistance of the tires and the resistance in the transmission, when a typical passenger car is cruising at a speed higher than approximately 90km/h (50mph). Similar conclusions can be found in Hucho's book [71], too.

In [74], Wong summarized that the vehicle aerodynamic resistance is mainly generated by two factors: the air flow over the exterior of the vehicle body; the air flow through the engine radiator and the interior of the vehicle for purposes of cooling and ventilating. The former factor is the dominant one in most cases and accounts for more than 90% of the total aerodynamic resistance of a passenger car.

In practice, the aerodynamic resistance, which is also called the drag force, is usually expressed in the following form:

$$R_a = \frac{\rho}{2} C_D A_f V_r^2 \tag{4.17}$$

where ρ represents the air density, C_D represents the drag coefficient that combines all the factors described above, which may also be affected by the yaw angle of the vehicle considerably. A_f represents the surface area, and V_r represents the relative velocity of the vehicle to the wind.

It should be mentioned that the air density ρ would be greatly affected by the atmospheric conditions. For examples, an increase in ambient temperature from $0^o C$ to $38^o C$ will cause a 14% reduction in aerodynamic resistance, and an increase in altitude of 1219m will lead to a decrease in aerodynamic resistance by 17%, [74].

The coefficient of aerodynamic resistance C_D and the front area A_f may significantly vary with different body shapes of vehicles. In [71], Hucho showed some typical drag coefficient values in his book for a variety of commercial vehicles.

A typical case that has often been addressed is the heavy tractor-semi-trailer. Hucho studied the influence of drag on fuel consumption of a 38-tonne semi-trailer in [71]. Cresswell and Hertz estimated in [66] that the fuel consumption L for a gasoline powered truck is

$$L \approx 1.264\, C_D \tag{4.18}$$

while for a diesel-engine powered truck it is,

$$L \approx 0.8051\, C_D \tag{4.19}$$

Considering a 35% power train efficiency for a diesel engine, estimates in [66], indicate 800 liters/yr fuel savings between the M1 and M3 mirror, for 150,000 km annual highway driving. Fuel savings estimated for a 25% power train efficiency for a gasoline fueled truck under similar estimation expects is 1250 liters.

In [73], Englar investigated those approaches for aerodynamic resistance reduction including: physically providing a smooth surface, such as a boat tail to prevent flow separation and turbulence at the rear of the trailer; completely sealing the gap between the tractor and the trailer; and sealing the underbody of the vehicles. He made estimation by personnel of the American Trucking Associations (ATA) and concluded that approaching 35% of these drag reductions could result in roughly 1.2 billion gallons of fuel saved per year if they are applied to today's US heavy vehicle fleet only. With fuel prices rising dramatically in 1999-2000, drag reduction naturally becomes a vital concern to the trucking industry. Nevertheless, People also found that inappropriate design will raise accident risk [67]. Therefore, with growing emphasis on fuel economy and environment protection, vehicle body structure design has gained more attention recently.

In order to optimize the vehicle's aerodynamic performance at moderate and higher speed, considerable efforts have been expended especially for Indy racers, trucks, and other high speed/sided vehicles. Let's still the tracker-semi-trailer for examples. Considerable techniques haven been tested to reduce Heavy Vehicle drag coefficients to improve highway operating efficiency.

Usually, C_D and A_f are obtained by wind tunnel testing of scale model or full size sample vehicle. In United States, 3/8 scale model is widely used for passenger car, while in Europe 1/4 scale is the most common. However, the wind tunnel testing method simply leads to considerable money and time consumption. Gradually recognizing the limitations of conventional wind tunnel testing, designers preferred considerable alternative efforts in the last decade to study vehicle aerodynamics computationally. These efforts range from studying the window profile of the vehicle [69] to drag implications of truck mirrors [66].

The computational fluid dynamics (CFD) method is a frequently used tools in this field. For instance, Takeuchi and Kohri described a numerical field flow analysis method in [70] for predicting aerodynamic drag and engine cooling performance for trucks and buses. In their paper, an adequate method was developed to accurately obtain the wake flow behind the body. Now, more and more researchers use CFD in vehicle shape design.

4.4 Advanced Vehicle Tracking and Braking Control

The braking subsystem is an essential part of vehicle to guarantee driving safety. Different vehicles may have different implementation of braking subsystem. In the last twenty years, related research can be divided into two aspects:

(1) enhanced vehicle tracking and braking control which aims to make vehicle powertrain and braking subsystems cooperate with each other [77]-[91];

(2) anti-lock braking system design [92]-[119].

However, these two parts cannot be thoroughly separated since the integrated tracking/braking control needs to consider the kinematics of the ABS system that has been applied.

4.4.1 Advanced Vehicle Tracking and Braking Control

When starting on slippery roads and slopes, excessive traction torque will cause the wheels to spin, which leads to tire slip and thus the decrease of traction performance, directional stability and steerability. In order to solve this problem, research on advanced tracking and braking control that maintains slip ratios of driven wheels low by controlling the traction torque using various vehicle parts have been conducted. Conventional commercial automotives often use empirical methods such as lookup table and gain scheduling to achieve proper performance under various road and vehicle conditions. Recently some non-empirical methods were presented.

It is naturally impossible to give a general model for cooperative tracking and braking control; because various engines, transmission systems and braking systems had been designed, while each of them has its unique kinematics. Thus, this section addresses a typical longitudinal powertrain model for heavy duty vehicle proposed by Lu and Hedrick in [88] as shown in Fig.4.13. The meanings of the symbols are listed as:

M - vehicle mass, ω - engine speed, ω_{idle} - engine idle speed,

$\omega_p = \omega$ - torque converter pump speed, ω_t - torque converter turbine speed,

ω_{tb} - turbocharger speed, ω_{tr} - transmission output speed,

ω_{dr_1} - propeller-shaft speed, ω_{dr_2} - driver-shaft speed,

ω_{dr} - drive-line speed (considered as lump sum), final driver end,

ω_w - wheel angular speed, v - vehicle wheel speed (longitudinal),

a - acceleration, α - throttle angle, α_f - fueling rate,

I_e - engine inertia, I_{dr_1} - drive line inertia (before final gear),

I_{tr} - transmission inertia, I_{dr_2} - drive line inertia (after final gear),

I_{dr} - lump sum drive line inertia, where $I_{dr} = I_{dr_1} + I_{dr_2}$,

I_w - wheel inertia,

P_m - intake manifold pressure or turbocharger booster pressure,

T_{net} - engine net output torque, T_p - torque converter pump torque,

T_t - torque converter turbine torque, T_b - service brake torque,

T_{jk_i} - engine Jake brake torque when i cylinder is on, here $i = 2,4,6$

T_{jk} - engine Jake brake torque, T_{tr} - transmission output torque,

T_{dr_1} - final gear input torque or equivalently propeller shaft final end torque,

T_{dr_2} - final gear input torque, T_{fric} - engine friction torque,

T_w - engine torque passed onto wheel, T_{rtd} - transmission retarder torque,

T_{e-brk} - engine braking effect torque when $T_{net}<T_{net\_des}$,

F_a - aeroynamic resistance force, F_r - rolling resistance force,

F_{eng_brk} - engine braking force transmitted to wheels when throttle is release,

F_w - engine brake force when clutch is engaged and fueling is in idle,

F_f - friction force, h_r - effective tire radius,

θ - road grade, $\theta > 0$ means ascending,

r_g - transmission gear ratio,

R_g - gear ratio, where $R_g = r_g r_d$, $\omega_t = vR_g / h_r$,

P_b - brake pressure force at brake drum or output pressure of diaphragm,

P_{app} - applied brake pressure, A_b - area of diaphragm,

τ_a - air brake activation time delay (pure time delay),

τ_r - air brake release time delay (pure time delay),

τ_{ba} - air brake time parameter for activation,

τ_{br} - time parameter for release,

τ_d - air brake drum radius,

V_b - applied brake voltage of transmission retard,

σ_0 - constant coefficient determined by manufacturer,

σ_b - friction coefficient between brake drum and brake pad.

The engine driving mode can be written as

$$T_{net} = I_e\dot{\omega} + T_p \tag{4.20}$$

and

$$T_t = \begin{cases} T_p, & lockup \\ \eta(T_p, \omega_p, \omega_t), & tq\_converter \end{cases} \tag{4.21}$$

In the reference of Lu and Hedrick 2003, they used static Kotwicki model which is an example of torque converter model for η-function.

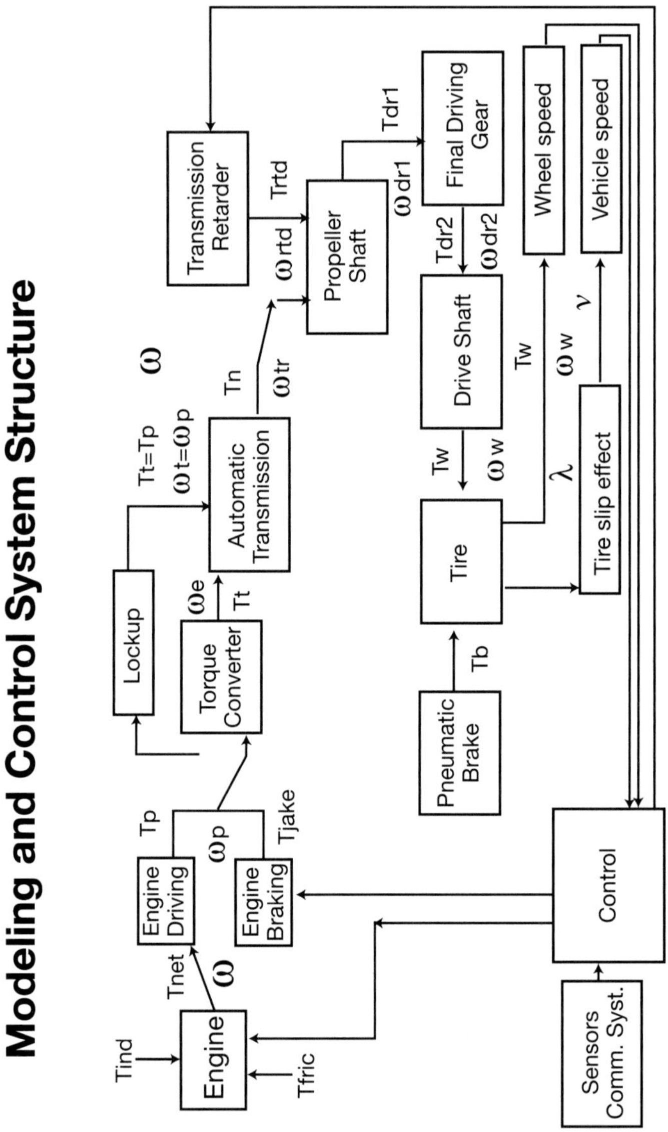

Fig.4.13 Schematic of powertrain system, from [88]. (© [2003] IEEE)

Notice for the following relationships

$$\omega_w = \omega_{dr2} = \frac{v}{h_r},\ \omega_{tr} = \omega_{dr1} = \frac{v}{h_r} r_d,\ \omega = \omega_t = \frac{v}{h_r} r_d r_g \tag{4.22}$$

The longitudinal vehicle dynamics can be modeled as

$$\dot{v} = \frac{r_d r_g T_{net} - (r_d T_{rtd} + T_b + F_a h_r + F_r h_r + Mgh_r \sin\theta)}{\bar{I}} \tag{4.23}$$

where

$$\bar{I} = r_d^2 r_g^2 \frac{I_e}{h_r} + r_d^2 \frac{I_{tr}}{h_r} + r_d^2 \frac{I_{dr1}}{h_r} + \frac{I_{dr2}}{h_r} + \frac{I_w}{h_r} + Mh_r \tag{4.24}$$

and

$$F_a = 0.5 C_a \rho_{air} A V_a^2 \tag{4.25}$$

where T_b is modeled separately.

It is obvious that the overall dynamics is intrinsically high dimensional and highly nonlinear, because the vehicle model includes varied components: vehicle dynamics, engine, braking systems, transmission and tire.

In the model shown above, the braking system for automatic control is composed of three parts: Jake brake, pneumatic brake and transmission retarder. And each part has its own characteristics. Therefore, the total braking torque on wheel can be written as

$$T_{total} = \begin{cases} T_b + T_{tr}^{(w)} + T_{jk}^{(w)}, & \omega \geq 800[rpm] \\ T_b + T_{tr}^{(w)}, & \omega < 800[rpm] \end{cases} \tag{4.26}$$

The Jake brake provides discrete and limited braking torque on the driving wheels with fast response. Its retarding force mainly depends on engine speed. Besides it, the engine braking effect can also be caused by the mismatch between engine speed and wheel speed when the throttle is released and drive-line is engaged [87]. It was recognized that this effect can be considered as a special case of Jake Brake when there is no valve is open. From this point of view, the braking torque on driving wheels provided by engine can be modeled in a variable structure model as

$$T_{jk} = \begin{cases} T_{jk\_0}, & \text{0 cyliner on} \\ T_{jk\_2}, & \text{2 cyliner on} \\ T_{jk\_4}, & \text{4 cyliner on} \\ T_{jk\_6}, & \text{6 cyliner on} \end{cases} \tag{4.27}$$

where

$$T_{jk\_0} = T_{eng\_brk} = \frac{C_{eng\_brk}(\omega - \omega_{idle})}{R_g} \tag{4.28}$$

and

$$T_{jk\_i} = \frac{C_{jk\_i}(\omega - \omega_{idle})}{R_g}, \ \omega \geq 800[rpm], \ i = 2,4,6 \tag{4.29}$$

together with

$$R_g = r_g r_d \tag{4.30}$$

It is clear that

$$T_{jk\_0} < T_{jk\_2} < T_{jk\_4} < T_{jk\_6} \tag{4.31}$$

And the pneumatic (air) brake provides continuous and large braking torque on all wheels. Its response is slow. Pneumatic brake model for control design was proposed as

$$\dot{P}_b = \begin{cases} \dfrac{1}{\tau_{ba}}(-P_b + A_b P_{app}(t - \tau_a)), & \text{activation} \\ \dfrac{1}{\tau_{br}}(-P_b + A_b P_{app}(t - \tau_r)), & \text{release} \end{cases} \tag{4.32}$$

The applied brake pressures $P_{app}(t - \tau_a)$ and $P_{app}(t - \tau_r)$ can be expends as follows using Taylor series approach

$$P_{app}(t - \tau_a) = P_{app}(t) - \dot{P}_{app}(t)\tau_a \tag{4.33}$$

and

$$P_{app}(t - \tau_r) = P_{app}(t) - \dot{P}_{app}(t)\tau_r \tag{4.34}$$

Substitute (4.33) and (4.34) to (4.32), it has

$$\dot{P}_b = \begin{cases} \dfrac{1}{\tau_{ba}}[-P_b + A_b(P_{app}(t) - \dot{P}_{app}(t)\tau_a)], & \text{activation} \\ \dfrac{1}{\tau_{br}}[-P_b + A_b(P_{app}(t) - \dot{P}_{app}(t)\tau_r)], & \text{release} \end{cases} \tag{4.35}$$

To implement (4.35), $\dot{P}_{app}(t)$ is actually calculated from $P_{app}(t)$ on line by a difference method or an integral filter.

The transmission retarder provides braking torque on driving wheels with slow response and limited braking torque. In [87], Lu and Hedrick establish the model for transmission retarder as

$$\dot{T}_{tr} = \begin{cases} \dfrac{1}{\tau_{tr}}[-T_{tr} + \varsigma(V_{rtd}(t-\kappa_{tr}), \omega_{dr})], & t < \tau_{tr0} \\ \dfrac{1}{\tau_{tr}}[-T_{tr} + \varsigma(V_{rtd}(t), \omega_{dr})], & t \geq \tau_{tr0} \end{cases} \tag{4.36}$$

and

$$\kappa_{tr} = \tau_{tr0}[1 - \frac{t}{\tau_{tr0}}] = \tau_{tr0} - t \tag{4.37}$$

where τ_{tr0} is pure time delay parameter. V_{rtd} is voltage or transmission retarder pedal deflection. $\varsigma(V_{rtd}(t), \omega_{dr})$ is a nonlinear function and κ_{tr} is chosen as 1.5s.

It is noted that the pure time delay parameter κ_{tr} is time varying and it disappears after 1.5s. This time period is used for filling up the liquid into the retarder chamber.

Applied torque on wheel is calculated as

$$T_{tr}^{(w)} = (T_{tr} - \omega_{dr1} I_{dr1}) r_d - \omega_{dr2} I_{dr2} \tag{4.38}$$

To deal with the time delay part in the model, Lu and Hedrick 2003 approximated the $\varsigma(V_{rtd}(t-\kappa_{tr}))$ by its first order Taylor series

$$\varsigma(V_{rtd}(t-\kappa_{tr})) = \varsigma(V_{rtd}(2t-\tau_{tr0})) = \varsigma(\widetilde{V}_{rtd}(t - \frac{\tau_{tr0}}{2})) \tag{4.39}$$

where $\widetilde{V}_{rtd}(p) = V_{rtd}(2p)$.

Thus, it can approximately have

$$\varsigma(V_{rtd}(t-\kappa_{tr})) \approx \varsigma(\widetilde{V}_{rtd}(t)) - \dot{\varsigma}(\widetilde{V}_{rtd}(t))\frac{\tau_{tr0}}{2} \tag{4.40}$$

From (4.38) and (4.40), the following relationship exits as

$$\dot{T}_{tr} = \begin{cases} \dfrac{1}{\tau_{tr}}[-T_{trd} + \varsigma(\widetilde{V}_{rtd}(t)) - \dot{\varsigma}(\widetilde{V}_{rtd}(t))\dfrac{\tau_{tr0}}{2}], & t < \tau_{tr0} \\ \dfrac{1}{\tau_{tr}}[-T_{trd} + \varsigma(V_{rtd}(t)], & t \geq \tau_{tr0} \end{cases} \tag{4.41}$$

Lu and Hedrick designed the braking system control strategy as follows

(1) If $T_{brk\_total} \le T_{jk\_0}$, category, no pneumatic brake nor Jake brake is necessary but throttle is released;

(2) If $T_{jk\_0} \le T_{brk\_total} \le T_{jk\_2}$, $T_b^{(des)} + T_{rtd}^{(des)} = T_{brk\_total} - T_{jk\_0}$;

(3) If $T_{jk\_2} \le T_{brk\_total} \le T_{jk\_4}$, $T_b^{(des)} + T_{rtd}^{(des)} = T_{brk\_total} - T_{jk\_2}$;

(4) If $T_{jk\_4} \le T_{brk\_total} \le T_{jk\_6}$, $T_b^{(des)} + T_{rtd}^{(des)} = T_{brk\_total} - T_{jk\_4}$;

(5) If $T_{jk\_6} \le T_{brk\_total}$, $T_b^{(des)} + T_{rtd}^{(des)} = T_{brk\_total} - T_{jk\_6}$.

The distribution of the desired braking torque on air brake and transmission retarder is completely up to the designer.

Based on the relationship between applied brake pressure and brake torque,

$$T_b = r_b \sigma_b P_b, \quad P_{app} = \sigma_0 V_b, \quad P_{app} \le P_{res} \tag{4.42}$$

the desired brake pressure should be

$$P_b^{(des)} = \frac{T_b^{(des)}}{r_b \sigma_b} \tag{4.43}$$

To make the drive error system

$$\dot{P}_b - \dot{P}_b^{(des)} = \begin{cases} \dfrac{1}{\tau_{ba}}[-P_b + A_b(P_{app}(t) - \dot{P}_{app}(t)\tau_a)] - \dot{P}_b^{(des)}, & \text{activation} \\ \dfrac{1}{\tau_{br}}[-P_b + A_b(P_{app}(t) - \dot{P}_{app}(t)\tau_r)] - \dot{P}_b^{(des)}, & \text{release} \end{cases} \tag{4.44}$$

When converging to zero, it should have

$$P_{app} = \begin{cases} \dfrac{-\tau_{ba}}{\tau_a A_b}\exp(\dfrac{t}{\tau_a})\displaystyle\int_0^t \left[\frac{1}{\tau_{ba}} P_b^{(des)} + \dot{P}_b^{(des)}\right] e^{\sigma/\tau_a} d\sigma, & \text{activation} \\ \dfrac{-\tau_{br}}{\tau_r A_b}\exp(\dfrac{t}{\tau_r})\displaystyle\int_0^t \left[\frac{1}{\tau_{br}} P_b^{(des)} + \dot{P}_b^{(des)}\right] e^{\sigma/\tau_r} d\sigma, & \text{release} \end{cases} \tag{4.45}$$

Experimental work with the modeling conducted as in and control strategy had been carried out at Richmond Field Station at low speed and Crows Landing at high speed for a single Freightliner. The used transmission system is Allision Company produced 6 gear automatic transmission set. The power source is the Detroit 365hp turbocharged diesel engine equipped with 3 Jake brake modes: 2, 4 and 6 cylinders respectively. The truck is partially loaded with total weight of 18182kg.

A reference speed trajectories used as the virtual front vehicle. Fig.4.14 and Fig.4.15 show the distance tracking error can be restricted with 2m. But transient distance tracking error is 15-25m.

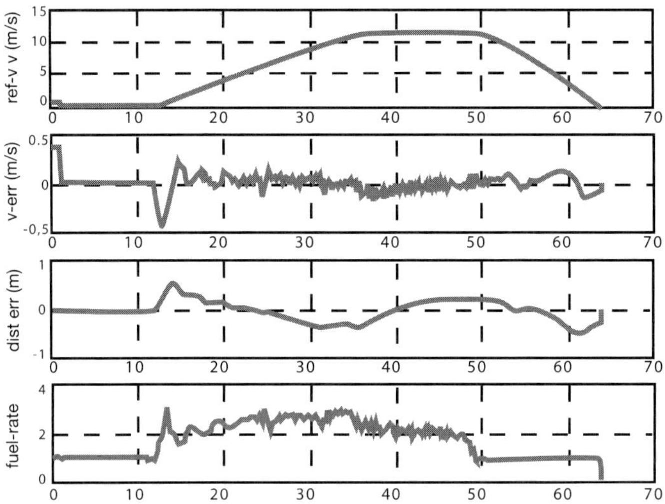

Fig.4.14 Low speed test: Max speed 25mph, from [88]. (© [2003] IEEE)

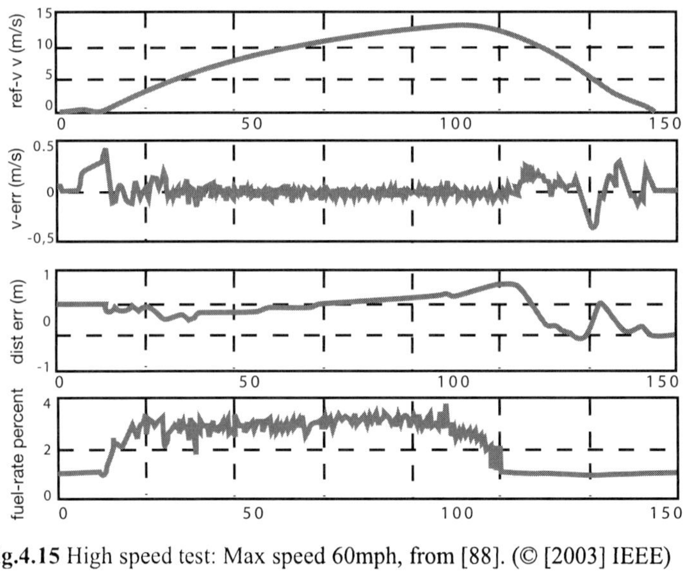

Fig.4.15 High speed test: Max speed 60mph, from [88]. (© [2003] IEEE)

4.4.2 Anti-Lock Braking System Design

Anti-lock braking system (ABS) for commercial vehicles first appeared on the market in 1960s and began its fast growth in 1970s with the emergences of microcomputers and advanced electronics technologies [92]-[95]. ABS had been recognized as an important contribution to road safety. It is now available in almost all types of vehicles. The automotive industry is continuously developing new generations of ABS. Now, ABS has appeared as a standard in nearly every new vehicle.

The advantage of ABS can be clearly seen when comparing the emergency braking situation of vehicles with and without ABS. Since the driver wants to reduce the speed of the vehicle as fast as possible, the wheels of the vehicle without ABS will lock and the vehicle will start sliding, which leads to some undesirable effects. One effect is that the tire wear will not be equally distributed over the whole tire. Another unexpected effect is that the vehicle becomes unsteerable as soon as the wheels lock. This might be quite dangerous in the case when the driver wants to avoid an obstacle during the braking maneuver. On the contrary, in a vehicle with ABS, the intelligent sensors can monitor the rotation of the wheels and reduced the brake pressure if the wheels are locked. Therefore, the wheels are still rolling steerable, and a high friction between the road and tire is simultaneously achieved, which stop the vehicle.

As been mentioned in *Chapter 2*, ABS design is tightly related to tire/road friction modeling. Thus, ABS controller can be divided into two kinds: ABS based on empirical tire/road friction models and analytical tire/road friction models.

In [108], Schinkel and Hunt 2002 formulated the quarter car braking model as

$$\dot{\lambda} = -\frac{1}{v}\left[\frac{1}{m}(1-\lambda) + \frac{r^2}{J}\right]F_z\mu(\lambda) + \frac{1}{v}\frac{r}{J}T_b \tag{4.46}$$

and

$$\dot{v} = -\frac{1}{m}F_z\mu(\lambda) \tag{4.47}$$

where v is the vehicle longitudinal speed, m is the vehicle mass, J is the wheel inertia, r is the wheel radius, λ is the tire slip, μ is the friction function between tire and road, F_z is the vertical force/dynamic load and T_b brake torque.

Schinkel and Hunt approximated $\mu(\lambda)$ with two piecewise linear functions

$$\mu = a\lambda\,,\ \lambda \le 0.1 \tag{4.48}$$

and

$$\mu = -\frac{1}{4}\lambda + \frac{3}{4} \pm 0.2\,,\ \lambda > 0.1 \tag{4.49}$$

where a was chosen as $a \in [5.75, 9.75]$ in [108]. And the notation ± 0.2 means that any arbitrary unfixed value should lies in the boundary interval $(-0.2, 0.2)$.

Linearizing the (4.50) and (4.51) through Taylor series expending, it has

$$\begin{cases} \dot{x} = A_q x + E_q + Bu^* \\ y = C_q x \\ q = f(x) \end{cases} \tag{4.50}$$

where A_q, B and C_q are the approximate system matrix, input matrix and output matrix, respectively. E_q is the affine terms and $f(\cdot)$ is the function telling which linearization is valid.

For each linearization point $q \in \{1, 2, \ldots, M\}$ with $\lambda \le 0.1$, it has

$$A_q = \begin{bmatrix} 0 & -a\dfrac{F_z}{m} \\ a\dfrac{F_z r^2 \lambda_{wp}}{v_{wp}^2 J} & -a\dfrac{F_z r^2}{v_{wp} J} \end{bmatrix},\ E_q = \begin{bmatrix} 0 \\ -a\dfrac{F_z r^2 \lambda_{wp}}{v_{wp} J} \end{bmatrix} \tag{4.51}$$

For each linearization point $q \in \{M+1, M+2, \ldots, N\}$ with $\lambda > 0.1$, it has

$$A_q = \begin{bmatrix} 0 & \dfrac{F_z}{4m} \\ (-\dfrac{\lambda_{wp}}{4} + \dfrac{3}{4})a\dfrac{F_z r^2}{v_{wp} J} \pm 0.2\dfrac{F_z r^2}{v_{wp} J}\dfrac{F_z r^2 \lambda_{wp}}{v_{wp}^2 J} & \dfrac{F_z r^2}{4 v_{wp} J} \end{bmatrix} \tag{4.52}$$

and

$$E_q = \begin{bmatrix} (-\dfrac{3}{4} + \dfrac{3}{4})\dfrac{F_z}{m} \\ (\dfrac{\lambda_{wp}}{4} - \dfrac{3}{2})a\dfrac{F_z r^2}{v_{wp} J} \pm 0.4\dfrac{F_z r^2}{v_{wp} J} \end{bmatrix},\ B = \begin{bmatrix} 0 \\ \dfrac{r}{J} \end{bmatrix} \tag{4.53}$$

and $u^* = u \cdot v$, $x^T = [v, \lambda]$ for $q \in \{1, 2, \ldots, N\}$.

Obviously, the braking objective is to design a controller which decelerates the vehicle as fast as possible and maintains steerability. Notice that we want to have $\dot{\lambda} > 0.1$, when $\lambda \leq 0.1$ and $v > 0$. Thus

$$\dot{\lambda} = -\frac{1}{v}\left[\frac{1}{m}(1-\lambda) + \frac{r^2}{J}\right]F_z\mu(\lambda) + \frac{1}{v}\frac{r}{J}T_b \geq 0 \tag{4.54}$$

Therefore

$$T_b \geq \frac{aJ}{r}\left[\frac{1}{m}(1-\lambda) + \frac{r^2}{J}\right]F_z\lambda$$

Or, approximately

$$T_b \geq aJF_z\lambda \tag{4.55}$$

However, as pointed out in [108], the admissible control inputs in the real situations, should satisfy $0 \leq T_b < 0.3rF_z$. Therefore, a more applicable sliding mode controller was introduced as

$$\dot{s} = -\frac{r^2 F_z\mu(\lambda)}{vJ} + \frac{1}{v}\frac{r}{J}T_b + Ke \tag{4.56}$$

where $s = (\frac{d}{dt} + K)\int_0^t ed\tau$ is the sliding surface and $e = \lambda - \lambda_d$.

$$T_b \geq aJF_z\lambda \tag{4.57}$$

And the control input is

$$\hat{T}_b = 10rF_z\lambda + \frac{vJ}{r}Ke \tag{4.58}$$

for $\lambda \leq 0.08$.

$$\hat{T}_b = -\frac{sK_p + K_i}{s}e \tag{4.59}$$

for $0.08 \leq \lambda \leq 0.12$.

$$\hat{T}_b = (-\frac{1}{4}\lambda + \frac{3}{4} - 0.2)rF_z - \frac{vJ}{r}Ke \tag{4.60}$$

for $\lambda \geq 0.12$.

There were several similar approaches based on piecewise line approximation of tire/road friction model including [94] and [109]. But recently, more and more approaches employed complex models. For instance, in [104], Yi and Alvarez applied the modified Burckhardt model. In their approach, the system dynamics are formulated as

$$\dot{v} = -c\mu - dv^2 \tag{4.61}$$

and

$$\dot{s} = -(a+c)\mu - b - dv^2 + eK_bP_b \tag{4.62}$$

where $a = R^2mg/4J$, $b = R\tau_d/J$, $c = g$, $d = C_{ax}/m$ and $e = R/J$. R is the effective rolling radius, τ_b is the braking torque and τ_d is the driving torque.

And the braking torque is approximated by K_bP_b, where K_b is an overall braking system gain and P_b the master cylinder pressure. During braking, it is assumed $\tau_d = 0$. The velocity and relative velocity are assumed to be uniformly continuous functions of time.

Assume that vehicles are equipped with ABS and that the longitudinal and angular velocities can be measured. In the case of AHS, the longitudinal velocity and acceleration can be derived from infrastructure devices designed to facilitate vehicle's position detection such as road magnets and accelerometer, respectively. Then the vehicle slippage can be estimated from the method presented below.

Define

$$\tilde{s} = s - s_m \tag{4.63}$$

where $s_m = \hat{\lambda}_m v$ is the peak relative velocity that corresponds to the estimated peak slip $\hat{\lambda}_m$ at velocity v.

The approximated tire/road model that was proposed is

$$\mu = p_1' e^{-p_2\lambda} \lambda^{(p_3\lambda + p_4)} e^{-p_5 v} \tag{4.64}$$

where p_1', p_2, p_3, p_4 and p_5 are parameters to be determined.

After applying logarithm to both sides of (4.64) and rearranging it in vector form, we obtain

$$y = \ln\mu = U\Theta \tag{4.65}$$

where $U = [1, -\lambda, \lambda\ln\lambda, \ln\lambda, -v]$, $\Theta = [p_1, p_2, p_3, p_4, p_5]$, and $p_1 = \ln p_1'$.

An estimation $\hat{\Theta}$ of the vector $\dot{\hat{\Theta}} = \Gamma U^T \hat{y}$ Θ can be obtained via a standard parameter adaptation algorithm (PAA)

$$\dot{\hat{\Theta}} = \Gamma U^T \hat{y} \tag{4.66}$$

where $\tilde{y} = y - \hat{y} = U(\Theta - \hat{\Theta}) = U\tilde{\Theta}$. Γ is a diagonal matrix of gains.

If the velocity is kept constant, the peak value of μ_m can be derived as:

$$\mu_m\Big|_{v=v_0} = p_1' e^{-p_2\lambda_m} \lambda^{(p_3\lambda_m + p_4)} e^{-p_5 v_0} \tag{4.67}$$

where λ_m the peak slip is given by the solution through

$$p_3\lambda_m(\ln\lambda_m + 1) = p_2\lambda_m - p_4 \tag{4.68}$$

Notice that although the peak friction value in (4.67) and (4.68) changes with velocity, the peak slip does not change with velocity in the same curves. This property of peak slip not depending on velocity is observed from the experimental data used to fit the pseudostatic friction models, like the "Magic Formula" or the Burckhardt model.

Therefore, s_m can be indirectly obtained from (4.67) and (4.68) as well as the current estimation $\tilde{\Theta}$. Correspondingly, the velocity error could be defined as $\tilde{v} = v - v_d$, with v_d as the desired velocity. However as for emergency braking $v_d = 0$, this definition of is not necessary.

With the estimation result, Yi et. al designed the braking pressure P_b as

$$P_b = \frac{\hat{M}_b}{e}\left[(a+c)\mu + b + dv^2 - \varsigma\hat{s} + \dot{\hat{\lambda}}_m v\right] \tag{4.69}$$

where $\varsigma > 0$ is a gain, $\hat{M}_b = 1/\hat{K}_b$ with $\hat{K}_b$ the estimated value of K_b and μ is derived from (4.61) under the assumption that the longitudinal acceleration can be measure.

Define $\tilde{M}_b = M_b - \hat{M}_b$, and $\tilde{M}_b$ is chosen as

$$\hat{M}_b = \xi\bar{s}\left[(a+c)\mu + b + dv^2 - \varsigma\hat{s} + \dot{\hat{\lambda}}_m v\right] \tag{4.70}$$

where $\xi > 0$ is a gain parameter. Yi, Alvarez and Horowitz proved that such an adaptive controller (4.69) and (4.70) can guarantee the stability of the system if

$$\hat{\lambda}_m \le \lambda_m ,\ \hat{\mu}_m \le \lambda_m \tag{4.71}$$

Then, they investigated the conditions on Θ , $\hat{\Theta}(0)$ and the adaptation gains which will guarantee Ineq.(4.71) to hold. This conclusion reveals that in this condition can be satisfied with appropriately selected initial parameters.

Fig.4.16 shows the simulation results for an emergency braking maneuver. Notice that underestimation of maximum friction coefficients and slip are always guaranteed, thus the system keeps stable.

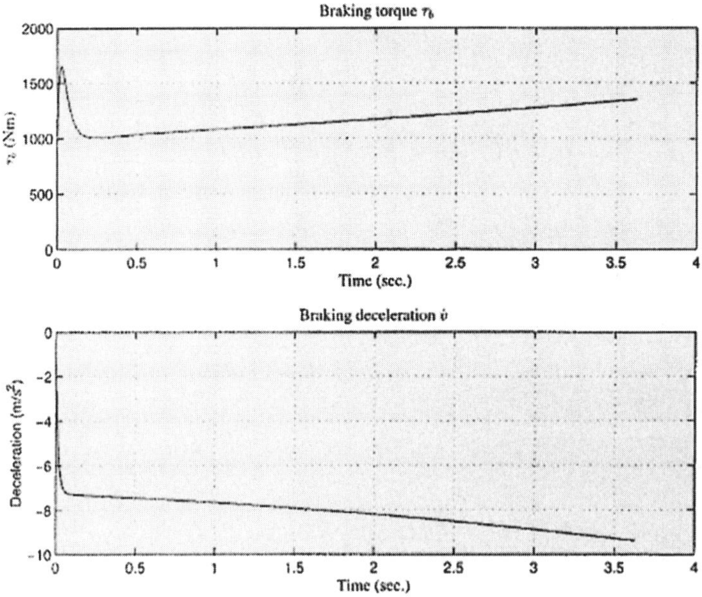

Fig.4.16 Braking torque and deceleration, from [104]. (© [2002] IEEE)

ABS control based on the famous "Magic" formula also received increasing attentions in the last decade. For example in [97], Tsiotras and De Wit formulated the equations of motion of the one-wheel dynamic model as Fig.4.17, where

$$F_r = m\dot{v}\,,\ J\dot{\omega} = -rF_r + u \tag{4.72}$$

where m denotes the 1/4 of the vehicle (suppose this is a four wheels vehicle and the gravity effect of the mass are equally assigned on four tires) and J , r are the inertia and radius of the wheel, respectively. u denotes the accelerating (or braking) torque.

For time optimality, the objective was selected as

$$J_{objective} = \int_0^{t_f} v(\tau)d\tau \tag{4.73}$$

subject to the boundary conditions

$$v(0) = v_0, \ \omega(0) = \omega_0, \ v(t_f) = \omega(t_f) = 0 \tag{4.74}$$

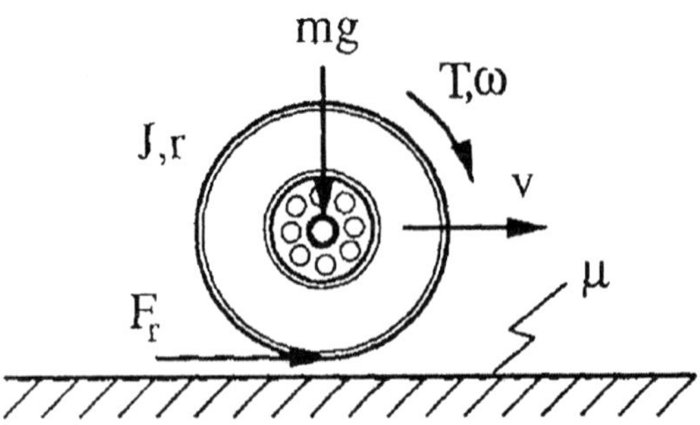

Fig.4.17 One-wheel dynamic model, from [97]. (© [2000] IEEE)

Introduce state variables $x_1 = v$, $x_2 = r\omega$, it gets

$$\begin{cases} \dot{x}_1 = f(x) \\ \dot{x}_2 = -\rho f(x) + u \end{cases} \tag{4.75}$$

where

$$f(x) = \frac{F_r}{m}, \ \rho = \frac{mr^2}{J}, \ u = \frac{rT}{J} \tag{4.76}$$

Therefore, the Hamiltonian for the previous optimal control problem is given by

$$H = x_1 + \lambda_1 f(x) + \lambda_2 u - \lambda_2 \rho f(x) \tag{4.77}$$

where λ_1 and λ_2 are the adjoint variables.

The adjoint system is then be given as

$$\dot{\lambda}_1 = -1 - (\lambda_1 - \lambda_2 \rho)\frac{\partial f}{\partial x_1}, \ \dot{\lambda}_2 = -(\lambda_1 - \lambda_2 \rho)\frac{\partial f}{\partial x_2} \tag{4.78}$$

Since the final time is not specified, the transversal condition yields

$$H(t_f) = 0 \tag{4.79}$$

Notice that the Hamiltonian is not an explicitly function of time, we can have

$$H(t) = 0, \ \forall t \in [0, t_f] \tag{4.80}$$

along the optimal trajectory. And the optimal control is expressed by

$$u_{opt} = \arg\min H(x,\lambda,u) = \begin{cases} u_{\min}, & H_1 > 0 \\ u_{\min}, & H_1 < 0 \\ u_{\min}, & H_1 \equiv 0 \end{cases} \tag{4.81}$$

where $H_1 = \lambda_2$ is the switching function of (4.78). And the Control input u is assumed to be bounded, that is $u_{\min} \leq u \leq 0$.

On the singular arc $\lambda_2 = 0$,

$$\ddot{H}_1 = \frac{d}{dt}\left(\frac{\partial f}{\partial x_2}\right) = 0 \tag{4.82}$$

Or further

$$\ddot{H}_1 = \alpha(x) - \beta(x)u_{\sin g} = 0 \tag{4.83}$$

where

$$\alpha(x) = -\frac{\partial^2 f}{\partial x_1 \partial x_2} f + \frac{\partial^2 f}{\partial x_2^2} f, \quad \beta(x) = \frac{\partial^2 f}{\partial x_2^2} \tag{4.84}$$

Then, under the implicit assumption that $\beta(x) \neq 0$, the singular control is given by

$$u_{sign} = \frac{\alpha(x)}{\beta(x)} \tag{4.85}$$

Based on Legendre-Clesch condion and Kelley-Contensou test, Tsiotras and De Wit pointed out that

$$\beta(x) \leq 0 \tag{4.86}$$

Further analysis shows that Eq.(4.86), which is the expression of singular control during braking, can be finally written as

$$u_{sign} = \frac{x_2 f(x_1, x_2) + x_1 \rho f(x_1, x_2)}{x_1} \tag{4.87}$$

The major drawback of such statistical approaches is transient phenomena are quite important for ABS. In [106]-[107], a control scheme for emergency braking of vehicles is designed utilizing LuGre dynamic friction model to estimate the tire/road friction. The control system output is the pressure to the braking system, and is calculated using only the wheel angular speed. The controller utilizes estimated state feedback control to achieve near maximum deceleration. The state observer gain is finally calculated by using linear matrix inequality (LMI) techniques. It is proven that it outperforms the similar optimal control based on static model given in [97].

Moreover, as pointed out in [97], static friction approach also requires a prior knowledge of the maximum friction force and the corresponding optimal slip, which may not be readily known in a realistic environment of changing road and tire conditions. This drawback can be avoided by implementing the optimal strategy via "extremum seeking" control schemes, such as the methods proposed in [111]-[114]. Besides, neural networks [115] and fuzzy theory [116]-[119] were also put into ABS design.

4.5 Adaptive Cruise Control

The cruise control means to keep the vehicle on the same speed. However, in overloaded roads, a cruise control is useless since the speed of the leading vehicle caries a lot. Thus, the concept of adaptive cruise control (ACC) is proposed to indicate intelligently maintain the distance with the leading vehicle [120]-[129].

ACC is a new technology that requires the use of forward-looking vision sensors to detect the speed and distance of the vehicle ahead of it, and then automatically adjusts your speed accordingly. As pointed out in [129], the major objectives of ACC are the following:

(1) the control of the speed between zero and the maximum vehicle velocity;

(2) the control of the relative distance and velocity from the preceding vehicle;

(3) the attainment of the best comfort for the driver and the passengers.

Fig.4.18 given in [127] provides a simplified technical portrayal of the basic one-on-one driving situation considering ACC systems, where V_P is the speed of the preceding vehicle, R is the range clearance between the two vehicles, V is the speed of the ACC vehicle. Thus it indicates

$$\dot{R} = V_p - V \tag{4.88}$$

The first difficulty is to correctly estimate the distance between two vehicles and predicate the future motion of the leading/preceding vehicle. The related research will be discussed in *Chapter 8* of this book.

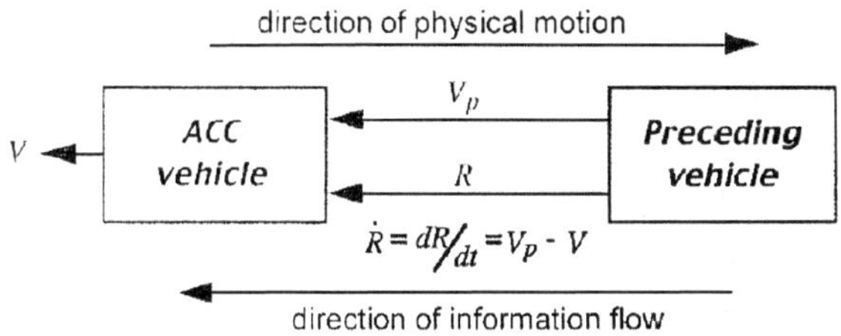

Fig.4.18 One-on-one driving, from [127]. (© [2003] IEEE)

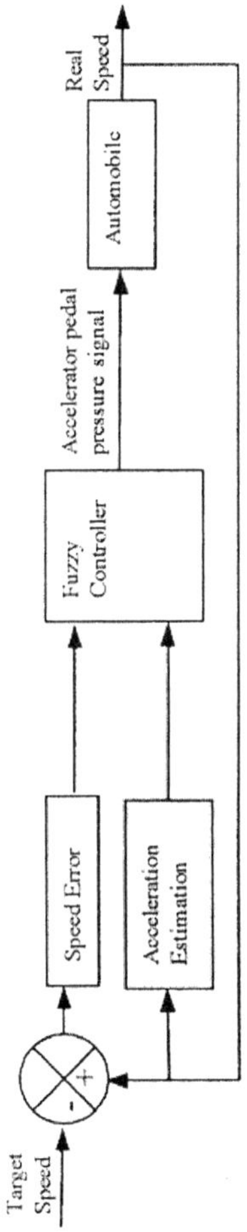

Fig.4.19 One-on-one driving, from [125]. (© [2003] IEEE)

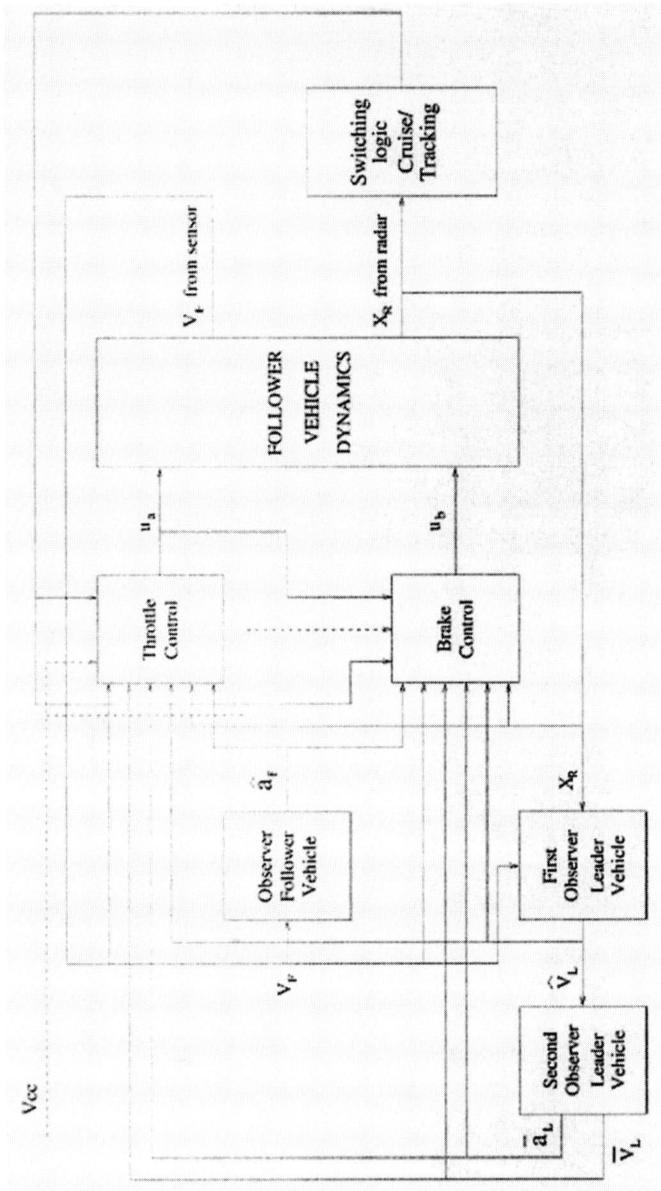

Fig.4.20 Diagram of integrated tracking, breaking, ACC control, from [129]. (© [2004] IEEE)

The second difficulty is how to apply proper control strategy since a vehicle cannot make linear acceleration/deceleration in real world because of its time variant nonlinear longitudinal dynamics. One simple approach is to directly set up approximate relationship between oil throttle input and vehicle speed output.

For example, an adaptive fuzzy control for inter-vehicle gap keeping was proposed in [125], see Fig.4.20. The control rules is quite straightforward, i.e.

"IF Speed Error MORE THAN null OR Acceleration MORE THAN null THEN Accelerator up

IF Speed Error LESS THAN null OR Acceleration LESS THAN null THEN Accelerator down

……

"

It is widely believed that such fuzzy control systems can yield acceptable results if correctly tuned, since human drivers ride in a similar way.

Another solving method is to establish complex but integrated model based on complete vehicle tracking and braking control models discussed in *Section 4.4.1*. For instance, Ferrara and Pisu considered a minimum sensor variable structure control strategy for cruise and tracking longitudinal control of vehicles, see Fig.4.21. Although they used a much simplified model, the control strategy is quite complex to analyze.

The third difficulty is how to describe driver/passenger's feelings of comfort. One frequently used way is to formulate a special objective function to measure driver/passenger's feelings [130]-[137]. However, it is believed that this method can provide driver-oriented service that meet different drivers' riding commands.

4.6 Summary

The recent trend of research on developments of vehicle dynamic components design and longitudinal motion control is reviewed in this Chapter. Although it is impossible to cover hundreds of publications in this area, the key findings and trends of research are included. However, due to length limits, some important contents leave untouched in this Chapter. For instance,

(1) The phenomenon shown Fig.2.12 inspires a motivation for new ABS brakes, since avoiding high longitudinal slip values will maintain high steerability and lateral stability of the vehicle during braking. Achieving this by manual control is difficult, because the slip dynamics are fast

and open loop unstable when operating at wheel slip values to the right of any peak of the friction curve. We expect a reasonable tradeoff between high longitudinal friction F_x and lateral friction F_y is achieved under all road conditions for longitudinal slip s_x close to its peak value on the longitudinal slip curve. There were some attempts in this direction using empirical models, i.e. [138]-[140]. For example, in [130], the above ABS control model (4.46) was modified as

$$\begin{cases} \dot{s}_x = -\dfrac{1}{v_x}\left[\dfrac{1}{m}(1-s_x)+\dfrac{r^2}{J}\right]F_z\mu(s_x,\alpha)+\dfrac{1}{v_x}\dfrac{r}{J}T_b \\ \dot{v} = -\dfrac{1}{m}F_z\mu(s_x,\alpha) \end{cases} \tag{4.89}$$

which address the effects of sideslip angle α. But the system dynamics become quite complex after considering α, and far more further efforts need to be carried out;

(2) vehicle fault-tolerant longitudinal motion control and fault detection begins to attract more attentions recently [141]. But comparing to vehicle lateral motion control, the related research in longitudinal part is more difficult, since much more components needs to be monitored.

4.7 Reference

1. U. Kiencke and L. Nielsen, Automotive Control Systems: for Engine, Driveline and Vehicle, Springer-Verlag, 2000.
2. T. Takeuchi, K.-G. Choi, and S. Kan, et. al, "Sensibility analysis of vibration transfer path and control of input force for reduction of acceleration and deceleration shock," Mitsubishi Motors Technical Review, 2003.
3. B. Jakoby, M. Scherer, and M. Buskies, et. al, "An automotive engine oil viscosity sensor," IEEE Sensors Journal, vol. 3, no. 5, pp. 562-568, 2003.
4. H.-P. Schoner and P. Hille, "Automotive power electronics: new challenges for power electronic," Proceedings of IEEE Annual Power Electronics Specialists Conference, vol. 1, pp. 6-11, 2000.
5. D. L. Harrington and J. A. Bolt, "Analysis and digital simulation of carburetor metering," SAE#700082, 1970.
6. J. D. Dobner, "A mathematical engine model for development of dynamic engine control," SAE#800054, 1980.
7. Y.-K. Chin and F. E. Coats, "Engine dynamics: time-based versus crank-angle based," SAE#860412, 1986.

8. J. D. Powell, "A review of IC engine models for control system design," Proceedings of World Congress on Automatic Control, vol. 3, 1987.
9. U. Kiencke, "A view of automotive control systems," IEEE Control Systems Magazine, vol. 8, no. 4, pp. 11-19, 1988.
10. J. B. Heywood, Internal Combustion Engine Fundamentals, New York: McGraw-Hill, 1988.
11. J. K. Hedrick and D. Cho, "A nonlinear controller design method for fuel-injected automotive engines," ASME Journal of Engineering for Gas Turbines and Power, vol. 110, no. 3, pp. 313-320, 1988.
12. E. Hendricks and S. C. Sorenson, "Mean value modelling of spark ignition engines," SAE#900616, 1990.
13. P. R. Crossley and J. A. Cook, "A nonlinear engine model for drivetrain system development," IEE Conference Publication 332, vol. 2, pp: 921-925, 1991.
14. J. J. Moskwa and J. K. Hedrick, "Modeling and validation of automotive engines for control algorithm development," ASME Journal of Dynamic Systems, Measurement, and Control, vol. 114, no. 2, pp. 278-285, 1992.
15. E. Hendricks, "Engine modelling for control applications: a critical survey," Meccanica, vol. 32, pp. 387–396, 1997.
16. H. Zhao, P. Calnan, and N. Ladommatos, et. al, "Development of an engine simulation program and its application to stratified charge SI engines," International Journal of Vehicle Design, vol. 22, no. 3/4, pp. 159-194, 2002.
17. H. Javaherian, D. Liu, and Y. Zhang, et. al, "Adaptive critic learning techniques for automotive engine control," Proceedings of American Control Conference, vol. 5, pp. 4066-4071, 2004.
18. S. R. Fozo and C. F. Aquino, "Transient A/F characteristics for cold operation of a 1.6 liter engine with sequential fuel injection," SAE#880691, 1991.
19. G. Almkvist and S. Eriksson, "An analysis of air to fuel ratio response in a multi point fuel injected engine under transient conditions," SAE#932753, 1993.
20. C.-F. Chang, N. P. Fekete, and J. D. Powell, "Engine air-fuel ratio control using an event based observer," SAE#930766, 1993.
21. C.-F. Chang, N. P. Fekete, and A. Amstutz, et. al, "Air-Fuel ratio control in Spark-Ignition engines using estimation theory," IEEE Transaction on Control Systems Technology, vol. 3, no. 1, pp. 22-31, 1995.
22. G. J. Vachtsevanos, S. S. Farinwata, and D. K. Pirovolou, "Fuzzy logic control of an automotive engine," IEEE Control Systems Magazine, vol. 13, no. 3, pp. 62-68, 1993.
23. H. Shiraishi, S. L. Ipri, and D. D. Cho, "CMAC neural network controller for fuel-injection systems," IEEE Transactions on Control System Technology, vol. 3, no. 1, pp. 32-38, 1995.
24. D. G. Coop, K. J. Burnham, and F. P. Lockett, "Fuzzy modeling techniques applied to an air/fuel ratio control system," Proceedings of IEE

Colloquium on Industrial Automation and Control: Applications in the Automotive Industry, pp: 3/1-3/7, 1998.

25. J. D. Powell, N. P. Fekete, and C.-F. Chang, "Observer-based air-fuel ratio control," IEEE Control Systems Magazine, vol. 18, no. 5, pp. 72-83, 1998.
26. J. K. Pieper and R. Mehrotra, "Air/fuel ratio control using sliding mode methods," Proceedings of American Control Conference, vol. 2, pp. 1027-1031, 1999.
27. C. Alippi, C. de Russis, and V. Piuri, "A neural-network based control solution to air-fuel ratio control for automotive fuel-injection systems," IEEE Transaction on Systems, Man and Cybernetics, Part C, vol. 33, no. 2, pp. 259-268, 2003.
28. V. I. Utkin, "Variable structure systems with Sliding modes," IEEE Transactions on Automatic Control, vol. AC-22, no. 2, pp. 212-222, 1997.
29. C. F. Aquino, "Transient A/F control characteristics of the 5 liters central fuel injection engine," SAE#810494, 1981.
30. E. Hendricks and T. Vesterholm, "Nonlinear transient fuel film compensation," SAE#930767, 1993.
31. Kjergaard, L., Nielsen, S., Vesterholm, T. and Hendricks, E., Advanced Nonlinear Engine Idle Speed Control Systems, SAE Technical Paper No. 940974, 1994.
32. M. Livshiz, D. J. Sanvido, and S. D. Stiles, "Nonlinear engine model for idle speed control," Proceedings of IEEE Conference on Decision and Control, vol. 3, pp. 2449-2451, 1994.
33. A. J. Bhatti, S. K. Spurgeon, and R. Dorey, et. al, "Rapid prototyping of a sliding mode controller for idle speed control," Proceedings of IEE Colloquium on System Control Integration and Rapid Prototyping in the Automotive Industry, pp. 3/1-3/7, 1997.
34. H.-G. Weil, G. Probst, and F. Graf, "Fuzzy expert system for automatic transmission control," Proceedings of IEEE Conference on Control Applications, vol. 2, pp. 716-721, 1992.
35. A. Bastian, S. Tano, and T. Oyama, et. al, "FATE: fuzzy logic automatic transmission expert system," Proceedings of IEEE International Conference on Fuzzy Systems, vol. 5, pp. 5-6, 1995.
36. K. Hayashi, Y. Shimizu, and Y., Dote, et. al, "Neuro fuzzy transmission control for automobile with variable loads," IEEE Transactions on Control Systems Technology, vol. 3, no. 1, pp. 49-53, 1995.
37. D.B. Maciuca, J.K. Hedrick, "Advanced nonlinear brake system control for vehicle platooning," Proceedings of European Control Conference, 1995.
38. Y. Lei, Y. Li, and H. Tian, et. al, "A research on gear position decision and combination control in braking condition," Proceedings of IEEE International Vehicle Electronics Conference, pp. 227-231, 2001.

39. J. Marco, R. Ball, and R. P. Jones, et. al, "A systems modelling and simulation approach to gear shift effort analysis," International Journal of Vehicle Design, vol. 25, no. 4, pp. 317-338, 2001.
40. B. Jacobson and M. Spickenreuther, "Gearshift sequence optimization for vehicles with automated non-powershifting transmissions," International Journal of Vehicle Design, vol. 32, no. 3/4, pp. 187-207, 2003.
41. S. E. Moon, M. S. Kim, and H. Yeo, et. al, "Design and implementation of clutch-by-wire system for automated manual transmissions," International Journal of Vehicle Design, vol. 36, no. 1, pp. 83-100, 2004.
42. J. Murin, "A controlled diesel drive line with hydrostatic transmission: Part 1 - mathematical model," International Journal of Vehicle Design, vol. 38, no. 2/3, pp. 109-122, 2005.
43. J. Murin, "A controlled diesel drive line with hydrostatic transmission: Part 2 - dynamic properties at periodic loading," International Journal of Vehicle Design, vol. 38, no. 2/3, pp. 123-138, 2005.
44. C. Chan, T. Volz, and D. Breitweiser, et. al, "System design and control considerations of automotive continuously variable transmissions," SAE#840048, 1984.
45. D. Cho and J. K. Hedrick, "Automotive powertrain modeling for control," ASME Journal of Dynamic Systems, Measurement, and Control, vol. 114, no. 4, pp. 568-576, 1989.
46. S. Sakaguchi, I. Sakai, and T. Haga, "Application of fuzzy logic to shift scheduling method for automatic transmission," Proceedings of IEEE International Conference on Fuzzy Systems, vol. 1, pp: 52-58, 1993.
47. S. Liu and B. Paden, "A survey of today's CVT controls," Proceedings of IEEE Conference on Decision and Control, pp. 4738–4743, 1997.
48. A. Beydoun, L. Y. Wang, and J. Sun, et. al, "Hybrid control of automotive powertrain systems: a case study," Proceedings of International Workshop on Hybrid Systems: Computation and Control, pp. 33-48, 1998.
49. S. Takahashi, "Fundamental study of low fuel consumption control scheme based on combination of direct fuel injection engine and continuously variable transmission," Proceedings of IEEE Conference on Control and Decision, pp. 1522–1529, 1998.
50. H. D. Lee and S. K. Sul, "Fuzzy-logic-based torque control strategy for parallel-type hybrid electric vehicle," IEEE Transactions on Industrial Electronics, vol. 45, pp. 625–632, 1998.
51. B. K. Powell, K. E. Bailey, and S. R. Cikanek, "Dynamic modeling and control of hybrid electric vehicle powertrain systems," IEEE Control Systems Magazine, vol. 18, no. 5, pp. 17–23, 1998.
52. I. Kolmanovsky, J. Sun, and L.Wang, "Coordinated control of lean burn gasoline engines with continuously variable transmissions," Proceedings of American Control Conference, vol. 4, pp. 2673–2677, 1999.
53. M. Yasuoka, M. Uchida, and S. Katakura, et. al, "An integrated control algorithm for an SI engine and a CVT," SAE#1999-01-0752, pp. 155–160, 1999.

54. H.-D. Lee, S.-K. Sul, and H.-S. Cho, et. al, "Advanced gear-shifting and clutching strategy for a parallel-hybrid vehicle," IEEE Industry Applications Magazine, vol. 6, no. 6, pp. 26-32, 2000.
55. P. Setlur, J. R. Wagner, and D. M. Dawson, et. al, "Nonlinear control of a continuously variable transmission (CVT)," IEEE Transactions on Control Systems Technology, vol. 11, no. 1, pp. 101-108, 2003.
56. S. Onoda and A. Emadi, "PSIM-based modeling of automotive power systems: conventional, electric, and hybrid electric vehicles," IEEE Transactions on Vehicular Technology, vol. 53, no. 2, pp. 390-400, 2004.
57. S. Shen and F. E. Veldpaus, "Analysis and control of a flywheel hybrid vehicular powertrain," IEEE Transactions on Control Systems Technology, vol. 12, no. 5, pp. 645-660, 2004.
58. R. W. Johnson, J. L. Evans, and P. Jacobsen, et. al, "The changing automotive environment: high-temperature electronics," IEEE Transactions on Electronics Packaging Manufacturing, Part C, vol. 27, no. 3, pp. 164-176, 2004.
59. J.-S. Won, R. Langari, M. Ehsani, "An energy management and charge sustaining strategy for a parallel hybrid vehicle with CVT," IEEE Transactions on Control Systems Technology, vol. 13, no. 2, pp, 313-320, 2005.
60. P. Stewart and P. J. Fleming, "Drive-by-wire control of automotive driveline oscillations by response surface methodology," IEEE Transactions on Control Systems Technology, vol. 12, no. 5, pp. 737-741, 2004.
61. K. Togai, K. Choi, and T. Takeuchi, "Vibration suppression strategy with model based command shaping: application to passenger car powertrain," Proceedings of SICE Annual Conference, vol. 2, pp. 941-943, 2002.
62. D. Lefebvre, P. Chevrel, and S. Richard, "An H-infinity-based control design methodology dedicated to the active control of vehicle longitudinal oscillations," IEEE Transactions on Control Systems Technology, vol. 11, no. 6, pp. 948-956, 2003.
63. K. Togai and T. Takeuchi, "CVT drive train vibration analysis and active control system design," Mitsubishi Motors Technical Review, 2005.
64. Y. Yu, S. M. Peelamedu, and N. G. Naganathan, "Automotive vehicle engine mounting systems: a survey," ASME Journal of Dynamic Systems, Measurement, and Control, vol. 123, pp. 186-194, 2001.
65. K. B. Kelly and H. J. Holcombe, "Aerodynamics for body engineers," Automotive Aerodynamics, Progress in Technology Series, Society of Automotive Engineers, vol. 16, 1978.
66. M. G. L. Cresswell and P. B. Hertz, "Aerodynamic drag implications of exterior truck mirrors," SAE#920202, pp: 29-34, 1992.
67. S. A. Coleman and C. J. Baker, "Reduction of accident risk for high sided road vehicles in cross winds," Journal of Wind Engineering and Industrial Aerodynamics, vol. 44, pp. 2685-2695, 1992.

68. R. H. Banard, Road Vehicle Aerodynamic Design - An Introduction, Addison Wesley, Longman Ltd, Harlow, 1996.
69. V. Eowsakul and T. J. Ortolani, "Improving the aerodynamic characteristics of a Dodge Ram pickup truck," ASME IMECE#97-WA/DE-18, 1997.
70. T. Takeuchi and I. Kohri, "Development of truck and bus aerodynamics using computational fluid dynamics," Journal of SAE Review, vol. 18, no. 2, pp. 188-188(1), 1997.
71. W.-H. Hucho, Aerodynamics of Road Vehicles: from Fluid Mechanics to Vehicle Engineering, Society of Automotive Engineers, Warrendale, 1998.
72. S. Roy and P. Srinivasan, "External flow analysis of a truck for drag reduction," SAE #2000-01-3500, 2000.
73. R. J. Englar, "Development of pneumatic aerodynamic devices to improve the performance, economics, and safety of heavy vehicles," SAW/TPS-2000-01-2208, 2000.
74. J.-W. Wong, The Theory of Ground Vehicle, 3rd ed., John Wiley & Sons, 2001.
75. W.-H. Hucho, "Aerodynamics of road vehicles - state of the art and tasks for the future," Proceedings of KULI Usermeeting, 2003.
76. H. Gotz and G. Mayr, "Aerodynamics of road vehicles," SAE International, pp. 415-488, 1998.
77. P. Kachroo, M. Tomizuka, and A. M. Agogino, "A comprehensive strategy for longitudinal vehicle control with fuzzy supervisory expert systems," Proceedings of IEEE International Conference on systems, Man and Cybernetics: Intelligent Systems for the 21st Century, pp. 765-770, 1995.
78. D. Yanakiev and I. Kanellakopoulos, "Longitudinal control of heavy-duty vehicles for automated highway systems," Proceedings of American Control Conference, vol. 5, pp. 3096-3100, 1995.
79. P. S. Fancher, H. Peng, and Z. Bareket, "Comparative analyses of three types of headway control systems for heavy commercial vehicles," Vehicle System Dynamics, Supplement, vol. 25, pp.139-151, 1996.
80. M. Shino, N. Miyamoto, and Y.-Q. Wang, et. al, "Traction control of electric vehicles considering vehicle stability," Proceedings of International Workshop on Advanced Motion Control, pp. 311-316, 2000.
81. M. Tai and M. Tomizuka, "Robust longitudinal velocity tracking of vehicles using traction and brake control," Proceedings of International Workshop on Advanced Motion Control, pp. 305-310, 2000.
82. G. Bartolini, A. Ferrara, and P. Pisu, "Longitudinal control design of passenger vehicles with second order sliding modes," Proceedings of American Control Conference, vol. 1, no. 6, pp. 120-124, 2000.
83. R. Outbib and A. Rachid, "Control of vehicle speed: a nonlinear approach," Proceedings of IEEE Conference on Decision and Control, vol. 1, pp. 462-463, 2000.

84. M. Druzhinina, A. G. Stefanopoulou, and L. Moklegaard, "Speed gradient approach to longitudinal control of heavy-duty vehicles equipped with variable compression brake," IEEE Transactions on Control Systems Technology, vol. 10, no. 2, pp. 209-220, 2002. Good paper!!!
85. R. Pusca, Y. Ait-Amirat, and A. Berthon, et. al, "Modeling and simulation of a traction control algorithm for an electric vehicle with four separate wheel drives, " Proceedings of IEEE Vehicular Technology Conference, vol. 3, pp. 1671-1675, 2002.
86. H.-C. Choi and S.-K. Hong, "Hybrid control for longitudinal speed and traction of vehicles," Proceedings of IEEE Conference of Industrial Electronics Society, vol. 2, pp. 1675-1680, 2002.
87. X. Y. Lu and J. K. Hedrick, "Modeling of heavy-duty vehicles for longitudinal control," Proceedings of International Symposium on Advanced Vehicle Control, 2002.
88. X. Y. Lu and J. K. Hedrick, "Longitudinal control design and experiment for heavy-duty trucks," Proceedings of American Control Conference, pp. 36-41, 2003.
89. H. Lee and M. Tomizuka, "Adaptive vehicle traction force control for intelligent vehicle highway systems (IVHSs)," IEEE Transactions on Industrial Electronics, vol. 50, no. 1, pp. 37-47, 2003.
90. P. Khatun, C. M. Bingham, and N. Schofield, et. al, "Application of fuzzy control algorithms for electric vehicle antilock braking/traction control systems," IEEE Transactions on Vehicular Technology, vol. 52, no. 5, pp. 1356-1364, 2003.
91. J. Sainte-Marie, S. Mammar, and L. Nouveliere, et. al, "Sub-optimal longitudinal control of road vehicles with capacity and safety considerations," ASME Journal of Dynamic Systems, Measurement, and Control, vol. 126, pp. 26-35, 2004.
92. R. Emig, H. Goebels, and H. J. Schramm, "Antilock braking systems (ABS) for commercial vehicles - status 1990 and future prospects," Proceedings of International Congress on Transportation Electronics, pp. 515-523, 1990.
93. D. B. Maciuca, J.K. Hedrick, "Advanced nonlinear brake system control for vehicle platooning," Proceedings of European Control Conference, 1995.
94. P. E. Wellstead and N. B. O. L. Pettit, "Analysis and redesign of an antilock brake system controller," IEE Proceedings - Control Theory and Applications, vol. 144, no. 5, pp. 413-426, 1997.
95. W. K. Lennon and K. M. Passino, "Intelligent control for brake systems," IEEE Transactions on Control Systems Technology, vol. 7, no. 2, pp. 188-202, 1999.
96. D. Zhang, H. Zheng, and J. Sun, et. al, "Simulation study for anti-lock braking system of a light bus," Proceedings of IEEE International Vehicle Electronics Conference, pp. 70-77, 1999.

97. P. Tsiotras and C. Canudas de Wit, "On the optimal braking of wheeled vehicles," Proceedings of American Control Conference, vol. 1, pp. 569-573, 2000.
98. F. Jiang and Z. Gao, "An application of nonlinear PID control to a class of truck ABS problems," Proceedings of IEEE Conference on Decision and Control, vol. 1, pp. 516-521, 2001.
99. L. Petersen, T. A. Johansen, and J. Kalkkuhl, et. al, "Wheel slip control in ABS using gain scheduled constrained LQR," Proceedings of European Control Conference, 2001.
100. R. Kazemi and K. J. Zaviyeh, "Development of a new ABS for passenger cars using dynamic surface control method," Proceedings of American Control Conference, vol. 2, pp. 677-683, 2001.
101. T. Kaneko, I. Kageyama, and H. Tsunashima, "Braking stability of articulated vehicle on highway," Vehicle System Dynamics, Supplement, vol. 37, pp. 1-11, 2002.
102. Y. Lee and S. H. Zak, "Designing a genetic neural fuzzy antilock-brake-system controller," IEEE Transactions on Evolutionary Computation, vol. 6, no. 2, pp. 198-211, 2002.
103. H. Yu and U. Ozguner, "Extremum-seeking control strategy for ABS system with time delay," Proceedings of American Control Conference, vol. 5, pp. 3753-3758, 2002.
104. J. Yi, L. Alvarez, and R. Horowitz, "Adaptive emergency braking control with underestimation of friction coefficient," IEEE Transactions on Control Systems Technology, vol. 10, no. 3, pp. 381-392, 2002.
105. C. Toy, K. Leung, L. Alvarez and R. Horowitz. "Emergency vehicle maneuvers and control laws for automated highway systems," IEEE Transactions on Intelligent Transportation Systems, vol. 3, no. 2, pp. 109-119, 2002.
106. J. Yi, L. Alvarez, and R. Horowitz, et. al, "Adaptive emergency braking control using a dynamic tire/road friction model," Proceedings of IEEE Conference on Decision and Control, vol. 1, pp: 456-461, 2000.
107. J. Yi, L. Alvarez, and X. Claeys, et. al, "Emergency braking control with an observer-based dynamic tire/road friction model and wheel angular velocity measurement," Vehicle System Dynamics, vol. 39, no. 2, pp. 81-97, 2003.
108. M. Schinkel and K. Hunt, "Anti-lock braking control using a sliding mode like approach," Proceedings of American Control Conference, vol. 3, pp. 2386-2391, 2002.
109. G. Celentano, R. Iervolino, and S. Porreca, et. al, "Car brake system modeling for longitudinal control design," Proceedings of IEEE Conference on Control Applications, vol. 1, pp. 25-30, 2003.
110. S. Armeni and E. Mosca, "ABS with constrained minimum energy control law," Proceedings of IEEE Conference on Control Applications, vol. 1, pp. 19-24, 2003.

111. S. Drakunov, U. Ozguner, and P. Dix, et. al, "ABS control using optimum search via sliding modes," IEEE Transactions on Control Systems Technology, vol. 3, no. 1, pp: 79-85, 1995.
112. M. Krstik and H.-H. Wang, "Design and stability analysis of extremum seeking feedback for general nonlinear systems," Proceedings of IEEE Conference on Decision and Control, pp. 1743-1748, 1997.
113. K. Park and S.-J. Heo, "A study on the brake-by-wire system using hardware-in-the-loop simulation," International Journal of Vehicle Design, vol. 36, no. 1, pp. 38-49, 2004.
114. S. Park, S. Bae, and J. M. Lee, "Numerical evaluation of braking feel to design optimal brake-by-wire system," International Journal of Vehicle Design, vol. 37, no. 1, pp. 1-23, 2005.
115. S. K. Mazumdar and C. C. Lim, "The application of neural networks to anti-skid brake system design," Proceedings of IEEE International Conference on Neural Networks, vol. 5, pp. 2409-2414, 1995.
116. G. F. Mauer, "A fuzzy logic controller for an ABS braking system," IEEE Transactions on Fuzzy Systems, vol. 3, no. 4, pp. 381-388, 1995.
117. M.-R. Akbarzadeh, K.-J. Emami, and N. Pariz, "Adaptive discrete-time fuzzy sliding mode control for anti-lock braking systems," Proceedings of Annual Meeting of the North American Fuzzy Information Processing Society, pp. 554-559, 2002.
118. C.-M. Lin and C.-F. Hsu, "Self-learning fuzzy sliding-mode control for antilock braking systems," IEEE Transactions on Control Systems Technology, vol. 11, no. 2, pp. 273-278, 2003.
119. C.-K. Chen, "Development of fuzzy controlled ABS systems for motorcycles," International Journal of Vehicle Design, vol. 34, no. 1, pp. 84-100, 2004.
120. S. Germann and R. Isermann, "Nonlinaer distance and cruise control for passenger cars," Proceedings of American Control Conference, pp. 3081-3085, 1995.
121. H. M. Kim, J. Dickerson, and B. Kosko, "Fuzzy throttle and brake control for platoon of smart cars," Fuzzy Sets and Systems, vol. 84, no. 3, pp. 209-234, 1996.
122. H. Holzmann, S. Germann, and C. Halfmann, et. al, "Intelligent fuzzy distance and cruise control for passenger cars," Journal of Intelligent and Fuzzy Systems, vol. 6, no. 3, pp. 315-327, 1998.
123. P. Seiler, B. Song, and J. K. Hedrick, "Development of a collision avoidance system," Proceedings of SAE Conference, pp. 97-103, 1998.
124. K. J. Hunt, T. A. Johansen, and J. Kalkkuhl, et. al, "Speed control design for an experimental vehicle using a generalized gain scheduling approach," IEEE Transactions on Control Systems Technology, vol. 8, no. 3, pp. 381-395, 2000.
125. J. E. Naranjo, C. Gonzalez, and J. Reviejo, et. al, "Adaptive fuzzy control for inter-vehicle gap keeping," IEEE Transactions on Intelligent Transportation Systems, vol. 4, no. 3, pp. 132-142, 2003.

126. B. J. Olson, S. W. Shaw, and G. Stepan, "Nonlinear dynamics of vehicle traction," Vehicle System Dynamics, vol. 40, no. 6, pp. 377-399, 2003.
127. Z. Bareket, P. S. Fancher, and H. Peng, et. al, "Methodology for assessing adaptive cruise control behavior," IEEE Transactions on Intelligent Transportation Systems, vol. 4, no. 3, pp. 123-131, 2003.
128. V. L. Bageshwar, W. L. Garrard, and R. Rajamani, "Model predictive control of transitional maneuvers for adaptive cruise control vehicles," IEEE Transactions on Vehicular Technology, vol. 53, no. 5, pp. 1573-1585, 2004.
129. A. Ferrara and P. Pisu, "Minimum sensor second-order sliding mode longitudinal control of passenger vehicles," IEEE Transactions Intelligent Transportation Systems, vol. 5, no. 1, pp. 20-32, 2004.
130. R. Cann, L. R. Pujara, and J. Lieh, "Influence of motor drive parameters on the robust stability of electric power steering systems," Power Electronics in Transportation, pp. 103-108, 1998.
131. C. Unsal, P. Kachroo and J. S. Bay, "Multiple stochastic learning automata for vehicle path control in an automated highway system," IEEE Transactions on Systems, Man, and Cybernetics, Part A, vol. 29, no. 1, pp. 120-128, 1999.
132. L. Wu, Y. Xu, and F.-Y. Wang, et. al, "Supervised learning of longitudinal driving behavior for intelligent vehicles using neuro-fuzzy networks: initial experimental results," International Journal of Intelligent Control and Systems, vol. 3, no. 4, pp. 443-463, 1999.
133. L. Wu, Y. Xu, and F.-Y. Wang, et. al, "Supervised learning of longitudinal driving behavior for intelligent vehicles using neuro-fuzzy networks: initial experimental results," International Journal of Intelligent Control and Systems, vol. 3, no. 4, pp. 443-463, 1999.
134. F.-Y. Wang, L. Li and P.g Li, "Adaptive vehicle driving control and driver assistance using Neuro-Fuzzy networks," International Journal of Intelligent Control and Systems, 2004.
135. J. Bengtsson, R. Johansson, and A. Sjogren, "Modeling of drivers' longitudinal behavior," Proceedings of IEEE/ASME International Conference on Advanced Intelligent Mechatronics, vol. 2, pp. 1076-1081, 2001.
136. S. I. Jeon, S. T. Jo and Y. I. Park, et. al, "Multi-Mode driving control of a parallel hybrid electric vehicle using driving pattern recognition," ASME Journal of Dynamic Systems Measurement and Control, vol 124, no. 1, pp. 141-149, 2002.
137. R. Iwaki, T. Kaneko, and I. Kageyama, ""A study on a driver model for longitudinal control on heavy duty vehicle," Vehicle System Dynamics, Supplement, vol. 41, pp. 332-340, 2004.
138. T. A. Johansen, I. Petersen, J. Kalkkuhl, J. Ludemann, "Gain-scheduled wheel slip control in automotive brake systems," IEEE Transactions on Control Systems Technology, vol. 11, pp. 799-811, 2003.
139. I. Petersen, T. A. Johansen, and J. Kalkkuhl, et. al, "Wheel slip control using gain-scheduled LQ - LPV/LMI analysis and experimental results," European Control Conference, 2003.

140. S. Taheri and E. H. Law, "Slip control braking of an automobile during combined braking and steering manoeuvres," Advanced Automotive Technologies, vol. 40, pp: 209–227, 1991.
141. B. Song, J. K. Hedrick, and A. Howell, "Fault tolerant control and classification for longitudinal vehicle control," ASME Journal of Dynamic Systems, Measurement, and Control, vol. 125, pp. 320-329, 2003.

Advanced Vehicle Vertical Motion Control

5.1 Introduction

Apart from vehicle lateral and longitudinal dynamics, vehicle vertical dynamics also received continuous research efforts during the last four decades. Generally, vehicle suspension control, vehicle rollover avoidance and road slope estimation are three main research directions in this field.

Vehicle suspension system is currently of great interest in both the academics and automobile industries worldwide, because of the growing demands for better ride comfort and controllability of road vehicles. In fact, it would be safe to say that the field has seen hundreds, if not thousands, of publications on the topic during its nearly 40-year history [1]-[8]. This Chapter will address on suspension controller design; vehicle rollover avoidance and road slope estimation will also be discussed.

As claimed by Appleyard and Wellstead in [2], developments of suspension control are mainly challenged by the range of performance characteristics that a good suspension system must achieve. Desirable characteristics include:

(1) regulation of body movement: ideally the suspension should isolate the body from road disturbances and inertial disturbances associated with cornering (causing body roll) and braking/acceleration (causing body pitch);

(1) regulation of suspension movement: excessive vertical wheel travel will result in non-optimum attitude of the tire relative to the road; the result will be poor handling and adhesion of vehicle and fatigue of driver [3];

(1) force distribution: to maintain good handling characteristics, the optimum tire/road contact must be maintained on all four wheels.

On average, suspension performance is primarily influenced by the type of suspension used. From the viewpoint of the control energy consumed, the vehicle suspension model could be divided into three types in an ascending order of improved performance: passive suspension, semi-active suspension and active suspensions.

Passive suspension systems could be found on most conventional vehicles. The term "passive" stems from the physics concept which means that the systems are absence of the external power sources. The passive suspension systems are normally implemented by springs and shock-absorbers, whose force-velocity relationship reflect power dissipation characteristic of passive components.

The ideal suspension system should be soft for the ride comfort, while it should be stiff for the drivability. However, passive suspension system could not reach this goal.

Different from passive suspensions, active suspension systems achieve better ride and handling performance. They require additional power sources such as compressors and pumps and sensors to measure the motions of the vehicle body suspension system and the un-sprung mass. The obtained information is then used in the online controller to command the actuator so as to provide the exact amount of desired force. Thus active suspension systems are more costly, more complex and less reliable than the passive ones, although active suspension systems can overcome the trade-off dilemma mentioned above. Now, the active actuator is usually secured in parallel with a spring and shock absorber in case of its failure.

More precise definition of active and passive suspension could be found in Hrovat et. al [4]-[7], which were based on similar definitions of passivity and passive operators as what is used in electrical networks.

Semi-active suspension proposed by Karnopp 1990 in [8] was based on the passive system in the structure, which guarantees the dissipation of vibration energy. It can only change the suspension damping coefficients by controlling the electromagnetic valve inside the absorber. So it can provide better performance than passive suspension, and does not consume as much energy as active suspension system. Because of it good property, semi-active suspension systems can be often found on today's high-end production sports cars. But it requires some measuring devices with a controller board in order to tune the damping properly, too.

Another division of suspension system is characterized by dynamic feature of the suspension components. From this aspect, all the frequently used suspension system could be distinguished into two parts: linear suspension system and nonlinear suspension system. The difference lies that the former one's dynamic behavior could be described by linear differential equations while that of the later one could not. The springs are assumed to have almost linear characteristics and most of the shock absorbers exhibit nonlinear relationship between force and velocity.

Since passive suspension cannot meet performance requirements in most driving cases, this Chapter mainly discusses active and semi-active suspensions in the rest.

5.2 Road Roughness Modeling

The vibration of vehicle mainly comes from three aspects: the roughness of the road, the vibration originated in various inertial and aerodynamic loadings and the wheel non-uniformities. Among them, the first one is the main factor. Thus, in order to analyze vehicle behavior for the purposes of evaluation the vibration reduction performance, it is necessary to investigate the road roughness input above all.

Extensive road roughness surveys have been performed in various countries [9]-[16]. Generally, road surfaces were taken as certain random processes with a given displacement power spectral density (P.S.D.).

The concept of P.S.D. comes from the Fourier analysis, in which random signals are described as a series of sine waves of different frequencies and amplitudes. Usually the so called spectral density is used to denote the square of the amplitude of the sine waves, divided by the frequency bandwidth. When the spectral density is plotted against frequency on linear scales, the area under the curve is equal to the mean–square value of the signal (square of the standard deviation). Thus the spectral density is also known as the power spectral density (P.S.D.) sometimes. Some typical spectral densities of various terrains obtained by Sevin and Philkey 1971 [9] are shown in Fig.5.1.

A frequently mentioned approximation of road displacement P.S.D. based on Fig.5.1 is given as

$$S(\Omega) = A\Omega^n \tag{5.1}$$

where Ω is the spatial frequency, typically in units of "radians per length", A and n are appropriate constants.

In the proposal given in [11], a more precise formulation is written as

$$S(\kappa) = \begin{cases} S(\kappa_0)(\kappa/\kappa_0)^{-n_1}, & \kappa \le \kappa_0 \\ S(\kappa_0)(\kappa/\kappa_0)^{-n_2}, & \kappa > \kappa_0 \end{cases} \tag{5.2}$$

where κ is the wave-number, κ_0 is the datum wave-number, $S(\kappa)$ denotes displacement spectral density, $S(\kappa_0)$ is the spectral density at κ_0.

Here, wave-number κ (cycles/m) denotes rate of change with respect to distance, while frequency f (cycles/s) denotes rate of change with respect to time. Noting that $T = 1/f$, the wavelength l (in meters) of a sinusoidal roughness component is related to the wave-number κ.

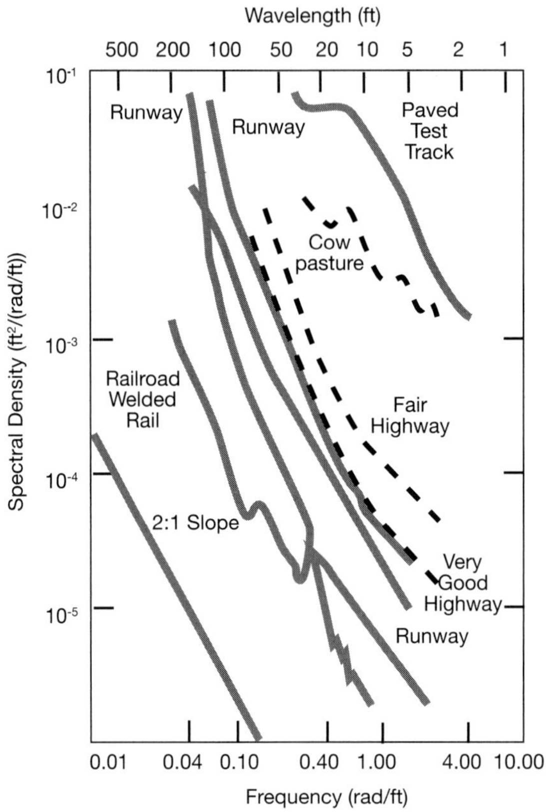

Fig.5.1 Power spectral density of various terrains, from [9].

Robson 1979 claimed that the above model fits the highway road well. A more general but also more complex road roughness model was proposed by Sayers 1988 [13]. It takes the form:

$$S(\kappa) = \frac{G_a}{(2\pi\kappa)^4} + \frac{G_s}{(2\pi\kappa)^2} + G_e \tag{5.3}$$

Because this model has more parameters: G_a, G_s and G_e, it provides the possibility to fit a wide range of measured road profile spectral densities.

In the suspension system simulation, there are some other road roughness models that were used to mimic road surface. One common approach is to model the road input as a Gaussian white noise [7]. Another frequently used approach is to measure and record some road profiles from typical road surfaces first, and then use the recorded data as road input in simulation.

Lu and DePoyster 2002 proposed a magnitude varying chirp signal as a new road input for nonlinear suspension system simulation in [17]. This signal is based on the following idea: if a system has significant nonlinearities, time responses are more appropriate for performance evaluation; and a time response with respect to a frequency-sweeping time sequences could provide certain frequency-response like characterization. From this consideration, a magnitude varying chirp signal road profile was proposed as shown in Fig.5.2, which has frequency contents from 0 to 15 Hz in order to excite the pitch and roll modes of the car body.

Caponetto 2003 proposed two new kinds of road profile for the suspension system optimization procedure [18]. The first, shown in Fig.5.3, is the so-called sleeper-plate test which consists of crossing a parallelepiped of fixed dimensions at a fixed speed. It is used as an impulse input to test the high frequency response of the car suspension system. The second, see Fig.5.4, consists of following the road profile modeled as half a sinusoid to test the low frequency features of the system, which is therefore called the sine-wave-hole test. These two new models could be used to appropriately mimic the unavoidable road fault that cannot be represented by classic road roughness models.

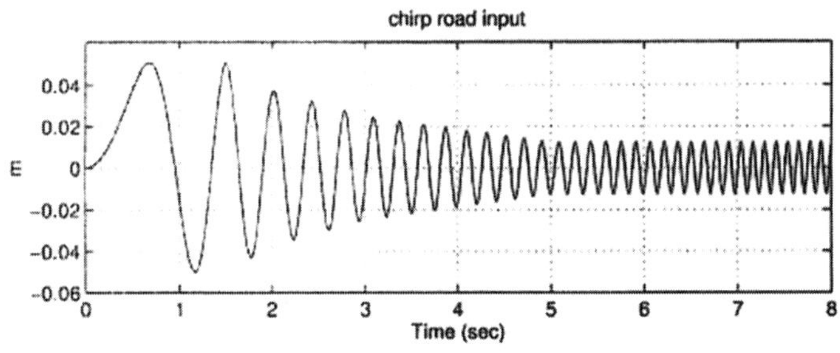

Fig.5.2 Time response of the simulated road inputs, from [17]. (© [2002] IEEE)

In some other discussions, the interactions between vehicle, especially heavy ones, and deformable soil [19]-[22] or bridges [23]-[24] were analyzed, since their responses are different from that on highway road surface. Constrained by book length, they are omitted here.

Fig.5.3 Plate test, from [18]. (© [2003] IEEE)

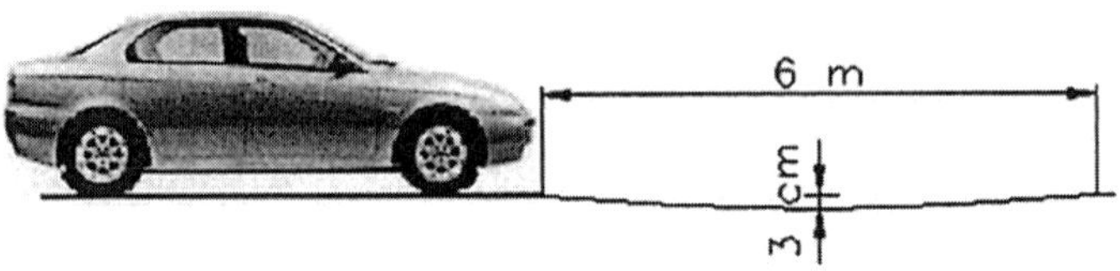

Fig.5.4 Sine-wave hole test, from [18]. (© [2003] IEEE)

5.3 Advanced Vehicle Suspension Systems

Generally, there are three simplified models that are used for vehicle suspension systems analysis: quart car suspension model, half car suspension model and full car suspension model. Usually, the vibrations originated in engine, driveline and aerodynamic loadings have been counted into road roughness input [25].

5.3.1 LTI Suspension Controllers

The linear time-invariant (LTI) active suspension controllers are the first widely discussed active suspension systems, which are based on linear optimal feedback control theory; [5]-[7], [26]. The core part of such method is to first formulate the state space model of the suspension system, and then use LTI optimal feedback method to design the controller.

For example, let's discuss 2 DOF quart car suspension model shown in Fig.5.5, which is widely used in suspension design. It consists of two masses, a sprung mass m_s and an un-sprung mass m_u, both of which are constrained to move vertically. The two degrees of freedom are the vertical sprung displacements $z_s(t)$ and un-sprung displacement $z_u(t)$. The road displacement input to the tire is denoted by $z_r(t)$. The stiffness and damping parameters of the model are represented as c_t, c_s, k_t and k_s, respectively. f_a denotes the control force, which is active suspension controller input.

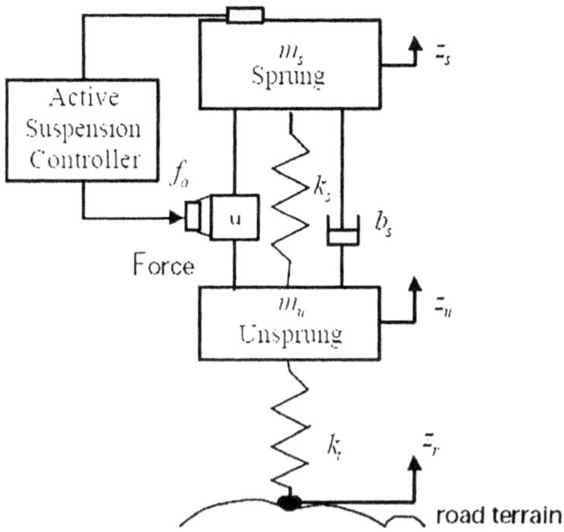

Fig.5.5 2 DOF quart car model, from [27]. (© [1999] IEEE)

By focusing on the first mode, the model dynamics could be given as

$$m_s \ddot{z}_s(t) = b_s[\dot{z}_u(t) - \dot{z}_s(t)] + k_s[z_u(t) - z_s(t)] + f_a \tag{5.4}$$

and

$$m_u \ddot{z}_u(t) = -b_s[\dot{z}_u(t) - \dot{z}_s(t)] - k_s[z_u(t) - z_s(t)] - f_a + k_t(z_r - z_u) \tag{5.5}$$

Introduce transformation as

$$x_1(t) = z_s(t) - z_u(t),\ x_2(t) = \dot{z}_s \tag{5.6}$$

and

$$x_3(t) = z_u(t) - z_r(t),\ x_4(t) = \dot{z}_u(t) \tag{5.7}$$

The system dynamics can be written in state space model format as

$$\dot{X}(t) = AX(t) + BU(t) + GZ(t) \tag{5.8}$$

where $X = [x_1 \quad x_2 \quad x_3 \quad x_4]^T$, $U = f_a$, and $Z = \dot{z}_r$. The matrix are defined as

$$A = \begin{bmatrix} 0 & 1 & 0 & -1 \\ -k_s/m_s & -b_s/m_s & 0 & b_s/m_s \\ 0 & 0 & 0 & 1 \\ k_s/m_u & b_s/m_u & -k_s/m_u & -b_s/m_u \end{bmatrix}, \quad B = \begin{bmatrix} 0 \\ 1/m_s \\ 0 \\ -1/m_u \end{bmatrix}, \quad G = \begin{bmatrix} 0 \\ 0 \\ -1 \\ 0 \end{bmatrix}.$$

Generally, assuming small vibration angle, the half car model and full car model could also be formulated into above format with $X(t)$, $U(t)$, and $Z(t)$ represent the general state variable, the control input and the road roughness input respectively, i.e. the half car model used in [26].

For the general model (5.8), a frequently used performance index J penalizes the state variables and the inputs can be introduced as:

$$J = \int_0^\infty \left(X^T QX + U^T RU\right)dt \tag{5.9}$$

where Q and R are positive definite matrix which is also called weighting matrices. Notice that acceleration which is an indicator of ride comfort is not being penalized here.

Based on Linear-Quadratic-Regulator (LQR) design theory, it is publicly known that the optimal state feedback controller of Eq.(5.9) in terms of (5.8) as

$$U = KX \tag{5.10}$$

where the gain matrix K is represented as

$$K = R^{-1}B^T P \tag{5.11}$$

where the matrix P is the solution of the following Algebraic Riccati Equation (ARE) written as

$$AP + A^T P - PBR^{-1}B^T P + Q = 0 \tag{5.12}$$

Fig.5.6, which given in Esmailzadeh and Taghirad 1998 [26] for a half-car model clearly, illustrates how the active suspension can effectively absorb the vehicle vibration in comparison to the passive system.

To improve the above conventional optimal state feedback method, it has been first proposed by Bender 1968 [28] that performance of active suspension can be further improved if knowledge of the road surface in front of the actively controlled axles, i.e., preview information is used in the control strategy. With preview information, the active suspension system can prepare the vehicle for a future road input and pass through abrupt road obstacles without severe impacts.

Usually the preview information is collected by look-ahead sensor as shown in Fig.5.7.

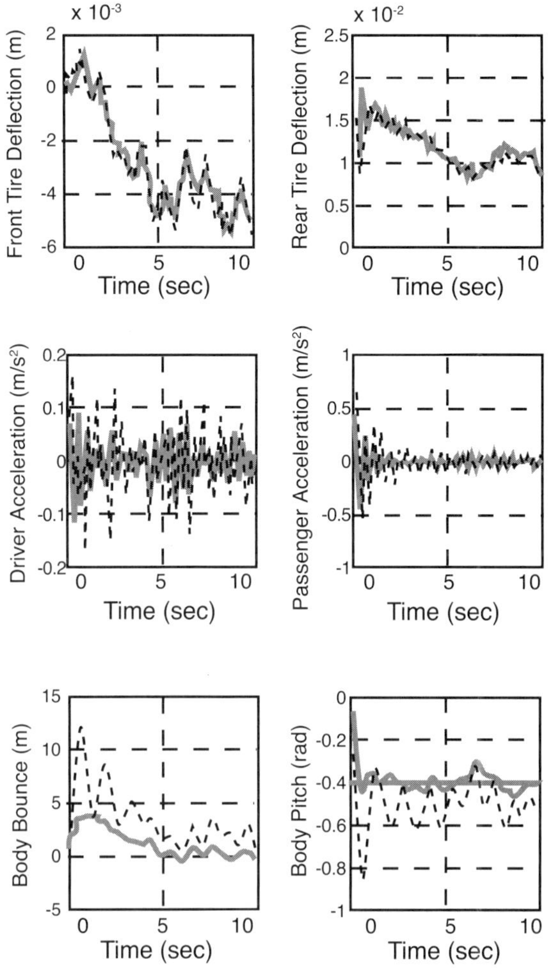

Fig.5.6 Comparison between passive and active systems: dotted - passive, solid - active, from [26].

In the stochastic preview optimal control method, the general dynamic model of a suspension model is discretized and formulated as

$$X(k+1) = A_d X(k) + B_d U(k) + E_d Z(k) \tag{5.13}$$

where $X(k)$, $U(k)$, and $Z(k)$ are sample value of $X(t)$, $U(t)$, and $Z(t)$ at time $t = k$. A_d, B_d and E_d are the associated regression matrix.

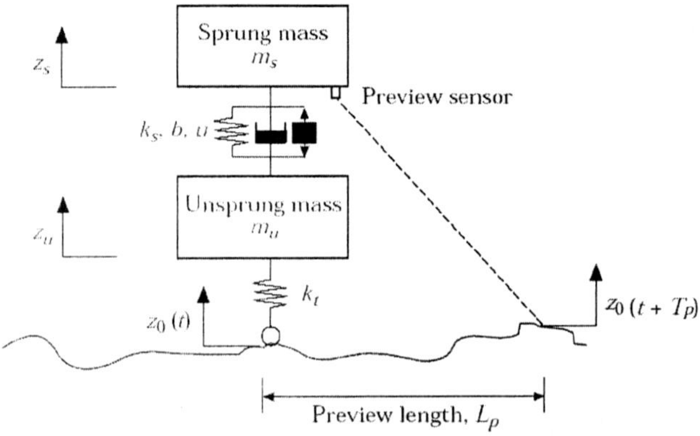

Fig.5.7 2 DOF quart car model with look-ahead sensor, Reprinted from [29] with permission from Elsevier.

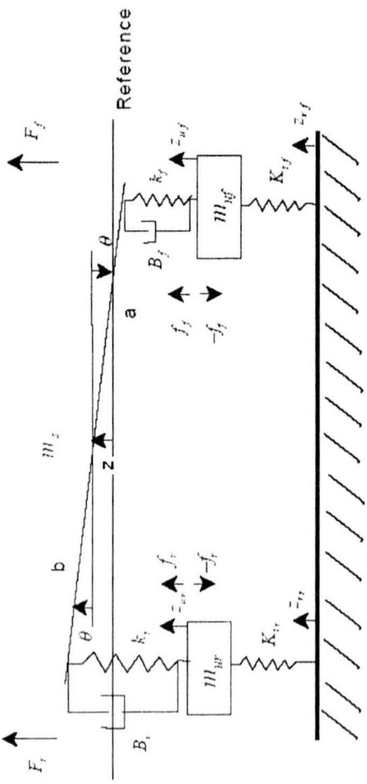

Fig.5.8 Heave and pitch/terrain mechanical subsystem, from [31]. (© [1999] IEEE)

And the performance index is also discretized as

$$J = \sum_{i=0}^{n} \left(X(k)^T Q X(k) + U^T(k) R U(k) \right) \tag{5.14}$$

By solving Eq.(5.14) with pre-assigned time range n, the stochastic preview optimal control input sequence $U(k)$ can be gotten. Because Eq.(5.13) contains more road roughness information than Eq.(5.8), it is usually assumed to provide better solution than conventional method.

There were other stochastic preview optimal control methods which do not use look-ahead sensor but predicate the coming road profile ahead based on the past road roughness input and system response. An illustrative example of such approach is shown in Fouskitakis and Fassois 1997 [30], which use long-memory Fractionally Integrated ARMA models.

Besides the above time domain approaches that are based on state space models, there are also some frequency domain approaches that aim to shape the vehicle vibration frequency response.

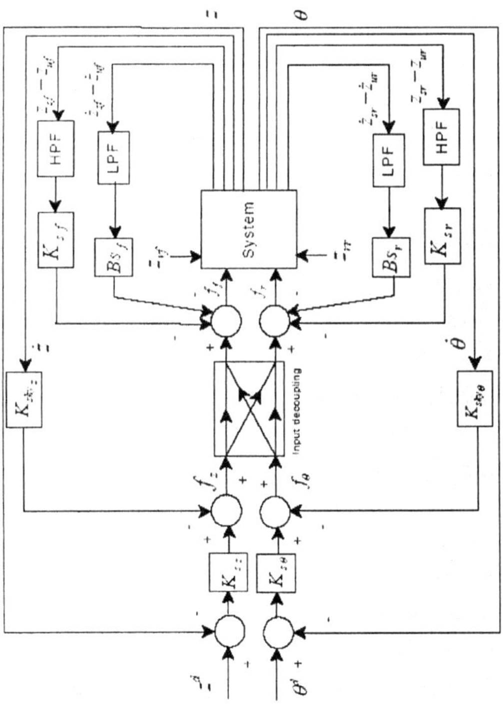

Fig.5.9 Generation of target strut force, from [31]. (© [1999] IEEE)

In [31], Campos et. al discussed a half car model shown in Fig.5.9. In the proposed model, the pitch angles are assumed to be small and the roll motion is neglected for simplicity. The mass of the body is m_s and its centroidal moment of inertia is J_y. Un-sprung masses on the front and rear wheels are denoted by m_{uf} and m_{ur}, respectively. z_{rf} and z_{rr} represent the road excitation on front and rear wheels. z_{uf}, z_{ur}, z_{sf} and z_{sr} represents the un-sprung/sprung displacements of the front and rear wheels, respectively.

The proposed control strategy is to introduce a low pass filter (LPF) and a high pass filter (HPF) for the feedback loop, which can be informally written as

$$x_{lpf} = \frac{\omega_0}{s+\omega_0}(\dot{z}-\dot{z}_{uf}),\ x_{lpr} = \frac{\omega_0}{s+\omega_0}(\dot{z}-\dot{z}_{ur}) \tag{5.15}$$

and

$$x_{hpf} = \frac{s}{s+\omega_0}(z_s-z_{uf}),\ x_{hpr} = \frac{s}{s+\omega_0}(z_s-z_{ur}) \tag{5.16}$$

where x_{lpf} and x_{hpf} are two feedback signals. $\omega_0 = \sqrt{k_u / m_u}$ denotes the natural frequency.

The object of this method is to use the low pass filter and high pass filter to eliminate the vibration to the maximum degree and allow only the natural vibration exits; because natural vibration cannot be totally gotten rid of. The above filters can be implemented through a careful selected feedback as Fig.5.10.

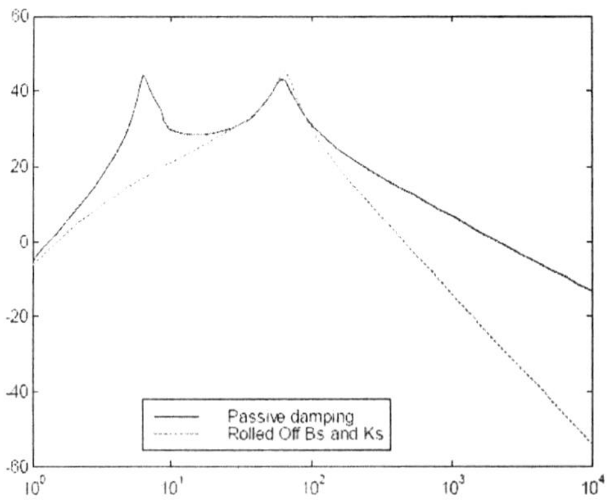

(a)

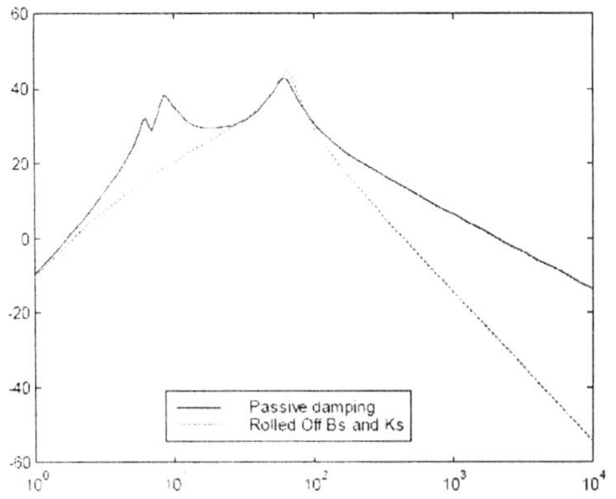

(b)

Fig.5.10 Effects of active damping: (a) heave acceleration from front wheel road disturbance, (b) pitch acceleration from front wheel road disturbance, from [31]. (© [1999] IEEE)

Fig.5.10 clearly illustrates the frequency response shaping results, where the vibration is concentrated to the neighborhood of natural frequency. The vibration is suppressed at every frequency though to different degrees.

Furthermore, Ikenaga et. al 2000 showed that this method can be successfully extended to full car models in [32].

5.3.2 Robust Suspension Controllers

One potential problem of the above methods discussed in *Section 5.2.1* is that they are all relied on explicit and accurate system models. However, the dynamic models for suspension systems are often unknown or partly-known. Moreover, the characteristics of the suspension components may change as times goes by. To handle this problem, many new controller design methods are proposed within the later decades [7], [33].

A possible way to solve this problem is to carry out system identification/estimation before we design the controller. The other methods include H_2, H_∞ design methods and Fuzzy rule based controllers which are supposed to be more robust and/or more flexible to the varied road condition and suspension status.

Robust control related theories have achieved significant success in the last two decades. Now, it is widely accepted that H_2 design method provides a convenient framework for discussing the system performance criterion and H_∞ design method yields a solid foundation for disturbance rejection analysis. The effectiveness of both H_2 and H_∞ design methods have been proved by thousands of different situations and varied application situations.

The early usage of robust control theory can be viewed as direct extension of the LQR theory [34]-[35]. Later, several multi-objective design frameworks are proposed in [17], [36]-[37]. These multi-objective approaches usually presented mixed H_2 / H_∞ synthesis for the suspension system to treat the standard H_2 and H_∞ optimal control problems as separate problems but in a unified state-space framework.

For example, Gaspar et. al 2002 analyze the half car model shown in Fig.5.9. The general dynamic models is still chosen as Eq.(5.8). The state, disturbance and input force vectors are respectively denoted by

$$X = \begin{bmatrix} z_s & \theta & z_{uf} & z_{ur} & \dot{z}_s & \dot{\theta} & \dot{z}_{uf} & \dot{z}_{ur} \end{bmatrix} \tag{5.17}$$

and

$$U = \begin{bmatrix} u_f & u_r \end{bmatrix}, W = \begin{bmatrix} z_{rf} & z_{rr} \end{bmatrix} \tag{5.18}$$

Introduce further measurement output for H_2 / H_∞ z_∞ and H_2 / H_∞ z_2, we could the integrated H_2 / H_∞ synthesis as (5.19) below, which is also shown in Fig.5.11.

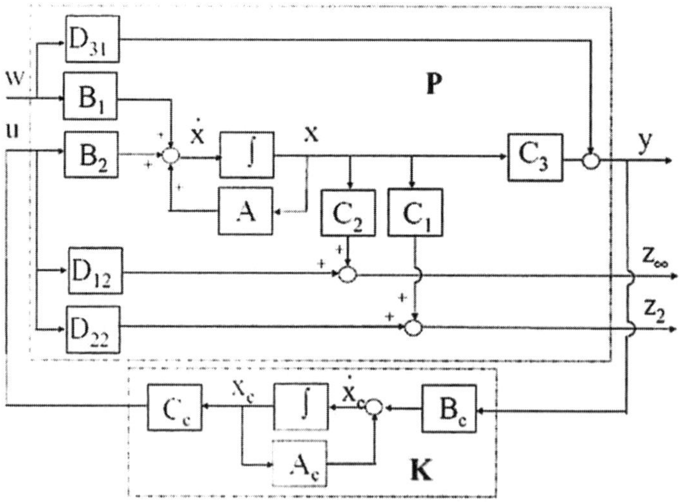

Fig.5.11 The closed-loop system for the mixed H_2 / H_∞ controller, from [37].

$$\begin{cases} \dot{X} = AX + BU + EZ \\ z_{\infty} = C_1 X + D_{12} U \\ z_{\infty} = C_2 X + D_{22} U \\ y = C_3 X + D_{31} U \end{cases} \tag{5.19}$$

The objective of mixed H_2 / H_∞ control is to minimize the H_2-norm of the closed-loop transfer function T_{z_2}, while constraining the H_∞-norm of the transfer function T_{z_∞} to be less than some specified levels.

More precisely, the objective to find an admissible controller which satisfies the following design criteria:

(1) the closed-loop system must be asymptotically stable;

(2) the closed-loop transfer function from z to z_∞ satisfies the constraint

$$\left\| T_{z_\infty}(s) \right\|_\infty < \gamma \tag{5.20}$$

where the sprung mass, the un-sprung mass, the suspension stiffness, the tire stiffness, and the suspension damping matrices are chosen as follows:

(3) there is no significant flow restriction between the pressure measurement and the actuator chambers.

$$\min \left\| T_{z_2}(s) \right\|_2 \tag{5.21}$$

Usually, the H_∞ objective will contradict with the H_2 objective, thus certain trade-off strategy should be used to obtain a satisfied solution.

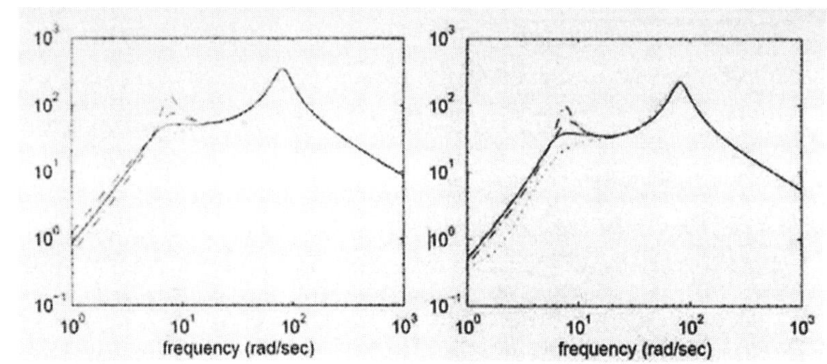

Fig.5.12 Effects of robust active damping: (left) heave acceleration from front wheel road disturbance and (right) pitch acceleration from front wheel road disturbance, from [37].

Fig.5.12 shows the robust active damping effects. The solid line corresponds to the mixed H_2 / H_∞ synthesis, the dashed line to the H_∞ synthesis, the dotted line to the LQG design, and the dashed-dotted line to the passive suspension system. Comparing Fig.5.10 and Fig.5.12, it is clearly that the robust design also eliminates the non-natural vibration. But it does not shape the high frequency part response too much due the robustness consideration. Similar discussions on full car models can be found in [38]-[41]. Besides frequency approaches, time domain approaches include LMI design and μ synthesis are also applied for robust suspension systems design [42]-[43].

5.3.3 Fuzzy Suspension Controllers

Fuzzy suspension controller is hottest topic in the recent decades. Till now, there are more than one hundred technique reports that discuss the feasibility of Fuzzy suspension controller, i.e. [1], [44]-[53].

$\dot{z}_s - \dot{z}_u$	$\dot{z}_s$	$\ddot{z}_s$	f_a	$\dot{z}_s - \dot{z}_u$	$\dot{z}_s$	$\ddot{z}_s$	f_a
PM	PM	ZE	ZE	PM	PM	P or N	NS
PS	PM	ZE	NS	PS	PM	P or N	NM
ZE	PM	ZE	NM	ZE	PM	P or N	NB
NS	PM	ZE	NM	NS	PM	P or N	NB
NM	PM	ZE	NB	NM	PM	P or N	NV
PM	PS	ZE	ZE	PM	PS	P or N	NS
PS	PS	ZE	NS	PS	PS	P or N	NM
ZE	PS	ZE	NS	ZE	PS	P or N	NM
NS	PS	ZE	NM	NS	PS	P or N	NB
NM	PS	ZE	NM	NM	PS	P or N	NB
PM	ZE	ZE	PS	PM	ZE	P or N	PM
PS	ZE	ZE	ZE	PS	ZE	P or N	PS
ZE	ZE	ZE	ZE	ZE	ZE	P or A	ZE
NS	ZE	ZE	ZE	NS	ZE	P or N	NS
NM	ZE	ZE	NS	NM	ZE	P or N	NM
PM	NS	ZE	PM	PM	NS	P or N	PB
PS	NS	ZE	PM	PS	NS	P or N	PB
ZE	NS	ZE	PS	ZE	NS	P or N	PM
NS	NS	ZE	PS	NS	NS	P or N	PM
NM	NS	ZE	ZE	NM	NS	P or N	PS
PM	NM	ZE	PB	PM	NM	P or N	PV
PS	NM	ZE	PM	PS	NM	P or N	PB
ZE	NM	ZE	PM	ZE	NM	P or N	PB
NS	NM	ZE	PS	NS	NM	P or N	PM
NM	NM	ZE	ZE	NM	NM	P or N	PS

Fig.5.13 The Propose Fuzzy rules, from [49].

The basic Fuzzy suspension controller is quite straight forward. In [49], Hyniova, Stribrsky and Honcu proposed a Fuzzy controller for the quart car model that is shown in previous Fig.5.5. It has three inputs: body acceleration $\ddot{z}_s$, velocity $\dot{z}_s$, body deflection velocity $\dot{z}_s - \dot{z}_u$ and one output: actuator force f_a.

The proposed Fuzzy rules is shown in Fig.5.13, where NV denotes Negative Very Big, NB denotes Negative Big, NM denotes Negative Medium, NS denotes Negative Small, ZE denotes Zero, PS denotes Positive Small, PM denotes Positive Medium, PB denotes Positive Big and PV denotes Positive Very Big. And the defuzzification stage uses center of gravity method. The proposed fuzzification Membership function is tightly related to the model parameters and thus not reproduced here.

In Fig5.14, Hyniova et. al 2001 showed compromising control performance of Fuzzy controller comparing to passive suspension system. Since the Fuzzy membership function may greatly affect the suspension performance, several optimization methods were developed to search the best Fuzzy membership function set. Among them, genetic algorithm (GA) is the most frequently used one to choose the optimal membership functions for Fuzzy suspension controllers. Some related research results can be found in [45], [47], [53], [50].

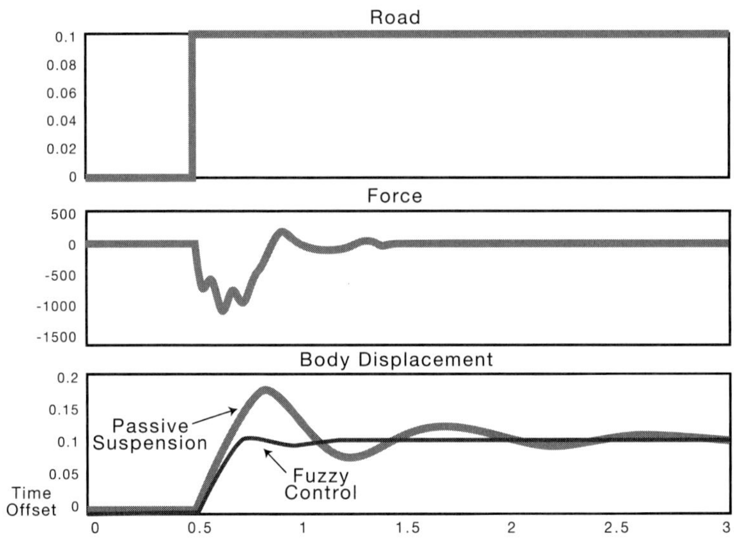

Fig.5.14 Active and passive suspension system under step response, from [49].

Besides these direct approaches, there are also some other suspension controllers that are based on complicated Fuzzy turning strategy. In [47], Kuo and Li proposed a composite control scheme which consists of a Fuzzy feedback controller and a Fuzzy feed-forward controller shown in Fig5.15. The Fuzzy feedback control is used to deal with the compromise between ride comfort and road holding ability according to the variations of the sprung mass and un-sprung mass velocities. The proposed Fuzzy feed-forward control is employed for the rejection of road disturbance. The theme of the propounded controller is to provide passenger ride comfort and maintain good road holding ability while vehicle running on a rough road. By using the merit of evolutionary programming (EP), the optimal decision making rules for both controllers are also constructed.

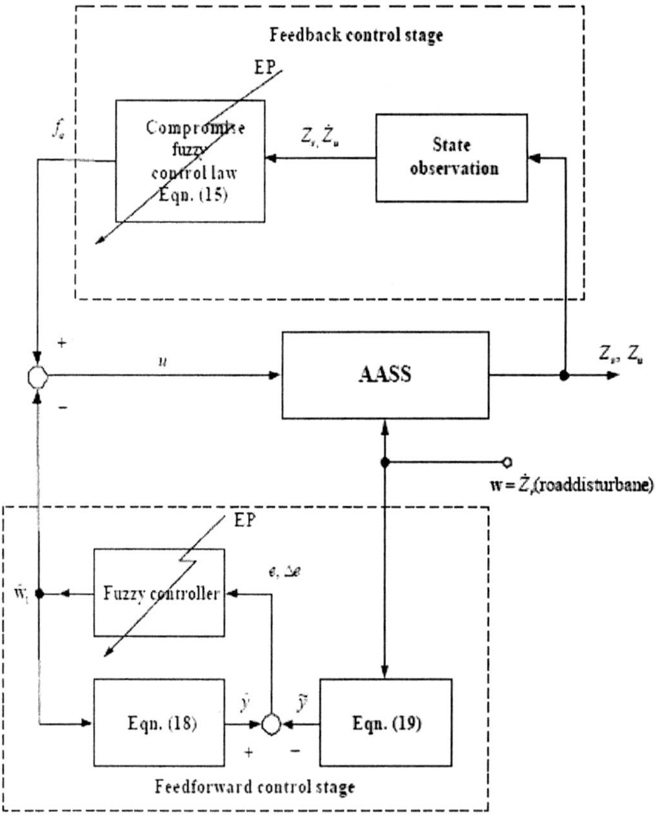

Fig.5.15 Structure of the composite Fuzzy suspension controller, from [47].

The real-time simulation results are illustrated in Fig5.16, where the solid, dashed, and dash-dot lines represent the performances of the composite Fuzzy control (CFC), the simple Fuzzy feedback control (SFFC) and the linear optimal control (LOC), respectively. Obviously, the proposed control strategy is effective and can provide much better performance compared to linear optimal control.

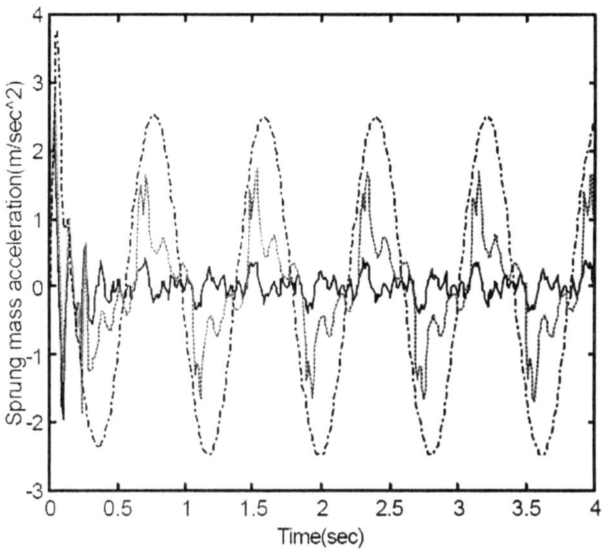

Fig.5.16 Time responses of sprung mass acceleration, from [47].

As the Fuzzy control method is extended from quart car model to half car and full car model, the Fuzzy rule computation cost emerges as a significant problem. Some researchers suggested that the less computationally complex Fuzzy system would be technically preferable in order to be more responsive in producing an output, as well as more economical due to the less number components used, thus implying smaller production costs. In [51], Rattasiri and Halgamuge discussed the possibility of using the so called hierarchical fuzzy system to solve this problem. However, it still needs further discussions.

Besides pure Fuzzy controller, some hybrid controllers were also developed recently to improve the flexibility of the suspension system. Normally such systems may combine integrate two or more intelligent controllers together, i.c. Fuzzy controller, sliding mode controller, neural network controller, i.e. [46].

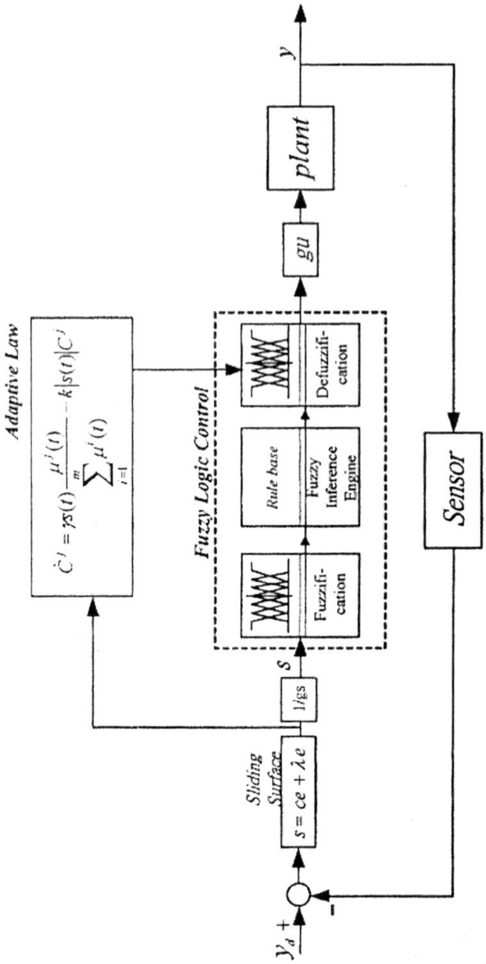

Fig.5.17 Block diagram of the control strategy proposed, from [52]. (© [2003] IEEE)

In [52], Huang and Lin proposed a hybrid suspension system which is shown in Fig.5.17. It has a special adaptive control rule with fuzzy and sliding mode control algorithms, which provides online learning ability to deal with the system time-varying and nonlinear uncertainty behaviors, and adjust the control rules parameters. Only eleven fuzzy rules are required for this active suspension system and these fuzzy control rules can be established and modified continuously by online learning. It is reported that the experimental results show that this intelligent control algorithm effectively suppresses the oscillation amplitude of the sprung mass with respect to various road surface disturbances. Similar online self adjusting mechanism was implemented by Foda in [48] through Neuro-Fuzzy suspension controller.

Besides the above suspension controller design approaches, there are several other methods employed [52]-[62]. For instance, adaptive suspension controllers were discussed in [54]-[56]; preview suspension controller was proposed in [57]; neural network suspension system modeling were mentioned in [58] and neural network suspension controller in [59].

5.4 Parameter Estimation and Fault Detection of Suspension Systems

In the previous *Section 5.2*, several different suspension control algorithms are studied, in which all the parameters and state variables in the model are assumed to be known or at least partly known. However, in the real applications, such model parameters are usually unknown and need to be identified before being used for analysis. And the existence of disturbance in the suspension components requires appropriate observers to achieve the value of some state variables that cannot be directly measured.

In [63], Tan and Bradshaw investigated the possibility of nonlinear hydraulic parameter identification and observer design. Particularly, their goal is to identify the effective bulk modules of the fluid in the system, the valve effective flow gain and the time history of the spool valve position. Then, the identified spool position can be further used for valve dynamics identification. The hydraulic actuator is modeled as a power cylinder with the piston which is controlled by a servo value. Fig.5.18 below illustrates the diagram of such servo value and hydraulic piston combination. The flows of the actuator are controlled by the position of the spool value x_{spv} that follows the commanded current i_{sv}. By assuming the servo valve orifices are matched and symmetrical, Tan and Bradshaw wrote the continuity equation to each piston chambers as

$$q_u - c_{ip}(P_u - P_l) - c_{ep}P_u = \frac{dV_u}{dt} + \frac{V_u}{\beta}\frac{dP_u}{dt} \tag{5.22}$$

and

$$q_l - c_{ip}(P_u - P_l) - c_{ep}P_l = \frac{dV_l}{dt} + \frac{V_l}{\beta}\frac{dP_l}{dt} \tag{5.23}$$

where q_u and q_l are the flows to upper and lower chamber respectively, V_u and V_l are the volumes of upper and lower chamber respectively, P_u and P_l are the pressures of upper and lower chamber respectively, c_{ip} and c_{ep} are the internal leakage coefficient across the piston and the external leakage coefficient of the piston respectively. β is the effective bulk modules of the hydraulic system (include hydraulic fluid, entrapped air, as well as mechanical compliance of the chambers and hoses).

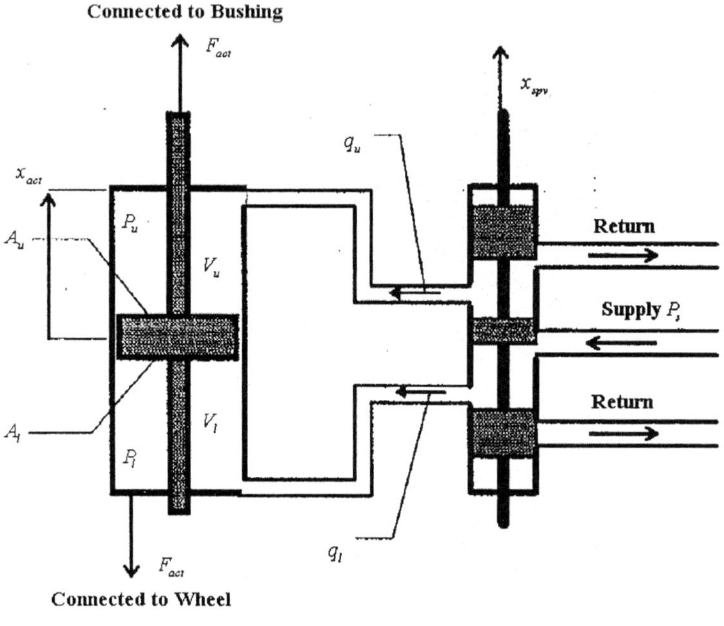

Fig.5.18 Hydraulic servo-valve and actuator combination, from [63]. (© [1997] IEEE)

From (5.22) and (5.23), we could see that the follow q_u and q_l are determined by the pressure differences across a sharp edge orifice whose size is a linear function of the spool position x_{spv}. The relationship between the supply pressure P_s and the unit flow resistance coefficient of the valve orifice k_q could be formulated as

$$q_u = -k_q x_{spv} \sqrt{\Delta P_u}\ , \ q_l = k_q x_{spv} \sqrt{\Delta P_l} \tag{5.24}$$

where

$$\Delta P_u = \begin{cases} \sqrt{P_u}\ , & if \quad x_{spv} \geq 0 \\ \sqrt{P_s - P_u}\ , & otherwise \end{cases} \tag{5.25}$$

and

$$\sqrt{\Delta P_u} = \begin{cases} \sqrt{P_s - P_l}\ , & if \quad x_{spv} \geq 0 \\ \sqrt{P_l}\ , & otherwise \end{cases} \tag{5.26}$$

Given the piston area of each chamber as A_l and A_u, we have

$$\frac{dV_u}{dt} = A_u \frac{dx_{act}}{dt}\ , \ \frac{dV_l}{dt} = -A_l \frac{dx_{act}}{dt} \tag{5.27}$$

where x_{act} is the actuator strut displacement with the convention. It is positive when the strut is compressed from its nominal position. The dynamics equations for the upper and lower chamber pressures could be further written as

$$\frac{dP_u}{dt} = \frac{\beta}{V_u}\left(-k_q x_{spv} \sqrt{\Delta P_u} - A_u \frac{dx_{act}}{dt} - c_{ip}(P_u - P_l) - c_{ep} P_u \right) \tag{5.28}$$

and

$$\frac{dP_l}{dt} = \frac{\beta}{V_l}\left(k_q x_{spv} \sqrt{\Delta P_l} + A_l \frac{dx_{act}}{dt} - c_{ip}(P_l - P_u) - c_{ep} P_l \right) \tag{5.29}$$

The actuator force F_{act} generated by the hydraulic strut is

$$F_{act} = A_l P_l - A_u P_u - F_{friction} \frac{dx_{act}}{dt} \tag{5.30}$$

where $F_{friction}$ is the friction force in the actuator strut.

Tan and Bradshaw also assumed that:

(1) there is no leakage in the system;

(2) the spool is line cut and symmetric;

(3) there is no significant flow restriction between the pressure measurement and the actuator chambers.

With these assumptions, Eqs. (5.28) and (5.29) could be rewritten as

$$\frac{V_u(t)\dot{P}_u(t)}{\beta} + k_q\sqrt{\Delta P_u(t)}x_{spv}(t) = -A_u\dot{x}_{act}(t) \tag{5.31}$$

and

$$\frac{V_l(t)\dot{P}_l(t)}{\beta} - k_q\sqrt{\Delta P_l(t)}x_{spv}(t) = -A_l\dot{x}_{act}(t) \tag{5.32}$$

where the postfix (t) is added here to emphasize the time sequence nature of variables.

Notice that the supply and return orifices are equal, Eq.(5.32) should be equal to the following formats

$$\left(\frac{V_u(t)\dot{P}_u(t)}{\sqrt{\Delta P_u(t)}} + \frac{V_l(t)\dot{P}_l(t)}{\sqrt{\Delta P_l(t)}}\right)\frac{1}{\beta} = \frac{A_l\dot{x}_{act}(t)}{\sqrt{\Delta P_l(t)}} - \frac{A_u\dot{x}_{act}(t)}{\sqrt{\Delta P_u(t)}} \tag{5.33}$$

The bulk modules can then be solved by using least squares with Eq.(5.33). To determine which flow path is open when computing $\sqrt{\Delta P_u(t)}$ and $\sqrt{\Delta P_l(t)}$ initially, the sign of the current command is used to determine whether the spool is on the positive or negative side.

With the identified valve of the bulk modules, Eq (5.33) can then be used to solve for the product of spool positive and flow gain. The process is repeated until convergence, which normally requires one or two iterations. The reason for iteration is to establish the correct sign of the spool position so that the correct identification equations can be used.

Besides obtaining the suspension model characteristics, parameter estimation technique is also helpful in detecting the fault of suspension system. Records show that the components, sensors and actuators often yield unexpected and/or un-permitted behavior due to fatigue, leakage or deterioration. As the suspension system is responsible for driving comfort and safety, certain fault detection process is essential for modern vehicles. Some recent reports proposed several dynamic models suitable for both system diagnosis and failure accommodation. And among them, the parameter estimation based fault detection method was most widely accepted and applied.

For example, the previous quart car model (5.4) and (5.5) can be discretized using the following approximations

$$\ddot{z}_u \approx \frac{\left(z_u(k) - 2z_u(k-1) + z_u(k-2)\right)}{2\Delta T} \tag{5.34}$$

and

$$\ddot{z}_s \approx \frac{(z_s(k) - 2z_s(k-1) + z_s(k-2))}{2\Delta T} \tag{5.35}$$

together with

$$\dot{z}_u \approx \frac{(z_u(k) - z_u(k-1))}{\Delta T}, \quad \dot{z}_s \approx \frac{(z_s(k) - z_s(k-1))}{\Delta T} \tag{5.36}$$

and

$$\dot{z}_r \approx \frac{(z_r(k) - z_r(k-1))}{\Delta T} \tag{5.37}$$

where ΔT represents the sample time span and normally should be a small positive number.

Then, we could have

$$z_u(k) = L_1(m_u, m_s, c_u, k_u, \Delta T)\begin{bmatrix} z_u(k-1) \\ z_u(k-2) \\ z_r(k) \\ z_r(k-1) \end{bmatrix} = \begin{bmatrix} L_{11} & L_{12} & L_{13} & L_{14} \end{bmatrix}\begin{bmatrix} z_u(k-1) \\ z_u(k-2) \\ z_r(k) \\ z_r(k-1) \end{bmatrix} \tag{5.38}$$

and

$$z_s(k) = L_2(m_s, c_s, k_s, \Delta T)\begin{bmatrix} z_s(k-1) \\ z_s(k-2) \\ z_r(k) \\ z_r(k-1) \end{bmatrix} = \begin{bmatrix} L_{21} & L_{22} & L_{23} & L_{24} \end{bmatrix}\begin{bmatrix} z_s(k-1) \\ z_s(k-2) \\ z_r(k) \\ z_r(k-1) \end{bmatrix} \tag{5.39}$$

where $L_1(m_u, m_s, c_u, k_u, \Delta T)$ and $L_2(m_s, c_s, k_s, \Delta T)$ are both 1×4 matrix in terms of the parameters that need to identify.

Notice the similarity of the above Eqs. (5.38) and (5.39) and the general linear parameter estimation Eq.(5.40), it is clear that we can identify and monitor the abstracted model parameter L_{ij}, $i = 1,2$, $j = 1,\ldots,4$ using almost all the known parameter estimation methods [64].

$$Y(k) = \begin{bmatrix} \phi_1 & \phi_2 & \cdots & \phi_n \end{bmatrix}\begin{bmatrix} Y(k-1) \\ \cdots \\ Y(k-m) \\ V(k) \\ \cdots \\ V(k-(n-m)) \end{bmatrix} \tag{5.40}$$

where $Y(k)$ denotes general output, $V(k)$ denotes general input and ϕ as the general parameters.

In [66]-[67], Borner, Zele and Isermann discussed the quart car parameter estimation problem. They compared several algorithms including non-recursive algorithm, prediction-error turning algorithm, recursive least square algorithm with forgetting factor, directional forgetting method, restricted exponential forgetting algorithm, exponential resetting and forgetting algorithm. They claimed that the estimations results have shown that the RLS with exponential forgetting factor and PEF algorithms received very good results. And it is able to adapt to the different damping configuration and to detect a fault in the shock absorber.

In [68]-[69], Fischer, Kaus and Isermann discussed a similar system which has an active controller shown in Fig.5.19. The complexity of this system is the nonlinearity of the active suspension controller. By using the so called local linear model tree (LOLIMOT) algorithm, they developed a new method which is quite different from the linearized estimation method used by Majjad 1997 for almost the same problem in [64].

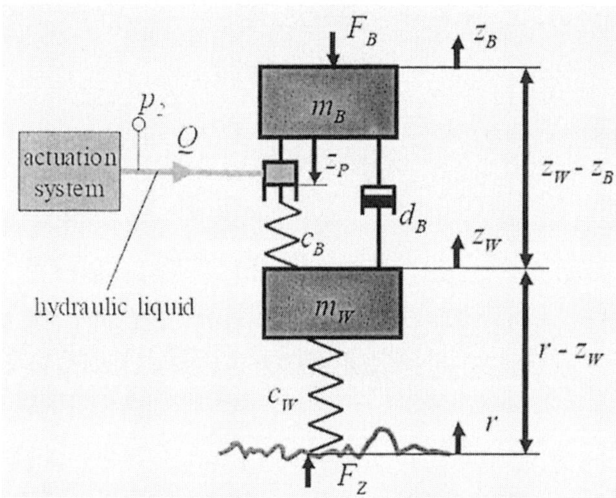

Fig.5.19 Schematic diagram of the active suspension model, from [69]. (© [2003] IEEE)

The LOLIMOT algorithm is based on the idea of approximating a nonlinear dynamic discrete function with m-dimensional inputs $u(k)$ and 1-dimensional output $y(k)$.

$$Y(k) = f(x(k)) \tag{5.41}$$

and

$$x(k) = [u_n(k-1) \quad \dots \quad u_m(k-n_{u_m}) \quad y(k-1) \quad \dots \quad y(k-n_y)]^T \tag{5.42}$$

by piece-wise linear models

$$y_i(k) = w_{0,i} + w_{1,i}x_1(k) + \dots + w_{n,i}x_n(k) \tag{5.43}$$

where $w_{j,i}$, $j = 0,1,\dots,n_i$ are the parameters of the i th linear regression model and x_i are the inputs. Each local model is valid in a sub-region of the input space. And the partitions of the local models are not crisp but Fuzzy.

The model output is calculated as the weighted sum of all M local models as

$$\hat{y} = \sum_{i=1}^{M}(w_{0,i} + w_{1,i}x_1 + \dots + w_{n,i}x_n)\cdot\Phi_i(x_i, c_i, \sigma_i) \tag{5.44}$$

where $\Phi_i(x_i, c_i, \sigma_i)$ is the normalized Gaussian weighting function for the i th model with the center c_i and standard deviation σ_i. This is indeed completes a nnonlinear interpolation between the local model outputs, see Fig.5.20.

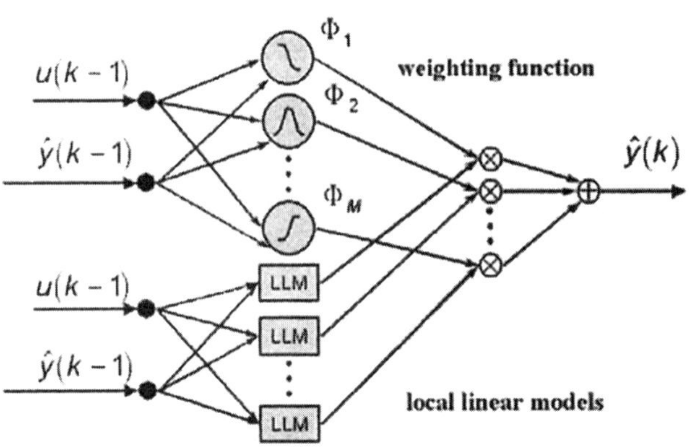

Fig.5.20 Concept of LOLIMOT, from [69]. (© [2003] IEEE)

Using the tree-like construction algorithm proposed by Nells and Isermann [65], c_i and σ_i can be determine through regression. Combine (5.40) and (5.43), we could get the estimation residual sequence as

$$r(k) = y(k) - \hat{y}(k) \tag{5.45}$$

It is apparent that when the estimation process converges, $r(k)$ would approach to zero. Thus, the residual sequence can be used to detect sensor faults associated with an appropriately assigned threshold. A simulation test is carried in [69] and the data obtained for every variable at $t = 0.3$ is used as the threshold. The corresponding detection results are shown in Fig.5.22. We can see that all the errors can be detected.

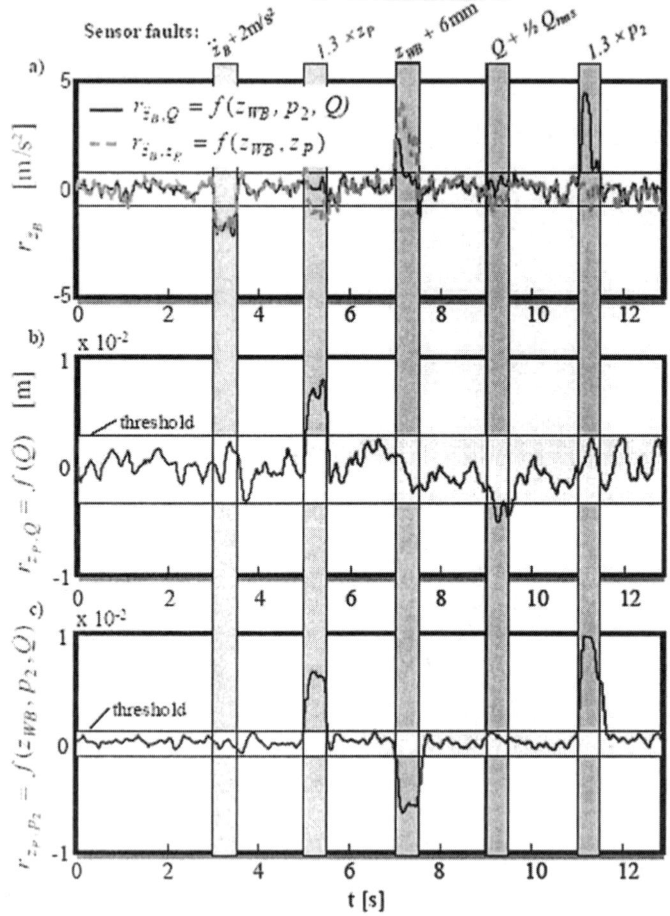

Fig.5.21 Residuals sequence, from [69]. (© [2003] IEEE)

5.5 Rollover Avoidance

Rollover avoidance using special steering control is gaining increasing attentions recently. The key idea of the rollover avoidance control concept is that in critical situations temporarily rollover avoidance is given priority over ideal lane keeping. As pointed out by Odenthal, Bunte and Ackermann in [76], vehicle rollover avoidance control aims at two goals. One is to improve the damping of the roll dynamics by active steering. This reduces the risk of rollover in dynamic steering maneuvers. The other task is to improve safety in emergency situations when the vehicle is already very close to rollover. They showed that applying active steering allows for an immediate and efficient impact on the vehicle's lateral, yaw and roll dynamics. Due to the fast and precise intervention of the active steering system which is in contrast to the limited reaction time of the driver there is a great potential for enhancing safety. Further, due to the small energy and actuator demand of the active steering system, it can be used not only in critical situations but also in continuous operation to improve ride comfort and handling. In addition to active steering, they studied braking actions considering rollover avoidance, too.

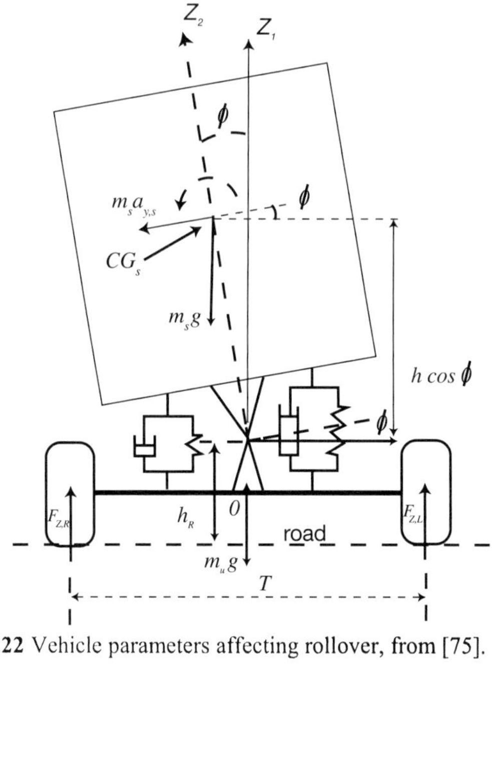

Fig.5.22 Vehicle parameters affecting rollover, from [75].

The first problem that needs to solve is how to determine the rollover of the vehicle. In the last ten years, great efforts had been put into rollover limit/index research [70]-[74]. For instance, the so called rollover coefficient R was used in [75], which depends on the vertical tire loads on the right and left hand side and is defined as

$$R = \frac{F_{Z,R} - F_{Z,L}}{F_{Z,R} + F_{Z,L}} \tag{5.46}$$

Fig.5.22 illustrates a simple model of the vehicle as seen from the front side during steady state. For straight driving the vertical tire loads on the left and right are equal and therefore R equals zero. If, at the beginning of rollover, the tires on one side are about to lift of the road, then, the absolute value of the rollover coefficient becomes one. For values of R within the range of -1 to 1 the vehicle is "rollover stable". R may be determined directly by measuring the vertical tire loads $F_{Z,R}$ and $F_{Z,L}$.

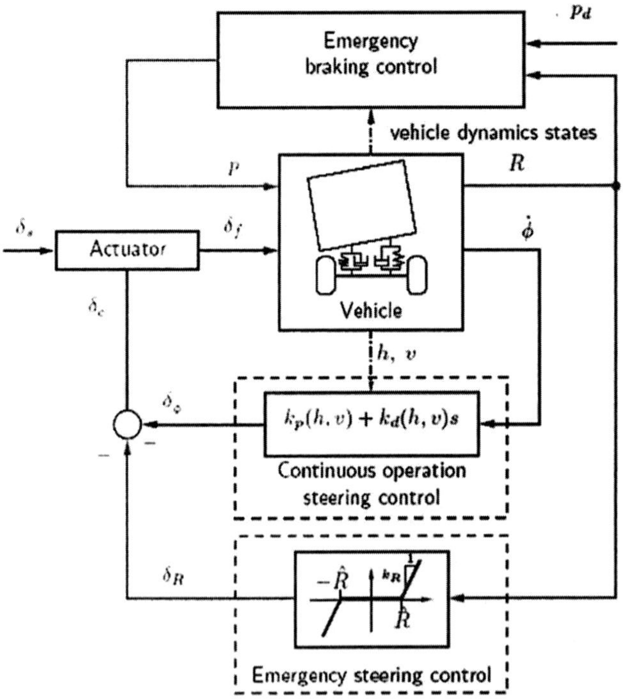

Fig.5.23 Vehicle parameters affecting rollover, from [75].

If this measurement is not available the simplified model shown in Fig.5.22 can be used to derive the following approximation of the rollover coefficient:

$$R \approx \frac{2(h_R + h)}{T} \cdot \frac{a_{y,s}}{g} \tag{5.47}$$

This approximation depends on the ratio of the roll body's CG height above the road $(h_R + h)$ and the constant track width T. It depends on the lateral acceleration $a_{y,s}$ at the roll body's *CG*. To provide this approximation of R for feedback the height of *CG* as well as the lateral acceleration must be determined. If the height h varies largely due to varying payloads its value can be estimated at the start of each ride. The lateral acceleration at the roll body's *CG* can e.g. be determined by interpolating between two accelerometers fixed at the top and bottom of the roll body. Note, that the signal $a_{y,s}$ should be available in body fixed coordinates such that road inclination can be considered.

In steady state cornering the lateral acceleration is $a_{y,s} = \rho v^2$, where ρ is the track curvature. From this formula it can be directly seen that, in case of rollover risk, the lateral acceleration has to be reduced either by reducing curvature (steering into a wider curve) or by reducing speed. In the present rollover control concept simultaneous emergency steering and braking is applied. The structure of the entire rollover avoidance control system is shown in Fig.5.23. It is implemented by nonlinear feedback of R to the front wheel steering angle δ_f. At the same time active braking by control of the braking pressure is applied to ensure precision of lane keeping and also to contribute to lateral acceleration reduction. As soon as $|R|$ falls below $\hat{R}$ the emergency control is deactivated. Additionally, in continuous operation the roll rate $\dot{\phi}$ is fed back to the front wheel steering angle δ_f. Gain scheduling of the controller with speed and CG height is used to achieve optimal performance for all operating conditions.

It had been proven that application of the vehicle dynamics control approach for rollover avoidance was especially suitable for the vehicles with an elevated center of gravity such as trucks, busses, sport and utility vehicles. The concept can be implemented in the vehicle with either steer by wire or a mechanical additional steering device. There were also some other rollover control approaches including shaped sliding mode control and neural network control [77]-[80]. Besides, in [81]-[82], rollover and suspension system integrated fault detection algorithms were studied. Constrained by the book length, this Chapter will not discuss them.

5.6 Summary

The recent developments of vehicle suspension control and rollover avoidance are studied in this Chapter. However, not all the vehicle vertical motion related topics are covered in this Chapter due to length limits. For instance,

(1) integrated 3D vehicle dynamics modeling, simulation and control gains more and more interest recently, since the vehicle's dynamics cannot be simply decomposed into three independent directions especially when the steering angle is relatively big [83]-[87];

(2) road slope estimation is another important research topic, since it is tightly linked to vehicle driveline control [88]-[89];

(3) Besides ordinary suspension systems installed in the chassis, seat suspension systems are proven to be capable to further improve ride performance in some luxury cars [90]-[91];

(4) Moreover, human feelings of vibrations are addressed in some recent literals [92]. Nevertheless, how to use human driver driving record to supervise active rollover avoidance systems is also under research now [93].

5.7 References

1. R. K. Jurgen, "Global '90 cars: electronics-aided," IEEE Spectrum, vol. 26, no. 12, pp. 45-49, 1989.
2. M. Appleyard and P. E. Wellstead, "Active suspensions: some background," IEE Proceedings - Control Theory and Applications, vol. 142, no. 2, pp. 123-127, 1995.
3. K. Bogsjo and A. Forsen, "Fatigue relevant road surgface statistics," Vehicle System Dynamics, Supplement, vol. 41, pp. 724-733, 2004.
4. D. Hrovat, Optimal Passive Vehicle Suspensions, Ph.D. thesis, University of California, Davis, CA, 1979.
5. D. Hrovat, D. L. Margolis, and M. Hubbard, "Suboptimal semi-active vehicle suspensions," Proceedings of Journalism Association of Community Colleges, 1980.
6. D. Hrovat and M. Hubbard, "Optimal vehicle suspension minimizing RMS rattle-space, sprung-mass acceleration and jerk," ASME Journal of Dynamic Systems, Measurement and Control, vol. 103, no. 2, pp. 228-236, 1981.
7. D. Hrovat, "Survey of advanced suspension developments and related optimal control applications," Automatica, vol. 33, no. 10, pp. 1781-1817, 1997.

8. D. Karnopp, "Design principles for vibration control systems using semi-active dampers," ASME Journal of Dynamics Systems, Measurements and Control, vol. 112, pp. 448-455, 1990.
9. R. P. La Barre, R. T. Forbes, and S. Andrews, "The measurement and analysis of road surface roughness," Motor Industry Research Association Report 1970/5, 1970.
10. E. Sevin and W. D. Pilkey, "Optimum shock and vibration isolation," The Shock and Vibration Information Center, United States Department of Defense, 1971.
11. Proposals for generalized road inputs to vehicles, ISO/DIS2631, 1972, pp. 1-7.
12. J. D. Robson, "Road surface description and vehicle response," International Journal of Vehicle Design, vol. 1, no. 1, pp. 25-35, 1979.
13. M. W. Sayers, "Dynamic terrain inputs to predict structural integrity of ground vehicles," University of Michigan Transportation Research Institute Report UMTRI-88-16, 1988.
14. M. W. Sayers, "Profiles of roughness," Transportation Research Record 1260, pp. 106-111, 1990.
15. V. Rouillard, "Using predicted ride quality to characterise pavement roughness," International Journal of Vehicle Design, vol. 36, no. 2/3, pp. 116-131, 2004.
16. K. Ahlin, J. Granlund, and F. Lindstrom, "Comparing road profiles with vehicle perceived roughness," International Journal of Vehicle Design, vol. 36, no. 2/3, pp. 270-286, 2004.
17. J. Lu and M. DePoyster, "Multiobjective optimal suspension control to achieve integrated ride and handling performance," IEEE Transactions on Control Systems Technology, vol. 10, no. 6, pp. 807-821, 2002.
18. R. Caponetto, O. Diamante, and G. Fargione, et. al, "A soft computing approach to fuzzy sky-hook control of semi-active suspension," IEEE Transactions on Control Systems Technology, vol. 11, no. 6, pp. 786-798, 2003.
19. I. Hostens and H. Ramon, "Descriptive analysis of combine cabin vibrations and their effect on the human body," Journal of Sound and Vibration, vol. 266, no. 3, pp. 453-464, 2003.
20. S. Park, A. A. Popov and D. J. Cole, "Vehicle suspension optimisation for heavy vehicles on deformable ground," Vehicle System Dynamics, Supplement, vol. 41, pp. 3-12, 2004.
21. D. Rubinstein and R. Hitron, "A detailed multi-body model for dynamic simulation of off-road tracked vehicles," Journal of Terramechanics, vol. 41, no. 2-3, pp. 163-173, 2004.
22. S. Park, A. A. Popov and D. J. Cole, "Influence of soil deformation on off-road heavy vehicle suspension vibration," Journal of Terramechanics, vol. 41, no. 1, pp. 41-68, 2004.
23. V. DeBrunner, D. Zhou, and M. Ta, "Adaptive vibration control of a bridge and heavy truck," Proceedings of IEEE Intelligent Vehicles Symposium, pp. 389-393, 2003.

24. M. Valask, J. Kejval, and J. Maca, et. al, "Bridge - friendly truck suspension," Vehicle System Dynamics, Supplement, vol. 41, pp. 13-22, 2004.
25. M. C. Smith and G. W. Walker, "Performance limitations and constraints for active and passive suspensions: a mechanical multi-port approach," Vehicle System Dynamics, vol. 33, no. 3, pp. 137–168, 2000
26. E. Esmailzadeh and H. D. Taghirad, "Active vehicle suspensions with optimal state-feedback control," International Journal of Modeling and Simulation, vol. 18, no. 3, pp. 228-238, 1998.
27. Y.-P. Kuo and T.-S. Li, "GA-based fuzzy PI/PD controller for automotive active suspension system," IEEE Transactions on Industrial Electronics, vol. 46, no. 6, pp. 1051-1056, 1999.
28. E. K. Bender, "Optimum linear preview control with application to vehicle suspension," Journal of Basic Engineering Series D90, pp. 213-221, 1968.
29. H.-S Roh and Y. Park, "Stochastic optimal preview control of an active vehicle suspension," Journal of Sound and Vibration, vol. 220, no. 2, pp. 313-330, 1999.
30. G. N. Fouskitakis and S. D. Fassois, "Long-memory modeling and prediction of automotive active suspension power consumption," Proceedings of the American Control Conference, pp. 3354-3358, 1997.
31. J. Campos, L. Davis, and F. L. Lewis, et. al, "Active suspension control of ground vehicle heave and pitch motions," Proceedings of IEEE Mediterranean Control Conference on Control and Automation, pp. 222-233, 1999.
32. S. Ikenaga, F. L. Lewis, and J. Campos, et. al, "Active suspension control of ground vehicle based on a full-vehicle model," Proceedings of American Control Conference, pp. 4019-4024, 2000.
33. C. Yue, T. Butsuen, and J. K. Hedrick, "Alternative control laws for automotive active suspensions," ASME Journal of Dynamic Systems, Measurement, and Control, vol. 111, pp. 286-291, 1989.
34. P. Michelberger, J. Bokor, and L. Palkovics, "Robust design of active suspension system," International Journal of Vehicle Design, vol. 14, pp. 145–165, 1994.
35. L. Palkovics, P. Gaspar, and J. Bokor, "Design of active suspension system in the presence of physical parameter uncertainties," Proceedings of American Control Conference, vol. 1, pp. 696-700, 1993.
36. B. B. Jager, "Multiobjective control: an overview," Proceedings of IEEE Conference on Decision and Control, pp. 440–445, 1997.
37. P. Gaspar, I. Szaszi, and J. Bokor, "Mixed H_2 / H_∞ control design for active suspension structures," Perodicapolytechnica Series of Transportation Engineering, vol. 28, no. 1-2, pp. 3-16, 2000.
38. S.-B. Choi, H.-S. Lee and Y.-P. Park, "H_∞ control performance of a full-vehicle suspension featuring magnetorheological dampers," Vehicle System Dynamics, vol. 38, no. 5, pp. 341–360, 2002.

39. L. Zuo and S. A. Nayfeh, "Structured H_2 optimization of vehicle suspensions based on multi-wheel models," Vehicle System Dynamics, vol. 40, no. 5, pp. 351–371, 2003.
40. D. Sammier, O. Sename and L. Dugard, "Skyhook an H_∞ control of semi-active suspensions: some practical aspects," Vehicle System Dynamics, vol. 39, no. 4, pp 279-308, 2003.
41. X. Shen and H. Peng, "Analysis of active suspension systems with hydraulic actuators," Vehicle System Dynamics, Supplement, vol. 41, pp. 143-152, 2004.
42. S.-H. Chen and J.-H. Chou, "Stability robustness of optimal active uncertain suspensions incorporating human sensitivity to vibration," International Journal of Vehicle Design, vol. 36, no. 4, pp. 303-319, 2004.
43. P. Gaspar, I. Szaszi, and J. Bokor, "Design of robust controllers for active vehicle suspension using the mixed μ synthesis," Vehicle System Dynamics, vol. 40, no. 4, pp. 193-228, 2003.
44. A. S. Cherry and R. P. Jones, "A fuzzy rule based approach to non-linear control of an automotive suspension system," Proceedings of IEE Colloquium on Active Suspension Technology for Automotive and Railway Applications, vol. 4, pp. 5/1-5/3, 1992.
45. T. Hashiyama, T. Furuhashi, and Y. Uchikawa, "Design of fuzzy controllers for semi-active suspension generated through the genetic algorithm," Proceedings of New Zealand International Two-Stream Conference on Artificial Neural Networks and Expert Systems, pp. 166-169, 1995.
46. N. Al-Holou, J. Weaver, and T. Lahdhiri, et. al, "Sliding mode-based fuzzy logic controller for a vehicle suspension system," Proceedings of American Control Conference, vol. 6, pp. 4188-4192, 1999.
47. Y.-P. Kuo and S. Li, "A composite EP-based fuzzy controller for active suspension system," International Journal of Fuzzy Systems, vol. 2, no. 3, pp. 183-191, 2000.
48. S. G. Foda, "Neuro-fuzzy control of a semi-active car suspension system," Proceedings of IEEE Pacific Rim Conference on Communications, Computers and signal Processing, vol. 2, pp. 686-689, 2001.
49. K. Hyniova, A. Stribrsky, and J. Honcu, "Fuzzy control of mechanical vibrating systems," Proceedings of International Carpathian Control Conference, pp. 393-398, 2001.
50. X. Peng, P. Vadakkepat, and T. H. Lee, "DNA coded GA for the rule base optimization of a fuzzy logic controller," Proceedings of Congress on Evolutionary Computation, vol. 2, pp. 1191-1196, 2001.
51. W. Rattasiri and S. K. Halgamuge, "Computationally advantageous and stable hierarchical fuzzy systems for active suspension," IEEE Transactions on Industrial Electronics, vol. 50, no. 1, pp. 48-61, 2003.
52. S.-J. Huang and W.-C. Lin, "Adaptive fuzzy controller with sliding surface for vehicle suspension control," IEEE Transactions on Fuzzy Systems, vol. 11, no. 4, pp. 550-559, 2003.

53. S. Y. Moon and W. H. Kwon, "Genetic-based fuzzy control for automotive active suspensions," Proceedings of IEEE International Conference on Fuzzy Systems, vol. 2, pp. 923-929, 1996.
54. A. G. Thompson, "An optimal suspension for an automobile on a random road," SAE#790478, 1979.
55. M. M. Oblak, A. S. Lesnika, and B. J. Butinar, "Optimum design of stochastically excited non-linear dynamic systems without geometric constraints," International Journal for Numerical Methods in Engineering, vol. 53, no. 11, pp. 2429-2443, 2002.
56. Y. Zhang and A. G. Alleyne, "A new approach to half-car active suspension control," Proceedings of American Control Conference, vol. 5, pp. 3762-3767, 2003.
57. J. Marzbanrad, G. Ahmadi, and H. Zohoor, "Stochastic optimal preview control of a vehicle suspension," Journal of Sound and Vibration, vol. 275, pp. 973-990, 2004.
58. A. Patel and J. F. Dunne, "NARX neural network modelling of hydraulic suspension dampers for steady-state and variable temperature operation," Vehicle System Dynamics, vol. 40, no. 5, pp. 285-328, 2003.
59. W. Kozukue and H. Miyaji, "Control of vehicle suspension using neural network," Vehicle System Dynamics, Supplement, vol. 41, pp. 153-161, 2004.
60. J.-S. Lin and C.-J. Huang, "Nonlinear backstepping control design of half-car active suspension systems," International Journal of Vehicle Design, vol. 32, no. 3/4, pp. 332-350, 2003.
61. M. Gobbi, F. Levi, and G. Mastinu, "Multi-objective robust design of the suspension system of road vehicles," Vehicle System Dynamics, Supplement, vol. 41, pp. 537-546, 2004.
62. H. Tokunaga, K. Misaji, and Y. Shimuzu, "On-Center steer feel evaluation based on non-linear vibration analytical method," Vehicle System Dynamics, Supplement, vol. 41, pp. 391-400, 2004.
63. H.-S. Tan and T. Bradshaw, "Model identification of an automotive hydraulic active suspension system," Proceedings of American Control Conference, vol. 5, pp. 2920-2924, 1997.
64. R. Majjad, "Estimation of suspension parameters," Proceedings of IEEE International Conference on Control Applications, pp. 522 - 527, 1997.
65. O. Nelles and R. Isermann, "Basis function networks for interpolation of local linear models," Proceedings of IEEE Conference on Decision and Control, vol. 1, pp. 470-475, 1996.
66. M. Borner, M. Zele, and R. Isermann, "Comparison of different fault detection algorithms for active body control components: automotive suspension system," Proceedings of American Control Conference, vol. 1, pp. 476-481, 2001.
67. M. Borner, H. Straky, and T. Weispfenning, et. al, "Model based fault detection of vehicle suspension and hydraulic brake systems," Mechatronics, vol. 12, no. 8, pp. 999-1010, 2002.

68. D. Fischer, E. Kaus, and R. Isermann, "Model based sensor fault detection for an active vehicle suspension," Proceedings of IFAC Symposium on Fault Detection, Supervision and Safety of Technical Processes, 2002.
69. D. Fischer, E. Kaus, and R. Isermann, "Fault detection for an active vehicle suspension," Proceedings of American Control Conference, pp. 4377-4382, 2003.
70. W.-H. Ma and H. Peng, "Worst-case manoeuvres for the roll-over and jackknife of articulated vehicles," Proceedings of American Control Conference, vol. 4, pp. 2263-2267, 1998.
71. B.-C. Chen and H. Peng, "A real-time rollover threat index for sports utility vehicles," Proceedings of American Control Conference, vol. 2, pp. 1233-1237, 1999.
72. I. Cech, "Anti-roll and active roll suspensions," Vehicle System Dynamics, vol. 33, no. 2, pp. 91–106, 2000.
73. V. Diaz, M. G. Fernandez, and J. L. S. Roman, et. al, "A new methology for predicting the rollover limit of buses," International Journal of Vehicle Design, vol. 34, no. 4, pp. 340-353, 2004.
74. A. Hac, T. Brown and J. Martens, "Detection of vehicle rollover," SAE#2004-01-1757, 2004.
75. J. Ackermann and D. Odenthal, "Damping of vehicle roll dynamics by gain scheduled active steering," Proceedings of European Control Conference, 1999.
76. D. Odenthal, T. Bunte, and J. Ackermann, "Nonlinear steering and braking control for vehicle rollover avoidance," Proceedings of European Control Conference, 1999.
77. S. Takano and M. Nagai, "Dynamics control of large vehicles for rollover prevention," Proceedings of IEEE International Vehicle Electronics Conference, pp. 85-89, 2001.
78. T. Acarman and U. Ozguner, "Rollover prevention for heavy trucks using frequency shaped sliding mode control," Proceedings of IEEE Conference on Control Applications, vol. 1, pp. 7-12, 2003.
79. D. J. M. Sampson and D. Cebon, "Active roll control of single unit heavy road vehicles," Vehicle System Dynamics, vol. 40, no. 4, pp. 229-270, 2003.
80. S. Takano, M. Suzuki, and M. Nagai, et. al, "Analysis of large vehicle dynamics for improving roll stability," Vehicle System Dynamics, Supplement, vol. 41, pp. 73-82, 2004.
81. B. P. Jeppesen and D. Cebon, "Real-time fault identification in an active roll control system," Vehicle System Dynamics, Supplement, vol. 37, pp. 360-372, 2003.
82. B. P. Jeppesen and D. Cebon, "Analytical redundancy techniques for fault detection in an active heavy vehicle suspension," Vehicle System Dynamics, vol. 42, no. 1/2, p75-88, 2004.

83. A. Alleyne, "Improved vehicle performance using combined suspension and braking forces," Proceedings of American Control Conference, vol. 3, pp. 1672-1676, 1995.
84. S. Rakheja, K. Wang, and R. Bhat, et. al, "Enhancement of ride vibration environment of tracked sidewalk snowploughs: vehicle modelling and analysis," International Journal of Vehicle Design, vol. 29, no. 4, pp. 193-222, 2002.
85. P.-E. Boileau, S. Rakheja, and Z. Wang, "Ride vibration environment of tracked sidewalk snowploughs: spectral classification," International Journal of Vehicle Design, vol. 30, no. 4, pp. 309-326, 2002.
86. S. Zetterstrom, "Electromechanical steering, suspension, drive and brake modules," Proceedings of IEEE Vehicular Technology Conference, vol. 3, pp. 1856-1863, 2002.
87. A. Trachtler, "Integrated vehicle dynamics control using active brake, steering and suspension systems," International Journal of Vehicle Design, vol. 36, no. 1, pp. 1-12, 2004.
88. P. Lingman and B. Schmidtbauer, "Road slope and vehicle mass estimation using Kalman filtering," Vehicle System Dynamics, Supplement, vol. 37, pp 12-23, 2002.
89. J. Ryu and J. C. Gerdes, "Estimation of vehicle roll and road bank angle," Proceedings of American Control Conference, vol.3, pp. 2110-2115, 2004.
90. X. Liu and J. Wagner, "Design of a vibration isolation actuator for automotive seating systems - Part I: modelling and passive isolator performance," International Journal of Vehicle Design, vol. 29, no. 4, pp. 335-356, 2002
91. X. Liu, J. Wagner, "Design of a vibration isolation actuator for automotive seating systems - Part II: controller design and actuator performance," International Journal of Vehicle Design, vol. 29, no. 4, pp. 357-375, 2002.
92. B.-C. Chen, "Human-in-the-loop optimization of vehicle dynamics control with rollover prevention," Vehicle System Dynamics, Supplement, vol. 41, pp. 252-261, 2004.
93. W. Choromanski and J. Kisilowski, "Human-vehicle system modelling - focus on heuristic modelling of driver-operator reactions and mechatronic suspension," Vehicle System Dynamics, Supplement, vol. 41, pp. 262-271, 2004.

Advanced Individual Vehicle Motion Control

6.1 Introduction

In previous *Chapter 3-5*, a single vehicle's dynamics is decomposed along three dimensions and studied. However, in many situations, the integrated 3-D vehicle motion control, which addresses the vehicle's short-term driving behavior under dynamic constraints, needs to be discussed.

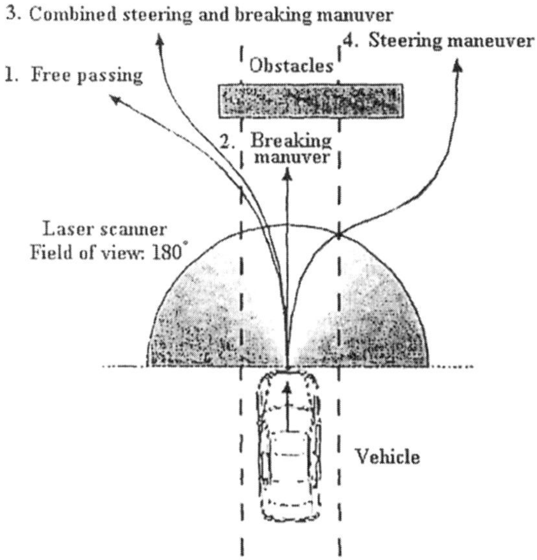

Fig.6.1 Diagram of vehicle driving behavior decision in emergences.

The most frequently mentioned integrated vehicle motion control problem is collision free trajectory/path planning. It consists of selecting the geometric path and vehicle speeds so as to avoid obstacles and to minimize some cost functions, such as time and energy. For example, Fig.6.1 shows four different strategies that were proposed by Lages 2001 in [6] to avoid collisions.

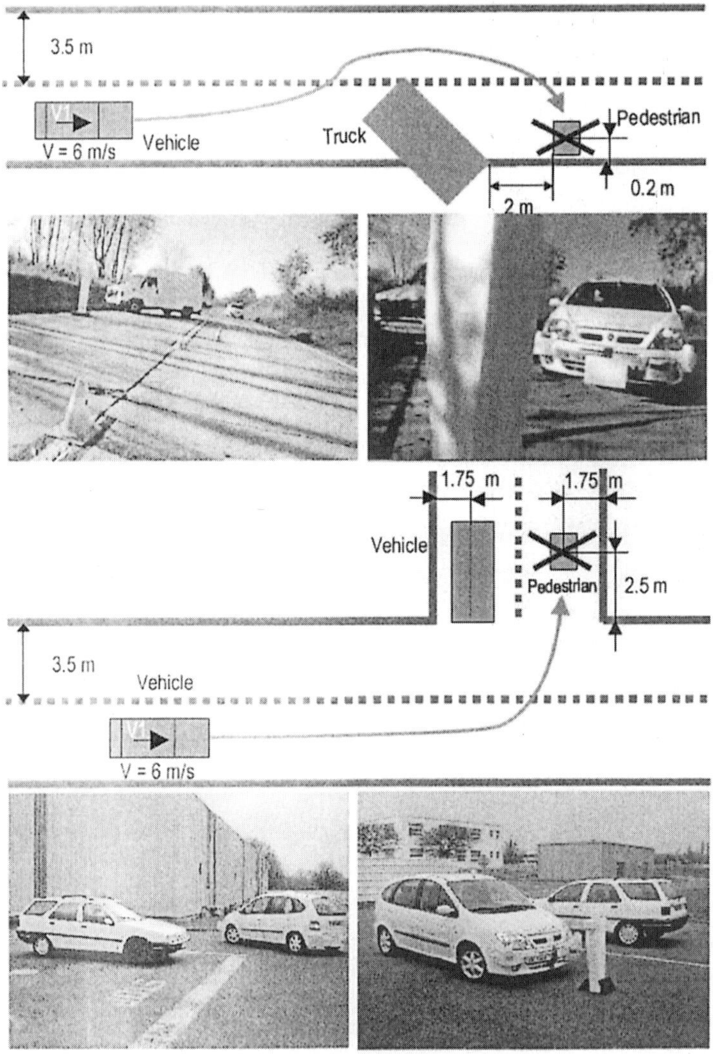

Fig.6.2 (upper) Emergency braking on a pedestrian hidden by a truck, (lower) emergency braking on a pedestrian located on a crossroad and hidden by a vehicle, from [8]. (© [2005] IEEE)

The first possible strategy is free passing which do not slow down vehicle speed but change the direction of motion. The second control strategy is to break down the vehicle without steering the vehicle. The third strategy is to slow down vehicle speed and steer the vehicle simultaneously. The fourth strategy is to steer the vehicle only without deceleration. Normally, not all four motion control strategies can be applied due to vehicle position, speed and mechanical constraints. But how to pick up the best strategy remains as a difficult problem to be thoroughly solved especially when computation time requirement is considered [1]-[8].

The first problem that needs to be carefully handled is how to achieve enough environment information to make correct decisions. For instance, Fig.2 shows two typical wrong driving strategies caused by vehicle occlusions. In the first case, the vehicle failed to detect a pedestrian hidden by a truck; in the second case, the vehicle failed to detect a pedestrian located on a crossroad and hidden by a vehicle. Constrained by book length, it assumes that information of road and traffic participators can be correctly obtained in the rest of this Chapter.

The second problem is to consider vehicle dynamics to determine whether to brake or steer or simultaneously brake and steer [60]-[84]. Trajectory planning is a main technique to test the validity of a generated driving strategy planning. Generally, the related works could be divided into three main categories: 1) path generation, 2) trajectory planning, and 3) path/trajectory tracking/following and corresponding controller design.

Path planning concerns how to attain an acceptable geometry path for a vehicle with certain requirements and constraints [9]-[12]. Trajectory planning problem focuses on how to generate the configuration of the vehicle motion as a function of time, as well as generate corresponding reference inputs to the trajectory tracking system. Comparing to path planning, it considers geometry path profile and associated time profiles for vehicle motion simultaneously [19].

It is remarkable that only few works had addressed trajectory generation problem for automatic vehicles. In many reports, the so called trajectory planning problems mentioned were actually about path planning, which only consider the geometry motion profile. While extensive work had focused on computing the geometric path, little attention had been given to selecting the optimal vehicle speeds. Selecting the wrong speeds can cause the vehicle to loose its path or to waste time or energy. Lane changing trajectory planning and parking trajectory planning are two discussed problems in this areas.

Path/trajectory tracking/following discusses how to guarantee the generated path/trajectory is feasible and how to track the generated path/trajectory [13]-[14]. Normally, it studies vehicle dynamics and steering/braking control according to pre-selected path/trajectory. For instance in [14], Chan and Tan emphasized the possibility of applying steering control to stabilize the trajectories of vehicles involved in a collision, which examined the descriptions of accident scenarios, limitations of applicability, and performance requirements. Thus, longitudinal and lateral combined vehicle motion control attracts notable interests recently.

It should be pointed out that trajectory planning can be viewed as an extension of the early proposed robot path planning problems, which aim to generate continuous curvature profile with certain performance requirement for robots [15]-[19]. Since the car-like robot's dynamics are similar to that of the vehicle, several planning methods, which were proven to be effective for intelligent robot, have been extended to vehicle path planning program. However, significant differences on mass, inertia, geometry size, driveline and navigation velocity of a vehicle lead to really different control methods that that of a robot.

6.2 Vehicle Path/Trajectory Planning

There are numerous methods that had been proposed for vehicle path planning. For instance GJK algorithm proposed by Bernabeu et. al 2001 [12], probabilistic roadmap approach by Lamiraux and Lammond 2001 [21], symmetrical clothoid method by Yossawee et. al 2002 [22]. But the most frequently used method is spline interpolation based method. For instance, Piazzi et. al [23]-[24] employed Quintic G^2-splines method to generate allowable path. In [25], Hilgert et. al applied elastic band theory, and Simon and Becker chose Bezier sub-spline as the interpolation base spline [26]-[27]. Generally, how to conveniently incorporate vehicle dynamic constraints into path/trajectory geometric constraints is the kernel part of the related research.

In [26], the desired trajectory was represented as a two dimensional Bezier spline. A sub-spline or trajectory segment consists of two fifth-order polynomials, which has the structure as

$$y(t) = a_5 t^5 + a_4 t^4 + a_3 t^3 + a_2 t^2 + a_1 t + a_0 \tag{6.1}$$

and

$$x(t) = b_5 t^5 + b_4 t^4 + b_3 t^3 + b_2 t^2 + b_1 t + b_0 \tag{6.2}$$

The boundary conditions can be written as

$$\begin{bmatrix} a_0 \\ a_1 \\ a_2 \\ a_3 \\ a_4 \\ a_5 \end{bmatrix} = \begin{bmatrix} 1 & 0 & 0 & 0 & 0 & 0 \\ 0 & 1 & 0 & 0 & 0 & 0 \\ 0 & 0 & 0.5 & 0 & 0 & 0 \\ -10 & -6 & -1.5 & 10 & -4 & 0.5 \\ 15 & 8 & 1.5 & -15 & 7 & -1 \\ -6 & -3 & -0.5 & 6 & -3 & 0.5 \end{bmatrix} \begin{bmatrix} y_{in} \\ \dot{y}_{in} \\ \ddot{y}_{in} \\ y_{fin} \\ \dot{y}_{fin} \\ \ddot{y}_{fin} \end{bmatrix} \tag{6.3}$$

and

$$\begin{bmatrix} b_0 \\ b_1 \\ b_2 \\ b_3 \\ b_4 \\ b_5 \end{bmatrix} = \begin{bmatrix} 1 & 0 & 0 & 0 & 0 & 0 \\ 0 & 1 & 0 & 0 & 0 & 0 \\ 0 & 0 & 0.5 & 0 & 0 & 0 \\ -10 & -6 & -1.5 & 10 & -4 & 0.5 \\ 15 & 8 & 1.5 & -15 & 7 & -1 \\ -6 & -3 & -0.5 & 6 & -3 & 0.5 \end{bmatrix} \begin{bmatrix} x_{in} \\ \dot{x}_{in} \\ \ddot{x}_{in} \\ x_{fin} \\ \dot{x}_{fin} \\ \ddot{x}_{fin} \end{bmatrix} \tag{6.4}$$

where a_i and b_i, $i = 0,1,\ldots 5$ are called the parameter variables that control the shape of the generated Bezier curve. Their meaning is shown in Fig.6.3.

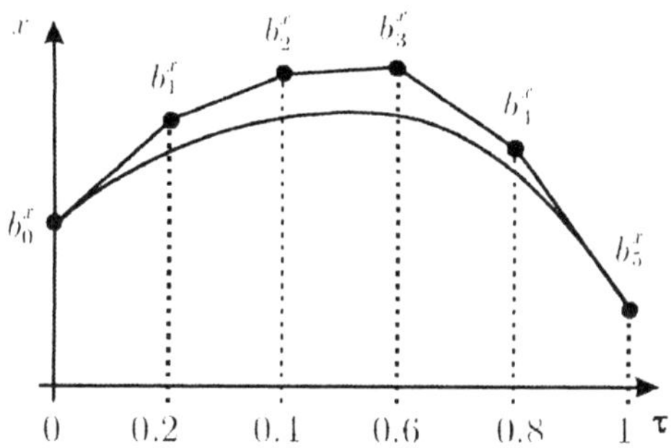

Fig.6.3 A fifth-point one-dimensional Bezier sub-spline, from [26]. (© [1999] IEEE)

In order to guarantee the generated the Bezier curve is a valid path for the vehicle, the corresponding steering angle constraints of the curve were further considered in [26].

Simon and Becker 1999 pointed out that curvature at the sub-spline boundary can be used to measure whether the designed trajectory is feasible or not. They considered a simplified vehicle dynamic model as shown in Fig.6.4, where the world coordinates of the vehicle is denoted by x and y. The vehicle's dynamic equation was written as

$$\dot{x} = v_F[\cos\phi\cos\theta - \frac{r}{l}\sin\phi\sin\theta] \tag{6.5}$$

and

$$\dot{y} = v_F[\cos\phi\sin\theta + \frac{r}{l}\sin\phi\cos\theta] \tag{6.6}$$

together with

$$\dot{\theta} = v_F\frac{\sin\phi}{l} \tag{6.7}$$

where θ is the angle between the vehicle's longitudinal direction and the x–axis. And ϕ is the steering angle. l is the length between the front tires and the rear tires. r is the length between the front tires and the reference point. v_F is the vehicle forward velocity.

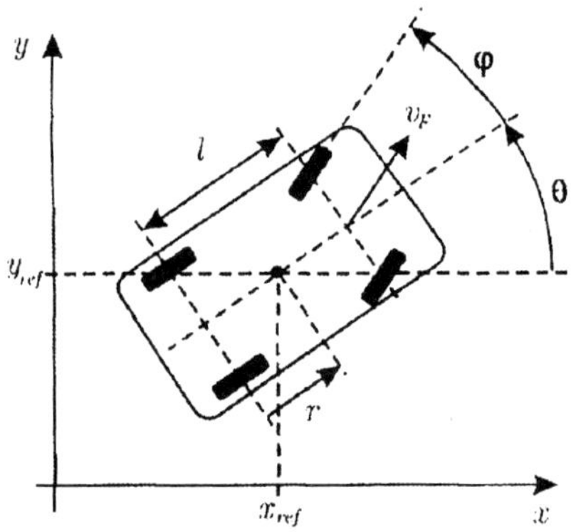

Fig.6.4 A one-dimensional Bezier sub-spline, from [26]. (© [1999] IEEE)

Suppose the lateral acceleration and steering velocity can be written as

$$a_{lat} = \kappa v^2 \tag{6.8}$$

and

$$\dot{\phi} = \frac{l}{1+(\kappa l)^2}\dot{\kappa}v \tag{6.9}$$

where κ is the real curvature of the vehicle and $\dot{\kappa}$ is the its derivative.

We can have the planning constraints as

$$\left| \kappa = \frac{4\lambda}{25\mu^2} \sin\alpha \right| \le \kappa_{\text{limit}} \tag{6.10}$$

where μ and λ are the lengths of two consequent polynomial edges and α is the angle between these two edges; see Fig.6.5.

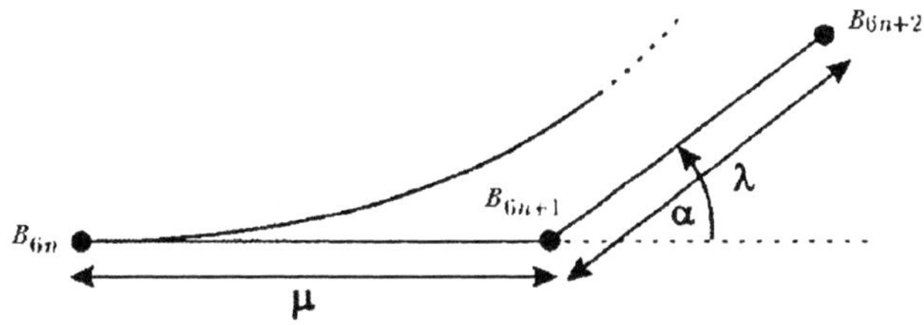

Fig.6.5 Curvature at the sub-spline boundary, from [26]. (© [1999] IEEE)

Taking the above constraints into account, Simon and Becker 1999 shown that the smoothness of trajectory decreases with an increasing sub-spline density. Fig.6.6 illustrates how to generate acceptable vehicle trajectory by segmentation and interpolation.

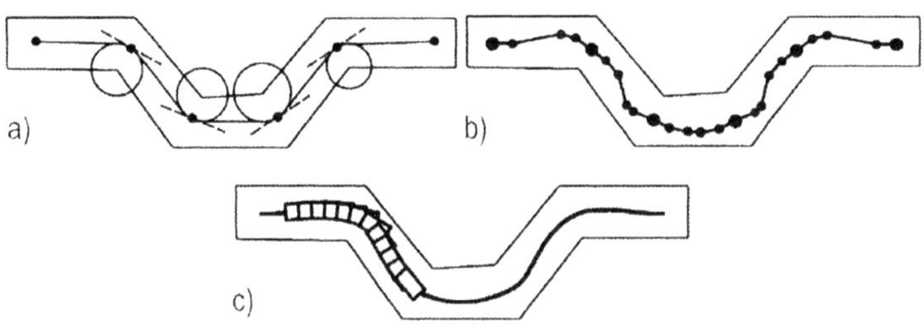

Fig.6.6 (a) determination of the trajectory interpolation points, (b) the remaining Bezire points, and (c) the final trajectory, from [26]. (© [1999] IEEE)

Lane change trajectory planning is a frequently analyzed trajectory planning problem. Also using Bezire interpolation method, Papadimitriou and Tomizuka proposed a fast lane changing path generation method in [27]. They considered the lane change process of a vehicle runs at longitudinal speed $\dot{x}$. If there is no obstacle, the time t_{lc} to complete the lane change obviously only depends on the passengers' comfort. Suppose t_{lc} and the width of the current and target lanes are assumed to be known in advance, and $\dot{x}$ can be measured by the vehicle speedometer. Papadimitriou and Tomizuka formulated the lane changing as a simple boundary condition problem as follows:

Find a smooth path that guides the vehicle from the initial state $(x_{in}, \dot{x}_{in}, \ddot{x}_{in}, y_{in}, \dot{y}_{in}, \ddot{y}_{in})$ to final state $(x_{fin}, \dot{x}_{fin}, \ddot{x}_{fin}, y_{fin}, \dot{y}_{fin}, \ddot{y}_{fin})$ shown in Fig.6.7.

To simplify the description as follows, let's define w_c as the width of the current path and w_t as the width of the target path. Here, x_{fin} corresponds to the longitudinal distance traveled if the speed was constant and y_{fin} corresponds to the sum of w_c and w_t.

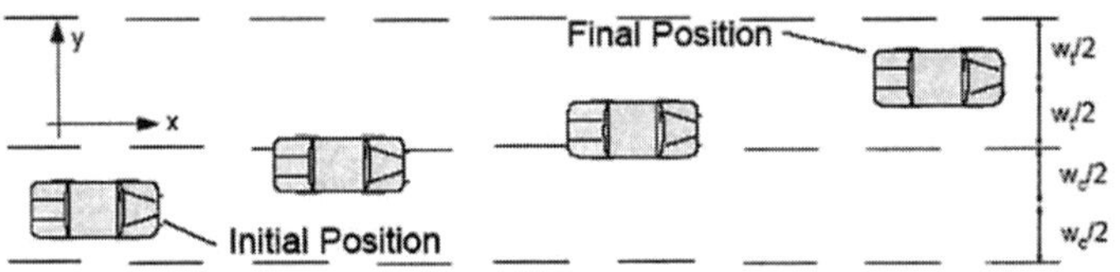

Fig.6.7 Lane changing path without obstacles, from [28]. (© [2003] IEEE)

Selecting fifth order polynomials for $x(t)$ and $y(t)$ position, the problem was then solved trivially through the following linear equations:

$$(a_0 \quad a_1 \quad a_2 \quad a_3 \quad a_4 \quad a_5)^T = A_{6\times6} \cdot (y_{in} \quad y_{fin} \quad \dot{y}_{in} \quad \dot{y}_{fin} \quad \ddot{y}_{in} \quad \ddot{y}_{fin})^T \tag{6.11}$$

and

$$(b_0 \quad b_1 \quad b_2 \quad b_3 \quad b_4 \quad b_5)^T = B_{6\times6} \cdot (x_{in} \quad x_{fin} \quad \dot{x}_{in} \quad \dot{x}_{fin} \quad \ddot{x}_{in} \quad \ddot{x}_{fin})^T \tag{6.12}$$

Papadimitriou and Tomizuka considered a typical scenario where the vehicle runs at 20m/s that starts a lane change at time $t_0 = 0$ s and completes it at $t_{lc} = 5$ s. The coordinate system is placed on the vehicle's geometric center at time $t_0 = 0$ s and remains fixed.

The lane change is considered complete when both longitudinal and lateral accelerations are 0. The width of each lane is 4m. Thus, the initial and final state of the vehicle is:

$$(x_{in}, \dot{x}_{in}, \ddot{x}_{in}, y_{in}, \dot{y}_{in}, \ddot{y}_{in}) = (0, 20, 0, 0, 0, 0) \tag{6.13}$$

and

$$(x_{fin}, \dot{x}_{fin}, \ddot{x}_{fin}, y_{fin}, \dot{y}_{fin}, \ddot{y}_{fin}) = (100, 20, 0, 4, 0, 0) \tag{6.14}$$

Solving Eqs. (6.1) and (6.2) yields the path shown in Fig.6.8. It is clear that such spline interpolation method will result in continuous curvature path profile.

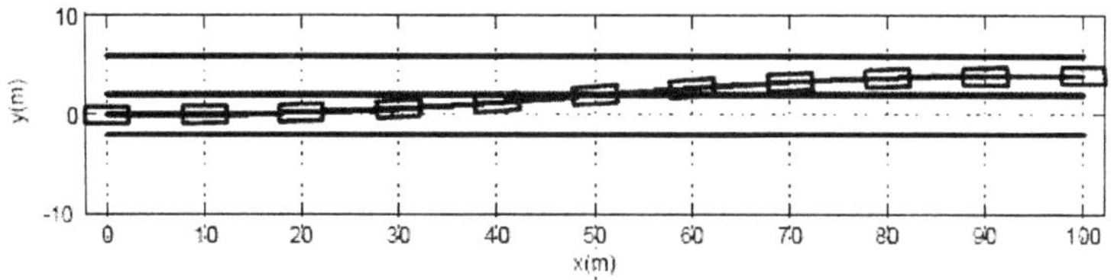

Fig.6.8 Lane changing path considering obstacles, from [28]. (© [2003] IEEE)

On the basis of above free lane changing path planning, Papadimitriou and Tomizuka took the obstacle into account. They claimed that the shape of the path changes can provide the vehicle with more freedom to avoid the obstacle if the order of one of the polynomials is increased, and the boundary conditions are still respected. Thus the problem was transformed into what polynomial should the order be increased of and what are the possible values of the extra coefficient that yield admissible trajectories.

However, as far as the former issue is concerned, increasing the order of $y(t)$ only, was not proven to be a wise choice. Indeed, shaping the lateral trajectory can yield only few admissible paths, since the vehicle is restricted to maneuvers in the current and target lane, which is not the case in the longitudinal direction. Moreover, the maximum achievable lateral accelerations were shown to be relatively small compared to the longitudinal ones (4-5 times smaller). If the order of both $x(t)$ and $y(t)$ is increased, then the complexity of the problem increases substantially because there are two extra coefficients to be determined. Thus, Papadimitriou and Tomizuka 2003 chose to increase the order of $x(t)$ only. More precisely, $x(t)$ is written as

$$x(t) = b_6t^6 + b_5t^5 + b_4t^4 + b_3t^3 + b_2t^2 + b_1t + b_0 \tag{6.15}$$

Substituting the boundary condition in (6.15), we can have the coefficients b_i in its first and second derivatives as

$$b_i = m_i + n_ib_6,\ i = 0,1,\ldots 5 \tag{6.16}$$

where $b_i = m_i$ and n_i are functions of the boundary conditions.

Substituting (6.16) into (6.15) will yield

$$x(t) = c_1(t) + c_2(t)b_6,\ i = 0,1,...5 \tag{6.17}$$

where $c_1(t)$ and $c_2(t)$ are functions of time the boundary conditions.

Besides the motion path of the vehicle center, the shape of the vehicle should also be considered. Because a fast way to represent objects in 2D space can be approximating them by an infinite number of circles, the vehicle was represented by the area swept by a dynamic circle of radius R as Fig.6.9, whose center obeys a linear function in $\lambda = [0,1]$ as

$$P = P_0 + \lambda(P_n - P_0) \tag{6.18}$$

where P_0 and P_n are the vectors of the end circle centers. This shape was called the oval shape in [28].

Fig.6.9 Vehicle representation, from [28]. (© [2003] IEEE)

Interpreter (6.18) into the following format

$$x = x_0 + \lambda(x_n - x_0) \tag{6.19}$$

and

$$y = y_0 + \lambda(y_n - y_0) \tag{6.20}$$

where x and y represent the line segment that connects P_0 and P_n.

And the criterion for collision avoidance is

$$(x_c - x_0)^2 + (x_c - x_0)^2 > (R_c - R_0)^2 \tag{6.21}$$

Assuming that the yaw angle of the controlled vehicle remains 0. taking the geometric characteristics, we can have

$$\begin{aligned}&[c_1 + c_2 b_6 - l_c/2 + \lambda_c l_c - \lambda_0(x_{on} - x_{o0})]^2 + \\ &[y_{c0} - y_{o0} - \lambda_0(y_{on} - y_{o0})]^2 > (R_c - R_0)^2\end{aligned} \tag{6.22}$$

where the subscript c corresponds to the controlled vehicle and subscript o corresponds to the obstacle.

Although the above assumption is not true in practice, it is still a fair approximation of reality. Further simplify (6.22) will gives

$$\alpha^2(t)b_6^2 + \beta(t)b_6 + \gamma(t) > 0 \tag{6.23}$$

where $\alpha(t)$, $\beta(t)$ and $\gamma(t)$ are functions of time, boundary conditions and vehicle geometry parameters. Since the coefficients of b_6^2 is larger than 0, the collision free condition can be expressed as

$$b_6 < \frac{-\beta - \sqrt{\Delta}}{2\alpha^2} \bigcup b_6 > \frac{-\beta + \sqrt{\Delta}}{2\alpha^2}, \ \Delta = \beta^2 - 4\alpha^2\gamma \tag{6.24}$$

Select a value for b_6 that satisfy this condition, the solution of (6.22) will ensure a collision free path. Indeed, Papadimitriou and Tomizuka 2003 found out that the number of combinations of $\lambda_c \in [0,1]$ and $\lambda_o \in [0,1]$ that are checked depends on the trade-off between accuracy and fast computation that exists in almost all trajectory planning methodologies. However, (6.23) is an affine function of λ_c so that the minimum and maximum acceptable values for b_6 are achieved at the minimum and maximum values of λ_c, hence, without loss of accuracy, the root plotting can be limited to the combinations corresponding to the possible collisions of the front and rear end circles of the controlled vehicle with the oval shape.

To illustrate this property, Papadimitriou and Tomizuka considered a typical scenario where a vehicle runs at 20m/s and a second vehicle running at 20m/s in the target lane exactly beside the controlled vehicle. It is assumed that the controlled vehicle will eventually adjust its speed to that of the obstacle, which is one of the principles in cooperative driving. Thus, the lane change is considered complete when both longitudinal and lateral accelerations of the controlled vehicle are 0, the longitudinal velocity is 20m/s and the x-position is 10m behind the obstacle. The total time of lane change is again $t_{lc} = 5$ s. The initial and final state of the vehicle is:

$$(x_{in}, \dot{x}_{in}, \ddot{x}_{in}, y_{in}, \dot{y}_{in}, \ddot{y}_{in}) = (0, 20, 0, 0, 0, 0) \tag{6.25}$$

and

$$(x_{fin}, \dot{x}_{fin}, \ddot{x}_{fin}, y_{fin}, \dot{y}_{fin}, \ddot{y}_{fin}) = (90, 20, 0, 4, 0, 0) \tag{6.26}$$

Applying the obstacle avoidance methodology results in the plot of the solutions of Eq.(6.24), corresponding to the rear and front-end circle of the controlled vehicle with the oval shape that represents the obstacle; see Fig.6.10.

Selecting $b_6 = 0.01$ yields the path shown in Fig.6.11. The corresponding positions, velocities and accelerations of the controlled vehicle versus time are shown Fig.6.12.

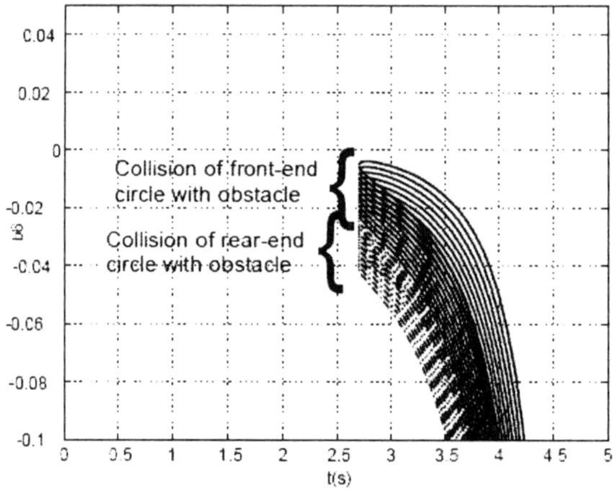

Fig.6.10 Map for selection of admissible trajectory, from [28]. (© [2003] IEEE)

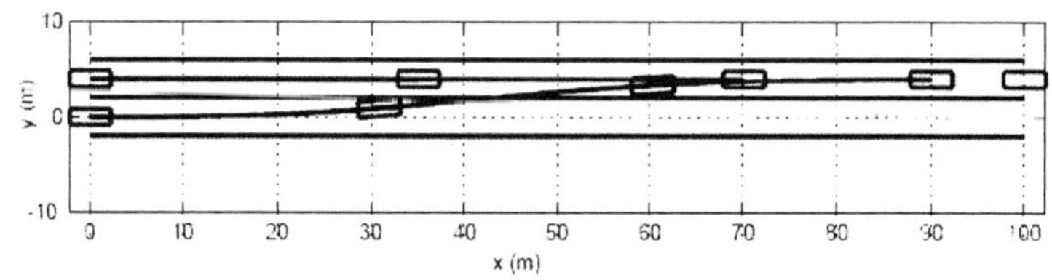

Fig.6.11 Vehicle representation, from [28]. (© [2003] IEEE)

The above results can be further modified by considering vehicle dynamic constraints. Assume the following constant threshold accelerations:

$$-|a_l| \leq \ddot{y} \leq |a_l| \tag{6.27}$$

and

$$-|a_{dac}| \leq \ddot{x} \leq |a_{ac}| \tag{6.28}$$

where $|a_l|$ represents the bound of vehicle lateral acceleration rate. $|a_{dac}|$ and $|a_{ac}|$ are the bound of vehicle longitudinal acceleration and deceleration rate respectively.

Differentiating Eq.(6.2), the longitudinal acceleration is

$$\ddot{x}(t) = 30b_6t^4 + 20b_5t^3 + 12b_4t^2 + 6b_3t + 2b_2 \tag{6.29}$$

Thus, we should guarantee that

$$-|a_{dac}| \le 30b_6t^4 + 20b_5t^3 + 12b_4t^2 + 6b_3t + 2b_2 \le |a_{ac}| \tag{6.30}$$

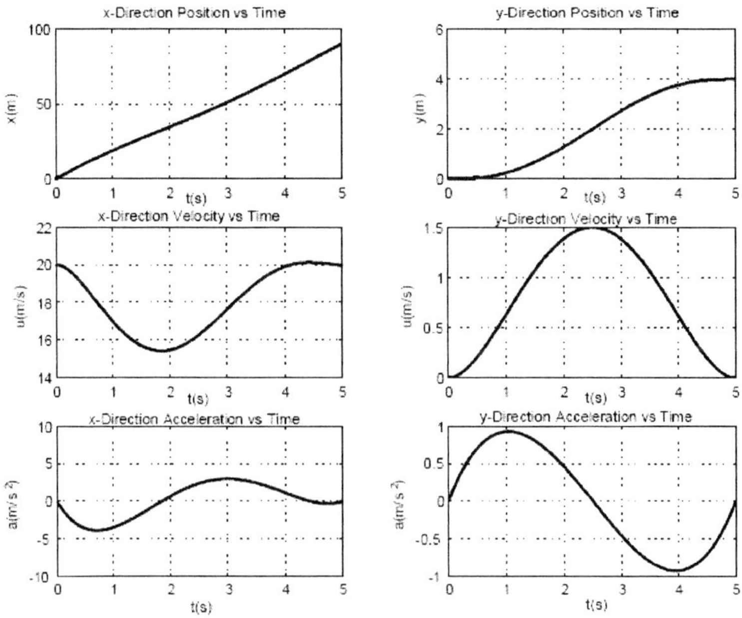

Fig.6.12 Positions, velocities and accelerations of the vehicle's geometric center, from [28]. (© [2003] IEEE)

Define $\ddot{x}(t)$ as an affine function of b_6, which is similar as (6.2), it gets

$$x(t) = c_3(t) + c_4(t)b_6,\ i = 0,1,\ldots 5 \tag{6.31}$$

where $c_3(t)$ and $c_4(t)$ are function of time and boundary conditions.

Substituting (6.31) to (6.30) yields

$$b_6 = \left[\pm |a_{dac}| - c_3\right] / c_4 \tag{6.32}$$

To illustrate this new result, let's consider the acceleration threshold as

$$-2m/s^2 \le \ddot{y} \le 2m/s^2,\ -10m/s^2 \le \ddot{x} \le 2.5m/s^2 \tag{6.33}$$

Solving the boundary condition problem for $y(t)$, we have the admissible range of b_6 as shown in Fig.6.13, in which the dynamic constraints are mapped on the same space as the virtual obstacles.

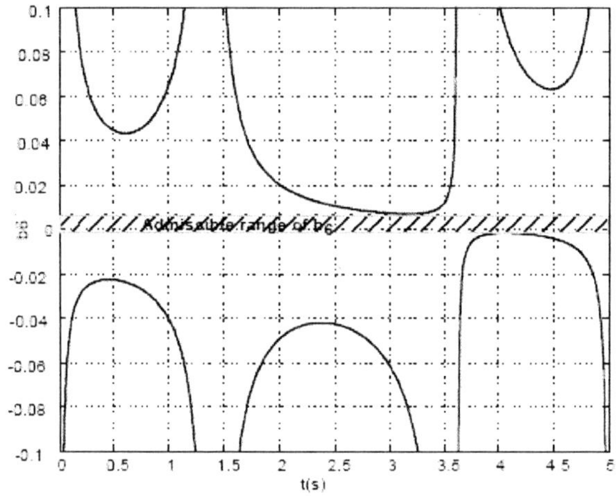

Fig.6.13 Dynamic constraints mapped on b_6 space, from [28]. (© [2003] IEEE)

Finally, Papadimitriou and Tomizuka considered the general lane changing problem where there is not only one vehicle on the road. In the general problems, the above one-to-one lane changing path generation method is carried out for all the vehicles nearby respectively. And the obtained path solution in terms of b_6 should exist in the common admissible region in the mapping space. Fig.6.14 shows a typical road traffic scenario, and the following Fig.6.15 demonstrates the corresponding the allowable solution in mapped on b_6 space.

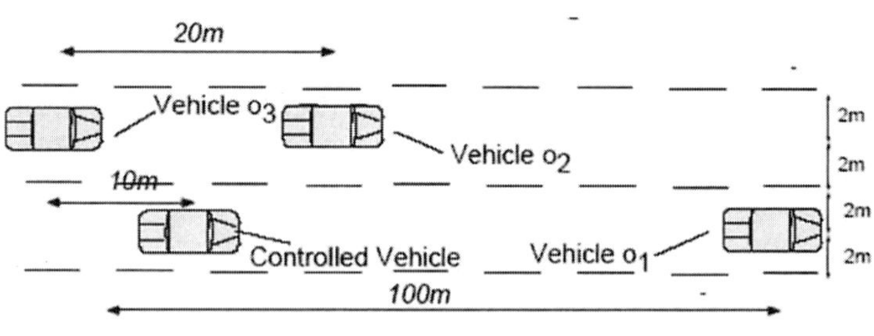

Fig.6.14 A lane changing scenario at initial state, from [28]. (© [2003] IEEE)

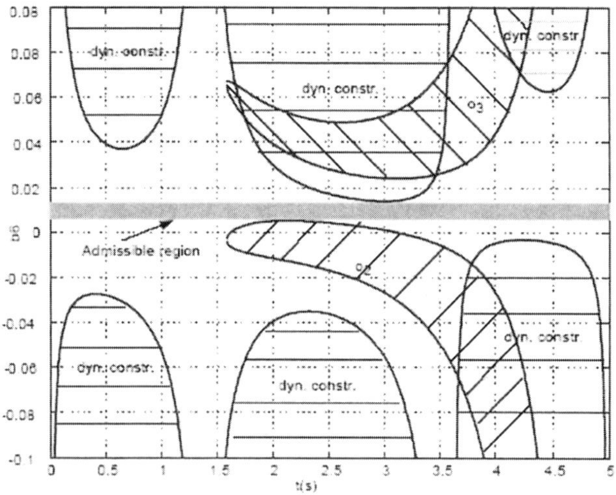

Fig.6.15 Region of admissible values mapped on b_6 space, from [28]. (© [2003] IEEE)

Besides driving safety constraints, how to obtain optimal driving trajectory is another important research direction in this field. For instance, how to get time-optimal trajectory for the vehicle were discussed in [29]-[31].

In [32]-[34], Li and Wang considered more general trajectory planning problem for the highway trip. In [34], the highway road was viewed as a series of several consequent road sections that are characterized by the curvature variations along the road respectively; see Fig.6.16. These sections are further divided into two types: approximately straight road sections (whose curvature variations were neglected) and curve road sections (whose curvature variations could not be neglected). It is obvious that the straight road sections and curve road sections always come alternatively along the road. The altitude variation and the width of the whole road are also neglected. And these road sections could be viewed as lines and curves on flat surface.

As shown in Fig.6.16, each straight road sections will be further divided into several sections that have the uniform length (50m or 100m). Merge the curve road section into the first straight road subsection that directly follows it, we could construct a new series of special road sections whose lengths are defined by a series length parameters l_1, ..., l_M. So the whole driving process could be characterized by M velocity variables u_1, ..., u_M that represented the entering velocity of each road sections. Here, we assume the start velocity u_1 and end velocity u_M could be arbitrary feasible velocity, since the starting and stopping driving process have little effect on the whole driving process and could be neglected.

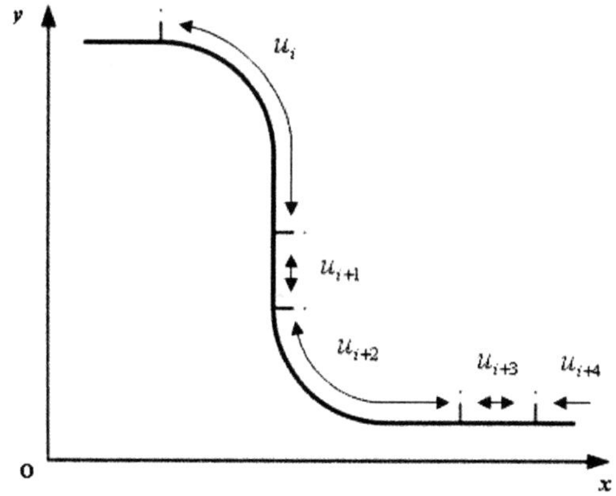

Fig.6.16 Definition of the road sections and associated speed variables.

To make the whole problem more tractable, let's assume that the velocity of a vehicle keeps constant in the curve road sections. And the entering velocity is used to represent the average speed along the road section.

These simplifications certainly introduce errors. However, ordinary persons do not change velocity too often as race-car driver. Indeed, they usually change the velocity when they drive on the straight roads and do not shift during steering. So the velocity could be assumed be constant in the curve road sections without lose of much generality.

Let Ω denotes the set of the feasible velocity. We will introduce the formulation of the longitudinal trajectory planning in the rest of this section and the trajectory constraints will be discussed in the next section.

Commonly, an optimal driving guidance generation is to determine the time history along a specified geometric path in terms of the path parameters that describes how the vehicle moves from its start point to its end point with an optimal objective function value under several constraints. A general form of the performance index can be defined as

$$\min_{u_1,\ldots,u_M \in \Omega} J(u_1,\ldots,u_M) = H(u_1,\ldots,u_M) \tag{6.34}$$

where $H(\cdot)$ represents a given objective function for the vehicle.

Wang etc. considered three basic objective functions and their weighted combination are developed to describe and model the normal behavior of the driver/passenger. The first is the minimum-time driving plan, which is obviously desired for those who prefer time saving objective. It is well known that this optimal driving plan can be achieved directly by always making the velocity to the allowable maximum value. For this situation, we will have the following dynamic objective function as

$$J_1 = \min_{u_1,\dots,u_M \in \Omega} \sum_{k=1}^{M} \frac{l_k}{u_k} \tag{6.35}$$

The second objective is the minimum-fuel-consumption driving planning, which can be used to save energy and protect the air quality. However this objective is somewhat difficult to handle because the relationship between the fuel-consumption and the speed is too complicated to be accurately modeled. In Khisty 1998, an approximate description for the fuel consumed in motion is represented as following

$$F(V) = \frac{h_1}{V} + h_2 V + h_3 V^2 \tag{6.36}$$

where F is the fuel consumed per unit distance and V is average speed measured over a distance. F h_1, F h_2 and F h_3 is coefficients determined by different vehicle weight, engine size and technological features. Though it's not accurate, we could take it as an acceptable estimate on the fuel-consumption performance index under certain conditions based on experience.

Therefore, a simple minimum-gasoline-consumption solution can be directly reached by simply setting the velocity to zero. Therefore, we instead consider the linear combination of both minimum-time and minimum-gasoline-consumption objectives as following

$$J_2 = \min_{u_1,\dots,u_m \in \Omega} \sum_{k=1}^{m} [Q_1 \cdot \frac{l_k}{u_k} + Q_2 \cdot F'(u_k) \cdot \frac{l_k}{u_k}] \tag{6.37}$$

where $F'(u_k) = \dfrac{h_1}{u_k} + h_2 u_k + h_3 u_{k_k}^2 \cdot Q_1$ and Q_2 are the positive weight coefficients that may from different groups of people.

The third objective is minimum-jerk driving plan. Because the driver will be apt to feel tired and uncomfortable if the acceleration/deceleration rate is too big and/or too often. So a minimum-jerk (the third derivative of the vehicle's position) performance index may be introduced as the objective function to make the acceleration/ deceleration smoothly. Since to accurately specify a minimum-jerk trajectory is too complicated and time consuming. To avoid involving the third derivatives directly, we choose the following approximate cost function

$$J_3 = \min_{u_1,\ldots,u_M \in \Omega} \sum_{k=1}^{M} [\frac{|u_{k+1} - u_k|}{l_{k1}} \cdot u_k] \tag{6.38}$$

Normally, drivers want to maintain a balance among timesaving, fuel saving and comfort feeling. Thus the optimal trajectory objective function should be a combination and compromise of the three basic planning strategies discussed above. There are numerous ways to deal with the multi-objectives problems, and the simplest way is to formulate it as a linearly weighted objective as follows

$$J_4 = \min_{u_1,\ldots,u_M \in \Omega} \sum_{k=1}^{M} [Q_1 \cdot \frac{l_k}{u_k} + Q_2 \cdot F'(u_k) \cdot \frac{l_k}{u_k} + Q_3 \cdot \frac{|u_{k+1} - u_k|}{l_{k1}} \cdot u_k] \tag{6.39}$$

where Q_1, Q_2 and Q_3 are some pre-selected positive weight coefficients. Naturally they vary greatly for the different drivers/passengers.

Generally, we have to consider three constraints for the longitudinal trajectory planning problem. The most important one is the maximum accelerate/decelerate rate of a vehicle. It is apparent that different vehicles have different maximum accelerate/decelerate rate. So we have

$$|u_k - u_{k+1}| \le E(u_k, l_k) \tag{6.40}$$

where $E(\cdot,\cdot)$ is an preset evaluation function to judge the maximum velocity determined by current velocity u_k and the length l_k of the specific road section. Some detailed discussion about $E(\cdot,\cdot)$ could be found in [35]].

The second constraint is the acceptable maximum velocity a vehicle could have. In the straight road sections, the acceptable maximum velocity will only be restricted to the power of the engine. In the curve road sections, the acceptable maximum velocity will be determined by the intrinsic dynamic property of the vehicle and the driver.

Here, those maximum velocity settings are assumed to be gotten as

$$u_k \le \bar{u}_k \tag{6.41}$$

where $\bar{u}_k$ denotes the maximum velocity for the k th road section

The third constraint is that the driver should keep a certain distance from the preceding vehicle, which leads to

$$u_k \le \min\{G(s_d, u_{pre}), u_k\} \tag{6.42}$$

where u_{pre} is the velocity of the preceding vehicle and s_d is the distance between the two vehicles, and $G(\cdot,\cdot)$ is an evaluation function to judge the safety distance.

From the discussion above, the whole problem has already been transformed into a nonlinear optimization problem. Since we could not predict u_{pre} and s_d for long distance, we will consider at most 50 special road sections ahead. And u_{pre} and s_d are assumed to be invariant at the same time. The optimizing process will be dynamically restarted after the vehicle passes a pre-selected sum of the specific road sections.

Embedding the constraints as the penalty functions into the objective function; the fitness function can be written as below, which was finally solved using genetic algorithm in [32]-[34].

$$\begin{aligned} J = \sum_{k=1}^{m} [Q_1 \cdot \frac{l_k}{u_k} + Q_2 \cdot F'(u_k) \cdot \frac{l_k}{u_k} + Q_3 \cdot \frac{|u_{k+1} - u_k|}{l_{k1}} \cdot u_k + \\ P_1 \cdot M(|u_k - u_{k+1} - E(u_k, l_k)) + P_2 \cdot M(u_k - \bar{u}_k) + \\ P_3 \cdot M(u_k - \min\{ G(s_d, u_{pre}), u_k \})] \end{aligned} \tag{6.43}$$

where the function $M(\cdot)$ is defined as $M(y) = \begin{cases} y, y \geq 0 \\ 0, y < 0 \end{cases}$. The P_1, P_2 and P_3 are positive penalty coefficients.

There were several other discussions on vehicle trajectory planning. In [36]-[37], the safety driving constraints was reversely mapped into the phase space of steering angle, and so does the trajectory planning objective. In [34], the so called cell mapping method [38]-[42] were applied to solved such kind of trajectory planning problems in terms of steering angle and speed.

6.3 Vehicle Parking Problems

Autonomous parking is another problem that had been addressed in the last ten years [43]-[45]. Comparing to lane changing trajectory planning, vehicle speed in autonomous parking problem is much lower. But the allowable moving area, i.e. parking bay, is relatively smaller with respect to that in lane change problem [46]-[48].

Ohata and Mio proposed a trajectory based approach for this problem in [44]. They formulated the vehicle dynamic model as shown in Fig.6.17, which indeed is equivalent to the bicycle model analyzed before when vehicle velocity is low.

In Fig.6.17, $[x, y]$ is the coordinates of the car position, P is the center between the rear wheels, θ is the posture angle of the car, z is the vehicle moving length, α is the steering angle of the front wheel, L is the wheel base, v is the moving velocity. Ohata and Mio focused on the control of α in [44].

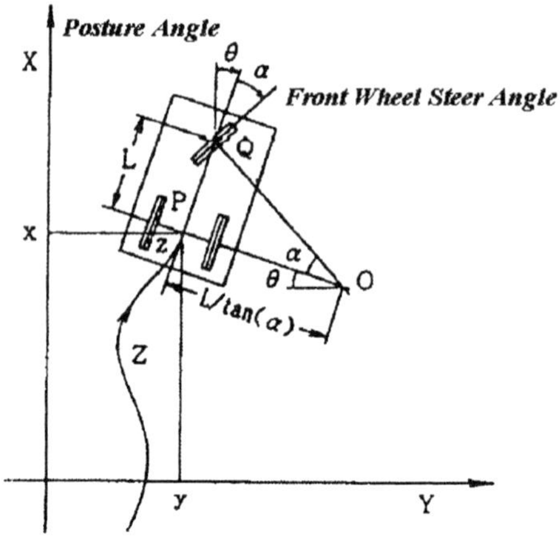

Fig.6.17 Tricycle behavior model, from [44].

And the dynamics of the system was written as

$$\frac{d\theta}{dt} = v\frac{\tan(\alpha)}{L}, \quad \frac{dY}{dt} = v\sin(\theta), \quad \frac{dX}{dt} = -v\cos(\theta), \quad \frac{dZ}{dt} = v \tag{6.44}$$

Thus,

$$\frac{d\theta}{dX} = -\frac{\tan(\alpha)}{L\cos(\theta)}, \quad \frac{dY}{dX} = -\tan(\theta) \tag{6.45}$$

Linearize (6.45) at the nominal state $Y = \theta = 0$, we can have the approximate state space model as

$$\frac{dx_l}{dX} = Ax_l + Bu_l \tag{6.46}$$

where $x_l = \begin{bmatrix} Y \\ \theta \end{bmatrix}$, $u_l = \frac{\theta}{L}$, $A = \begin{bmatrix} 0 & -1 \\ 0 & 0 \end{bmatrix}$, $B = \begin{bmatrix} 0 \\ -1 \end{bmatrix}$.

Suppose that the parking position is at $[X, Y] = [0,0]$ and the desired moving direction is equal to the X -axis; Fig.6.17. In [44], Ohata and Mio first designed the LQ regulator for the system as

$$\alpha = L(F_1 Y + F_2 \theta) \qquad (6.47)$$

where F_1 and F_2 are constants obtained by solving the corresponding Algebraic Riccati equation of (6.46).

Besides LQ controller, Ohata and Mio presented a nonlinear controller as

$$\alpha = \tan^{-1}\left[L^3(\theta)[F_1 Y + F_2 \tan(\theta)]\right] \qquad (6.48)$$

This controller will only be applied when $-\pi/2 < \theta < \pi/2$, because we have $|\tan(\theta)| \to \infty$ if $\theta \to \pi/2$ or $\theta \to -\pi/2$.

Finally, Ohata and Mio 1991 mimicked human driver decision process and designed a so called driver model control as

$$\alpha = -k\left[Y - (L_c - L_d)\theta\right] \qquad (6.49)$$

where k is a pre-selected feedback gain. L_c is the distance from the driver to the gazing point. And L_d is the length from the driver to the point p, which is the center between the rear wheels. As shown in Fig.6.18, it indicates that the driver reduces the order of inputs mapping from Y and θ to $\left[Y - (L_c - L_d)\theta\right]$.

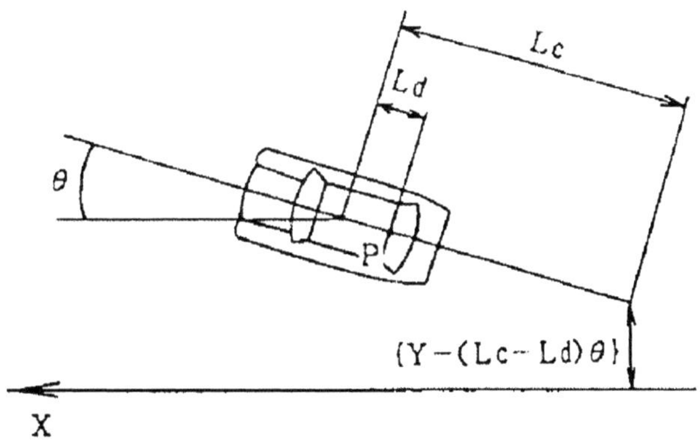

Fig.6.18 Human model during reverse motion, from [44].

Simulation results with each control are shown in Fig.6.19 where the car size is 1.5m in width and 4.0m in length. Fig.6.19(a) shows the trajectories from the initial state of $Y = 10.0m$ and $\theta = 0$. The nonlinear controller was shown to give the most rapid response that the car converges on the guideline. Fig.6.19(b) demonstrates the car trajectoriesare calculated from the initial state of $Y = 20.0m$ and $\theta = 0$. Apparently, the response of the nonlinear controller resemble its behavior in Fig.6.19(a). However, the response of LQ controller and the driver model control yield quite different response according to the initial state. Simulations proved that it becomes more conspicuous in the cases of an even farther initial position from the guideline.

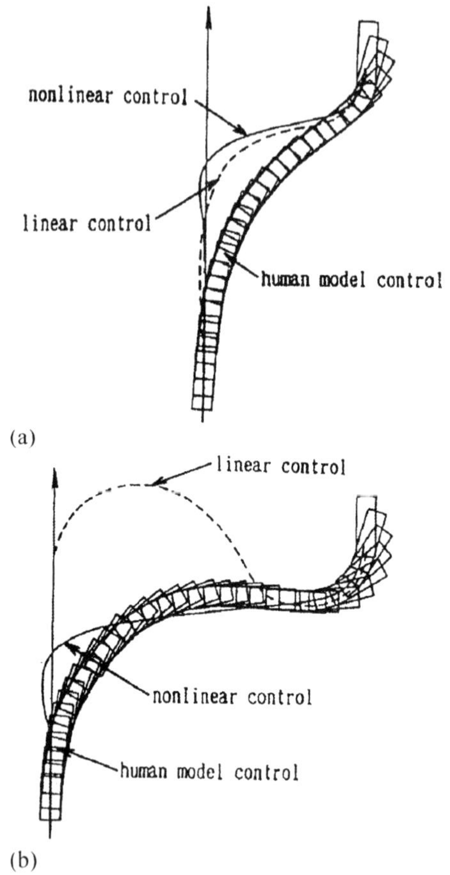

Fig.6.19 Comparisons of nonlinear controller, LQ controller and human model controller, from [44].

Further analysis shows that even the nonlinear controller yields unacceptable solution since the posture angle may becomes bigger than the right angle. Fig.6.20 shows a typical scenario in which the initial posture angle is near a right angle. Obviously, the trajectory has a large overshoot which often leads to collision with other cars parking nearby.

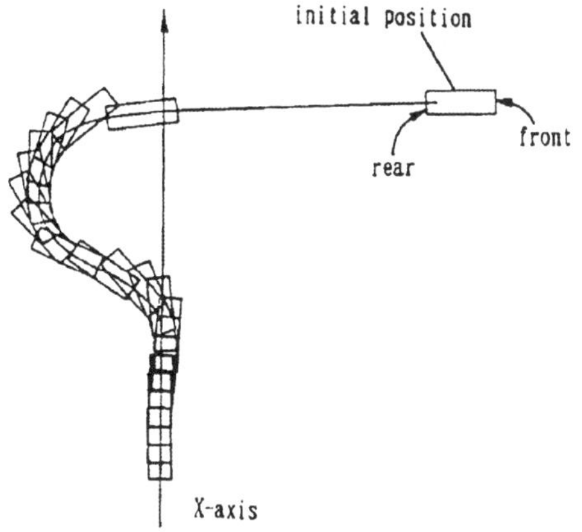

Fig.6.20 Unacceptable car behavior under nonlinear control, from [44].

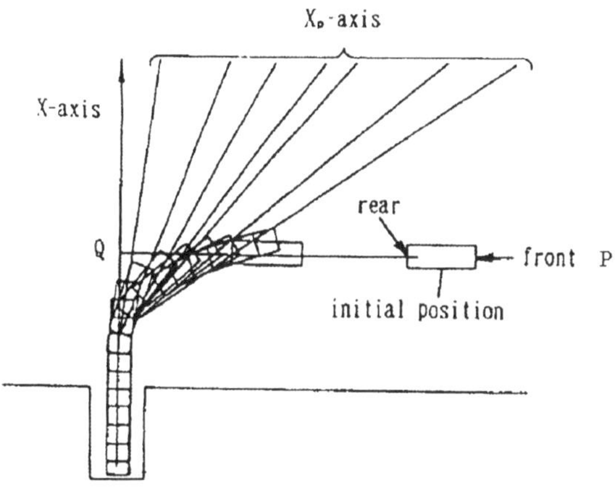

Fig.6.21 Simulation results using reference axle correcting method when $\theta = \pi / 2$, from [44].

To deal with such problems, Ohata and Mio proposed a method to change the straight guideline according to the car position and the posture angle that was introduced. This method was indeed a continuous sequence of transformation of reference axis. As shown in Fig.6.21, the guideline is firstly the one passing through the points P and Q and finally converges approximately to the X -axis when the car approaches near the parking position. Here X_I and X_{II} represent two intermediate reference axis that had been used.

In the case of Fig.6.22 where the posture angle from the line PQ is greater than a right angle, the guideline is set as the Y-axis parallel to the car at first and changes to the X-axis when the posture angle form the X -axis becomes sufficiently smaller than a right angle.

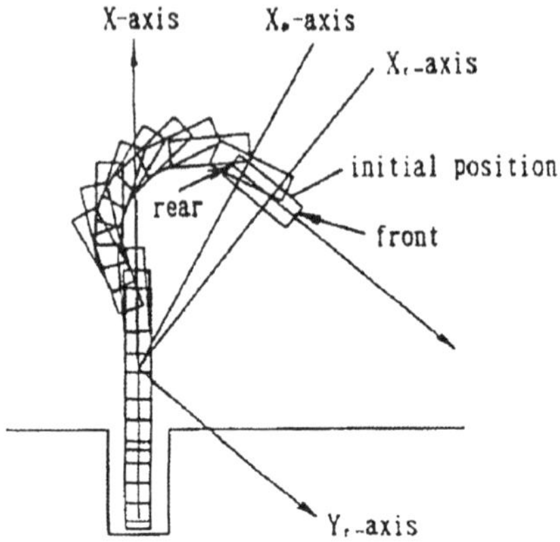

Fig.6.22 Simulation results using reference axle correcting method when $\theta = 3\pi / 2$, from [44].

However, there are still three problems need further discussion.

The first problem lies in that the ability of parking control is greatly limited by allowable moving area nearby. Thus, how to obtain the nearby environment information is of great importance. Zhao and Collins emphasized the parking space detection task in [51]. They used a laser ranger finder which is mounted in the front of vehicle with a 180° field of view and a maximum range of 25m. When the vehicle is passing past the parking space, a map of the vehicle's environment will be built. This map can provide a contour constructed by the obstacles on the right side of the vehicle. A sufficiently large rectangle can then be fitted into the contour. Finally, they used fuzzy logic controller to move the vehicle apart from the other cars and obstacles and keep a safety distance during the whole parking process. In [71], the GPS and GIS are combined to provide available parking lots for the driver.

Stepped even further than Zhao and Collins, Song and Goldenberg [45], Paromtchik and Laugier [46], Chang et. al [49], Gorinevsky et. al [53] developed several different parking monitoring and assisting systems. As a typical example, Wada et. al 1999, 2003 presented the concept of Advanced Parking Assistance System as Fig.6.23 in [55]-[56]. The proposed system is an integration of a parking assistance system, a parking administration system, and an employed sensor system.

In this integration system, the parking administration system is responsible for the parking lot management, assignment of parking bay, and transmission of data that is necessary in the assistance system. The sensor system provides real-time vehicle state estimation. The driver assistance system uses parking environment static data provided by the parking administration system, vehicles' external dimensions, and the estimated vehicle state to derive the commands necessary to perform the parking task. The information flow between these systems is shown in Fig.6.23.

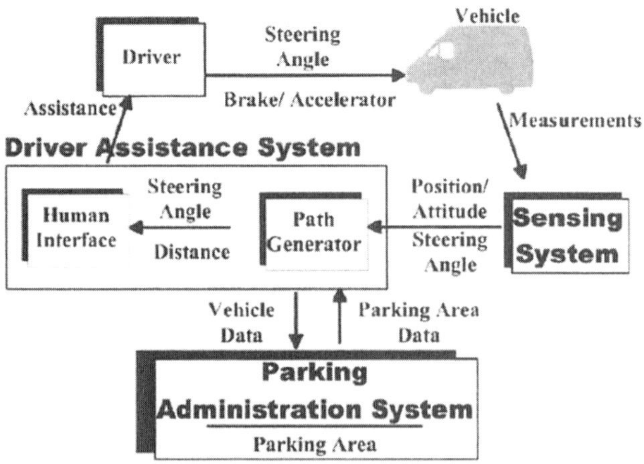

Fig.6.23 Advanced parking assistance steps, from [56]. (© [2003] IEEE)

In [56], Wada etc. designed another parking assistance system to maintain the driver in the vehicle control loop. In the system's design, they considered some factors including driver delay, driver errors, and driver insecurity. Driver delay was defined as the time interval between the task command display and the start of the commanded task. Driver errors emerge while the driver is following the system commands using the steering wheel, brake pedal, and accelerator pedal. A system that has an excessive dependence on an audio–visual interface to perform the driving task may cause driver insecurity. The necessity of addressing those problems makes a completely different design approach from that employed for similar autonomous systems.

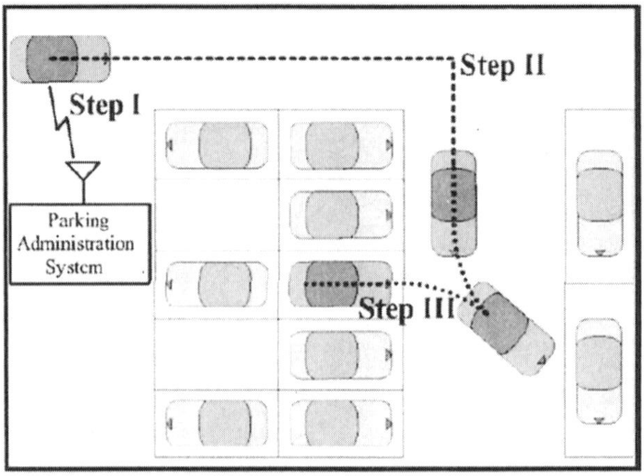

Fig.6.24 Advanced parking assistance steps, from [56]. (© [2003] IEEE)

As illustrated in Fig.6.24, Wada et. al built the assistance systems that steps from the parking lot entrance until parking is completed as follows:

(1) Data transmission performed by the parking assistance system, which contains two sub actions: Parking bay assignment; and Parking data (parking bay dimensions, parking bay front lane dimension) transmission.

(2) Guidance from the parking lot entrance to the entrance of the parking bay;

(3) Guidance until parking is completed.

The second problem is that we cannot only considered continuous movement of the vehicle like [44]. Many parking action consists of parking process with multiple movements. For example, Paromtchik and Laugier showed a typical two stage sequences of actions in autonomous parking process in [46] as Fig.6.25.

To achieve benefit from multi-stage parking, Wada et. al. presented the so called parking possibility region (PPR) based parking path planning. Following definition, there is a vehicle pose corresponding to each PPR. This pose will be called from now on the PPR angle. The row and lane parking possibility regions when the angle between the vehicle and the parking center line is are illustrated, respectively, in Fig.6.26.

The path planning algorithm using PPR is implemented through the use of the following five modules:

(1) PPR computation.

(2) path generation;

(3) collision danger judgment.

(4) fine parking guidance.
(5) parking conclusion judgment.

Fig.6.25 Sequence of motions in parking (left-top to right-bottom), from [46]. (© [1999] IEEE)

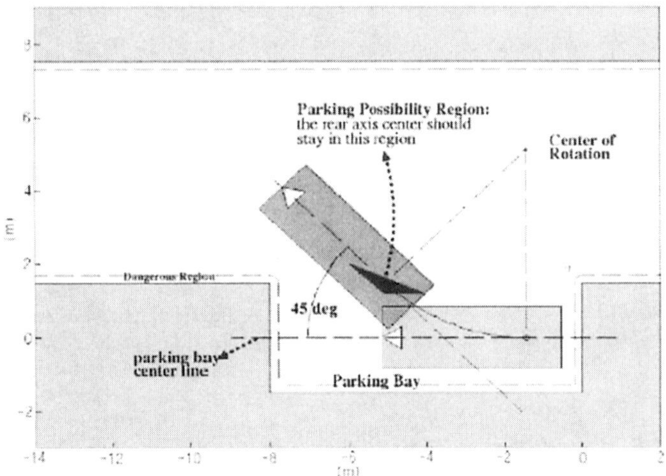

(a)

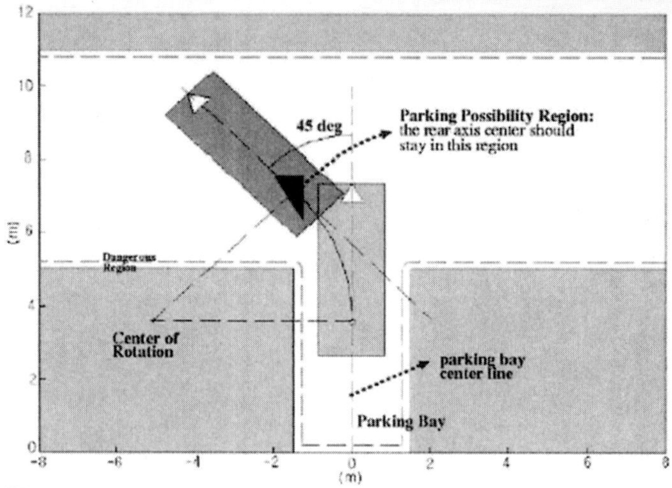

(b)

Fig.6.26 Concept of parking possibility region, from [56]. (© [2003] IEEE)

The path-generation module generates line and/or arc-composed paths given the departure and arrival poses/positions. Collision danger judgment is employed during the vehicle displacements. If necessary, the fine parking guidance module will guide the vehicle to its final parking position.

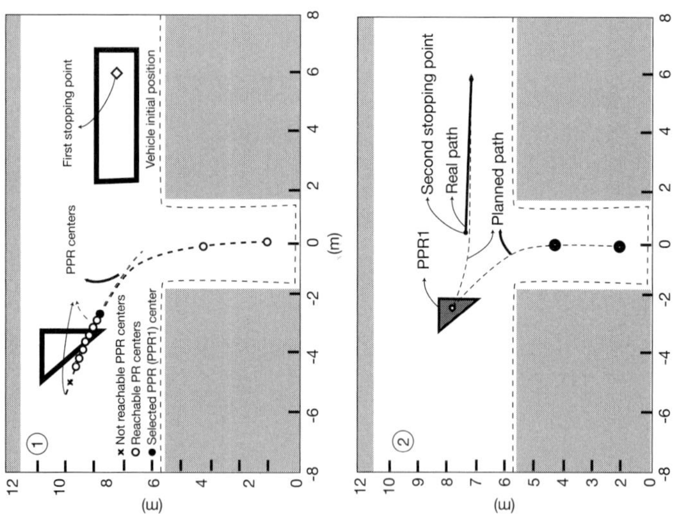

(a) (b)

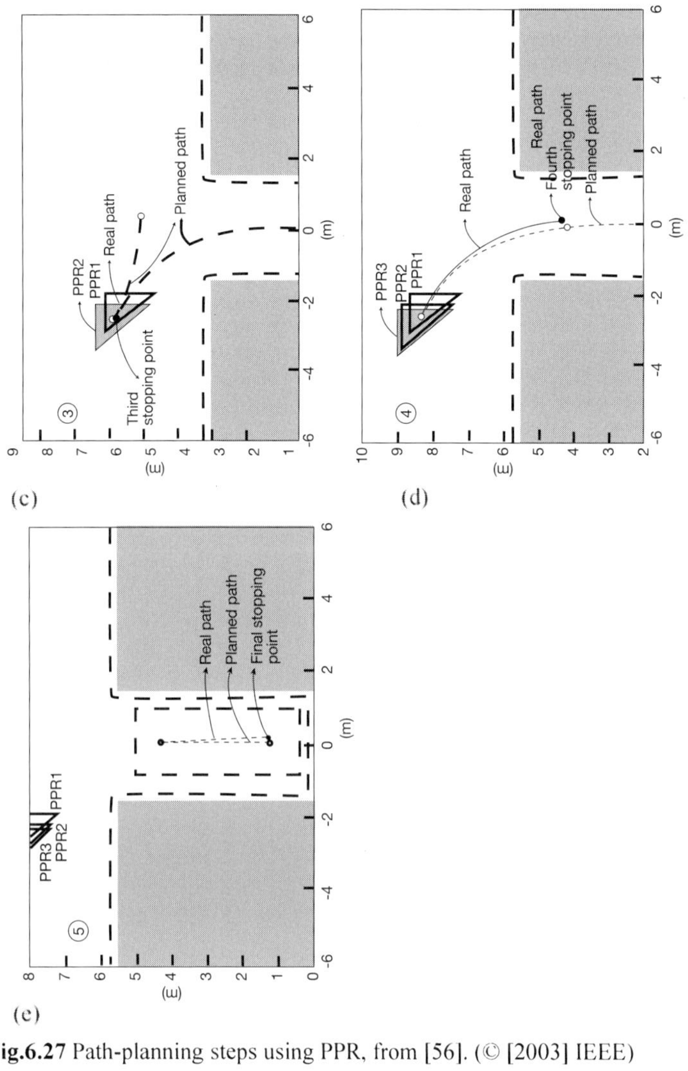

Fig.6.27 Path-planning steps using PPR, from [56]. (© [2003] IEEE)

Fig.6.27 shows how PPR is used in the path-planning algorithm steps. During these steps, it is necessary to evaluate if the vehicle is in a PPR or if a point in a PPR is reachable. The vehicle is said to be in a PPR-A if its position is inside PPR-A, and its pose is equal to the PPR-A angle. A point in PPR-A is said to be reachable if there is a path between the present vehicle (position) (pose) and (PPR-A point position) (PPR-A angle) exists (it should be stressed that the existence of a path between the present (position) (pose) and the PPR-A point position is not sufficient).

The algorithm first computes the PPR centers. Part of not reachable PPR centers, the reachable PPR centers, and the PPR (PPR1) center selected for path generation is shown in the portion of Fig.6.27(a). The selected PPR1 center, the planned path, and the vehicle traveled path is shown in the portion of Fig.6.27(b). If a driver error occurs during the subsequent line movement, a new point belonging to a PPR (PPR2) reachable with an arc is selected. The newly selected PPR2 point, PPR2, the regenerated path, and the path the vehicle has traveled is shown in the portion of Fig.6.27(c). If a driver error occurs during the subsequent arc movement, it is necessary to determine if the subsequent stopping point is in a PPR. The newly emerged PPR (PPR3), the regenerated path, and the path the vehicle has traveled is shown in the portion of Fig.6.27(d). Finally, the portion of the figure labeled shows the path generated to the final parking position and the path the vehicle has traveled. If it is not possible to park with one or two displacements, the PPR-based algorithm allows the path for parking the vehicle to be re-planned with one additional switchback maneuver. It was claimed the PPR-based path planning algorithm can generate paths that conform to the system constraints and result in a reduced number of maneuvers.

There were some other methods using the concept of PPR but not the same algorithms presented above. For example, fuzzy inference methods in [49]-[51] and neural networks in [52]-[54] were supervised by human driver experiences and/or driving records to generate allowable path from the possible parking area to the parking bay.

The third problem is how to improve the system robustness against model uncertainty and handle the nonlinearity and nonholonomic vehicle dynamics [44], [47]-[48], [50]. In [57]-[58], long distance trajectory planning was considered, which guide the vehicle from the entrance of the parking lot to the available parking lot. It naturally leads to more complex trajectory generation process. A typical approach will be discussed in the following Section 6.4.

Comparing to single track vehicle parking problems, track-trailers parking problems gained more interests in the last two decades [52], [59]. Because of the nonholonomic characteristics, track-trailers parking problems are even harder to analyze. Although some useful methods had been proposed, it seems there is still have a long way to pass before such problems are completely solved.

6.4 Longitudinal/Lateral/Vertical Vehicle Motion Control Synthesis

It is well known that vehicle dynamics are not completely independent in both directions, especially when the steering angle is large than a certain degree [60]-[85]. For instance, the coupling effect of longitudinal and lateral control method had also been analyzed in several literals.

In [69], Lim and Hedrick classified the coupling effects into three types:

(1) Kinetic coupling - steering angle affects longitudinal dynamics by changing the vehicle motion direction and the lateral cornering forces on the steered wheels; while lateral centripetal force is affected by the longitudinal velocity simultaneously;

(2) Tire force coupling - the tire friction force is distributed nonlinearly in lateral and longitudinal directions for each tire, see *Chapter 2*. And the tire forces coupling effect becomes increasingly significant as the friction forces on the tire approach saturation;

(3) Weight shift coupling - a third form of coupling which is a result of weight shift due to longitudinal acceleration which affects the lateral dynamics by redistributing the tire normal forces. In this manner, longitudinal acceleration has a significant effect on lateral dynamics. Moreover, lateral accelerations change the weight distribution between the left and right tires.

To solve the coupling problem, in [60], Pham, Hedrick and Tomizuka built a simplified model as

$$\dot{V}_x = f_1(V_x, V_y, \dot{\psi}) + k_1 T_{net} \tag{6.50}$$

and

$$\dot{V}_y = f_2(V_x, V_y, \dot{\psi}) + k_2 \delta_f \tag{6.51}$$

together with

$$\ddot{\psi} = f_3(V_x, V_y, \dot{\psi}) + k_3 \delta_f \tag{6.52}$$

where f_1, f_2 and f_3 are nonlinear functions of the longitudinal velocity V_x, the lateral velocity V_y and the heading angle ψ. The system inputs are the steering angle δ_f and the net engine torque T_{net}.

Suppose the control objective is to keep proper longitudinal spacing. Assume the spacing error between the desired and actual spacing is denoted by ε, and the lateral sensor error is denoted as y_s, which represents the offset between the guideline or the road center and the vehicle mounted sensor. Then, it yields

$$\dot{\varepsilon} = V_x - V_{lead} \tag{6.53}$$

and

$$\dot{y}_s = V_y + V_x \tilde{\psi} + d_s \dot{\tilde{\psi}} \tag{6.54}$$

where V_{lead} is the preceding vehicle velocity and $\tilde{\psi} = \psi - \psi_{road}$ is the error between the velocity's yaw and the road's yaw angles.

In order to accommodate ride quality issues, we can replace ψ_{road} in Eq.(6.54) with a desired yaw angle ψ_{des}, thus we get

$$\dot{y}_s = V_y + V_x(\psi - \psi_{des}) + d_s(\dot{\psi} - \dot{\psi}_{des}) \tag{6.55}$$

The desire yaw angle is obtained from a smoothing of ψ_{road}, and it assumes to have limited derivatives. Implicit in this is the need for a future values of ψ_{road} at time t.

Using sliding mode control skills, Pham, Hedrick and Tomizuka defined the desired plant dynamics of the sliding surface as

$$S_1 = \dot{\varepsilon} + 2\lambda_1 \varepsilon + \lambda_1^2 \int \varepsilon dt = 0 \tag{6.56}$$

and

$$S_2 = \dot{y}_s + 2\lambda_2 \dot{y}_s + \lambda_2^2 \int \dot{y}_s dt = 0 \tag{6.57}$$

where λ_1 and λ_2 are two positive constants. Note that since S is defined with relative degree one, the tracking problem is reduced to a first order problem in S.

In [60], Pham, Hedrick and Tomizuka introduced a continuous approximation of the switching term

$$\dot{S} = \frac{\eta S}{\sqrt{S^2 + \mu^2}} \tag{6.58}$$

Choose the inputs as

$$\begin{cases} T_{net} = -\dfrac{1}{b_1}[f_1 + \dot{V}_{lead} + 2\lambda_1 \dot{\varepsilon} + \lambda_1^2 \varepsilon] - \dfrac{\eta_1}{b_1} \dfrac{S_1}{\sqrt{S_1^2 + \mu_1^2}} \\ \delta_f = -\dfrac{1}{b_2}[f_2 + V_x \dot{\psi} + d_2(f_3 - \ddot{\psi}_{des}) + 2\lambda_2 \dot{y}_s + \lambda_2^2 y_s] - \dfrac{\eta_2}{b_2} \dfrac{S_2}{\sqrt{S_2^2 + \mu_2^2}} \end{cases} \tag{6.59}$$

it leads to

$$\dot{S}_i \le -\frac{\eta_i S_i}{\sqrt{S_i^2 + \mu_i^2}} + |f_{i\max}| \tag{6.60}$$

Thus, it guarantees that the S_i stay within a "sausage" area defined by

$$\dot{S}_i \le \left| \frac{f_{i\max}\eta_i}{\sqrt{\eta_i^2 - f_{i\max}^2}} \right| \tag{6.61}$$

where the η_i 's compensation gain were carefully chosen to assure

$$\eta_i \ge |d_{i\max}| \tag{6.62}$$

In [60], the proposed surface control law against a lateral frequency shaped linear quadratic (FSLQ) optimal controller utilizing the full 18-sate model were compared. The simulation results prove such control strategies can guarantee the stability of the vehicle. In [69], Lim and Hedrick used sliding mode methods to control a more complicated longitudinal/lateral 2D vehicle model.

Integrated longitudinal/lateral problems also received significant interests in trajectory planning situation. For instance, several researchers tried to incorporate lane changing profile design and vehicle dynamic control models together [71].

In [62], [72], the vehicle dynamic model was formulated as a Bicycle model as

$$\ddot{y} = \frac{A_1}{V}\dot{y} - A_1\varepsilon + \frac{A_2}{V}\dot{\varepsilon} + B_1\delta - \frac{K_y}{m}\{V_{wy} - V\varepsilon + \dot{y}\}|V_{wy} - V\varepsilon + \dot{y}| \tag{6.63}$$

and

$$\ddot{\varepsilon} = \frac{A_3}{V}\dot{y} - A_3\varepsilon + \frac{A_4}{V}\dot{\varepsilon} + B_2\delta \tag{6.64}$$

where y and ε are the lateral position and the yaw angle of the vehicle, as Fig.6.28. δ is the front steering wheel angle. V is the speed of the vehicle. V_{wy} is the lateral wind speed and K_y is lateral air drag coefficients, respectively. C_s denotes the cornering stiffness.

And $A_1 - A_4$, $B_1 - B_2$ were defined as follows

$$A_1 = \frac{-4C_s}{m}\ A_1 = \frac{-4C_s}{m},\ A_2 = \frac{-2C_s(l_1 - l_2)}{m},\ A_3 = \frac{-2C_s(l_1 - l_2)}{I_z} \tag{6.65}$$

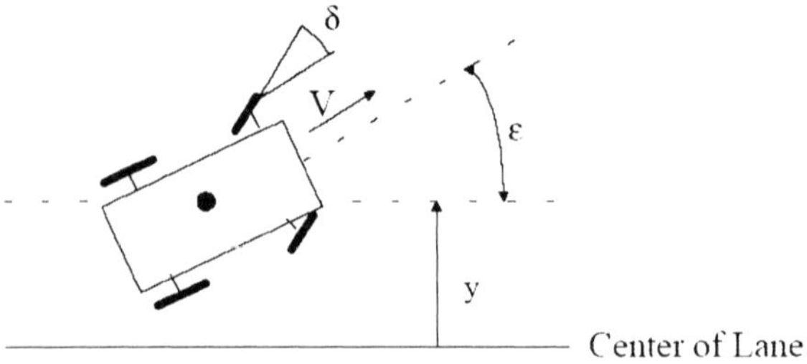

Fig.6.28 Lane changing diagram, from [62]. (© [1995] IEEE)

$$A_4 = \frac{-2C_s(l_1^2 - l_2^2)}{I_z},\ B_1 = \frac{2C_s}{m},\ B_2 = \frac{2C_s l_1}{I_z} \tag{6.66}$$

where m is the mass and I_z is the moment of inertia of a vehicle. l_1 and l_2 are the distances from center of gravity to front and rear axles. Furthermore, let's define $a_{max} > 0$ as lateral acceleration limit and J_{max} as jerk limit.

In [62], three different acceleration profiles: circular trajectory, trapezoidal acceleration trajectory, and fifth order polynomial trajectory, were used. The trapezoidal acceleration trajectory was defined as Fig.6.29, where the lateral jerk is specified by selecting the slope of the trapezoid, and the lateral acceleration by choosing the height of the trapezoid.

The desired lateral profile $\ddot{y}_d$ is written as

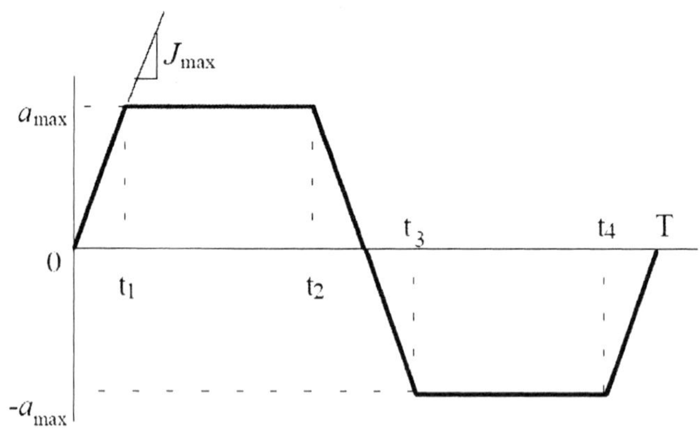

Fig.6.29 Trapezoidal acceleration profile, from [62]. (© [1995] IEEE)

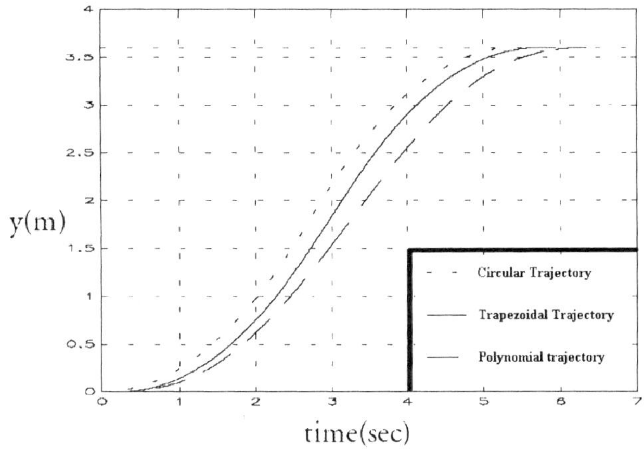

Fig.6.30 Virtual desired trajectories, from [62]. (© [1995] IEEE)

$$\begin{aligned}\ddot{y}_d &= J_{max}(t)\cdot u(t) - J_{max}(t-t_1)\cdot u(t-t_1) \\ &\quad - J_{max}(t-t_2)\cdot u(t-t_2) + J_{max}(t-t_3)\cdot u(t-t_3) \\ &\quad + J_{max}(t-t_4)\cdot u(t-t_4) - J_{max}(t-T)\cdot u(t-T)\end{aligned} \tag{6.67}$$

where $u(t)$ is a unit step function. Temporal parameters for this trajectory t_1, t_2, t_3, t_4 and T were given as

$$t_1 = \frac{a_{max}}{J_{max}},\ t_1 = \frac{-t_1^2 + \sqrt{t_1^4 + 4t_1 \dfrac{d}{J_{max}}}}{2t_1} \tag{6.68}$$

and

$$t_3 = 2t_1 + t_2,\ t_4 = t_1 + 2t_2,\ T = 2t_1 + 2t_2 \tag{6.69}$$

where $d = y(T)$ is the width of the lane.

Fig.6.30 shows the three trajectories for $J_{max} = 0.1g/s$, $a_{max} = 0.05g$, and $V = 31.3\ m/s$ (70 mph) for three trajectories, where the circular trajectory has the smallest transition time and the polynomial trajectory has the largest one.

In [62], two controllers were designed to control the vehicle to follow the virtual desired trajectory under the system uncertainties and the disturbances. In addition to the robustness to the parameter uncertainties and disturbances, ride comfort, such as jerk limit and acceleration limit, was also considered as the controller design objectives.

The first controller, Frequency Shaping Linear Quadratic (FSLQ) optimal controller, is based on the linear quadratic optimal control theory with a quadratic performance index involving those frequency dependent weighting functions. It was shown that frequency-dependent ride quality index can be explicitly included in the cost function.

The second controller is a sliding mode controller, where the tracking error is

$$e(t) = \{y(t) - y_d(t)\} + \psi\{\varepsilon(t) - \varepsilon_d(t)\} \tag{6.70}$$

where y_d and ε_d are the desired lateral position and yaw angle, respectively. ψ is a dimension conversion factor and a weighting factor at the same time. Here, ψ is assumed to be 1 for the convenience of the calculation.

Then, the sliding surface can be set as

$$S(t) = \left(\frac{d}{dt} + \lambda\right)^2 v(t) = 0 \tag{6.71}$$

where $v(t)$ is the filter tracking error satisfying

$$\frac{dv(t)}{dt} - \ln\gamma \cdot v(t) = e(t), \quad 0 < \gamma < 1 \tag{6.72}$$

where γ is the filter coefficient to reduce the unexpected vibration that appears frequently in sliding mode controller.

In [62], γ was implemented in the following development. Therefore, $v(t)$ can be viewed as the weighted integral of the tracking error

$$v(t) = \int_0^t \gamma^{(t-\tau)} e(\tau) d\tau \tag{6.73}$$

The impulse response of the filter is shown in Fig.6.31. We can see that, with $0 < \gamma < 1$, the impulse response places more weights on near-past errors than fast-past errors. Thus, $v(t)$ implies the weighted integration of the tracking error with emphasis on near-past errors. These relatively light weights on far-past errors imply a forgetting action.

The stability of the sliding surface $S(t) = 0$ can be satisfied by the condition

$$\frac{1}{2}\frac{d}{dt}\left(S^2\right) < -\eta \cdot |S| \tag{6.74}$$

Considering the bound property of $S(t)$, let introduce a upper bound as

$$|S| < \Phi \tag{6.75}$$

it leads to

$$\frac{1}{2}\frac{d}{dt}\left(S^2\right) < -\frac{\eta}{\Phi}\cdot\left(S^2\right) \tag{6.76}$$

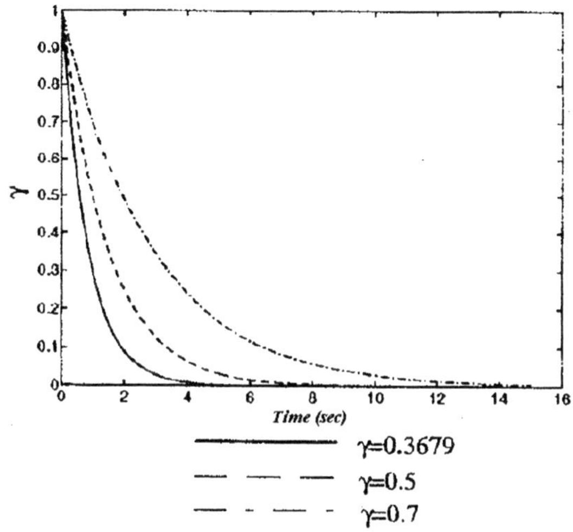

Fig.6.31 Impulse response of filter, from [62]. (© [1995] IEEE)

A control input that satisfies this condition (6.76) was proposed in [62] as

$$\delta = \frac{1}{B_1 + B_2}\left\{-\frac{A_1 + A_3}{V}\dot{y} + (A_1 + A_3)\varepsilon + \frac{K_y}{m}(\dot{y} - V\varepsilon)|\dot{y} - V\varepsilon| + \ddot{y}_d + \ddot{\varepsilon}_d - (2\lambda + \ln\gamma)\dot{e} - (\lambda + \ln\gamma)^2 e - \ln\gamma\cdot(\lambda + \ln\gamma)^2 v - K\,\mathrm{sgn}(S) - \frac{A_2 + A_4}{V}\dot{\varepsilon}\right\} \tag{6.77}$$

and

$$K = \eta + 2\alpha\left|\frac{A_1 + A_3}{V}\dot{y} - (A_1 + A_3)\varepsilon + \frac{A_2 + A_4}{V}\dot{\varepsilon}\right| + \tag{6.78}$$

$$\frac{K_y}{m}\left\{V_{wm}^2 + [2V_{wm} + \alpha|\dot{y} - V\varepsilon|]|\dot{y} - V\varepsilon|\right\} +$$

$$\alpha\left|\ddot{y}_d + \ddot{\varepsilon}_d - (2\lambda + \ln\gamma)\dot{e} - (\lambda + \ln\gamma)^2 e - \ln\gamma \cdot (\lambda + \ln\gamma)^2 v\right|$$

where α is the system uncertainty bound obtained from equation. The choice of γ is bounded by ride comfort limit and allowable steady state tracking error.

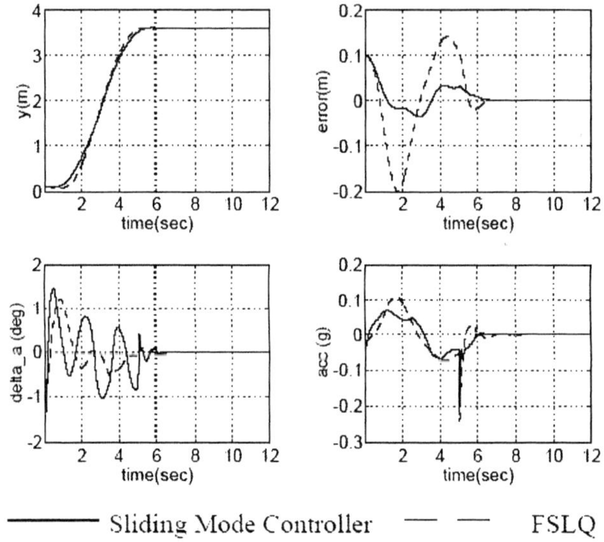

Fig.6.32 Comparison between FSLQ and sliding mode controller, from [62]. (© [1995] IEEE)

In the simulation, only the on-board sensors, a yaw rate sensor and lateral accelerometer, were used in lane change maneuvers. These sensors were assumed to be noise free and the state variables that are not directly measured are obtained by integrating these two signals. It was assumed that the vehicle speed was maintained at 70MPH during the lane change maneuver. The trapezoidal acceleration trajectory with 0.05g-lateral acceleration limit and 0.1g/sec-lateral jerk limit is selected as the VDT and the corresponding transition time(T) is 5.92 sec. Fig.6.32 shows the simulation results of the two controllers' performances. For the road surface condition, cornering stiffness was reduced to 20% of the nominal value from 0 to 5 sec, and, after 5 sec, the nominal road surface condition was assumed. A 55MPH wind disturbance is assumed to present from 1.5 sec to 5 sec. From the simulation results, we can find that the sliding mode controller can control the vehicle with smaller tracking errors while maintaining smaller lateral accelerations, as compared to the FSLQ controller.

Chee et. al also proven that, if the model uncertainty can be given as the bounds of the system coefficients as follows

$$A_1 = \hat{A}_1(1+\Delta A_1),\ A_2 = \hat{A}_2(1+\Delta A_2),\ A_3 = \hat{A}_3(1+\Delta A_3) \tag{6.79}$$

and

$$A_4 = \hat{A}_4(1+\Delta A_4),\ B_1 = \hat{B}_1(1+\Delta B_1),\ B_2 = \hat{B}_2(1+\Delta B_2) \tag{6.80}$$

where

$$|\Delta A_i| \le \alpha,\ |\Delta B_j| \le \alpha,\ j = 1,\dots,4 \tag{6.81}$$

the nominate controller can still guarantee the stability of the system

$$\begin{aligned}\delta = \frac{1}{\hat{B}_1+\hat{B}_2}\Bigg\{ & -\frac{\hat{A}_1+\hat{A}_3}{V}\dot{y} + (\hat{A}_1+\hat{A}_3)\varepsilon + \frac{K_y}{m}(\dot{y}-V\varepsilon)|\dot{y}-V\varepsilon| \\ & + \ddot{y}_d + \ddot{\varepsilon}_d - (2\lambda+\ln\gamma)\dot{e} - (\lambda+\ln\gamma)^2 e \\ & - \ln\gamma\cdot(\lambda+\ln\gamma)^2 v - K\,\mathrm{sgn}(S) - \frac{\hat{A}_2+\hat{A}_4}{V}\dot{\varepsilon}\Bigg\}\end{aligned} \tag{6.82}$$

and

$$\begin{aligned}K = \eta + 2\alpha & \left|\frac{\hat{A}_1+\hat{A}_3}{V}\dot{y} - (\hat{A}_1+\hat{A}_3)\varepsilon + \frac{\hat{A}_2+\hat{A}_4}{V}\dot{\varepsilon}\right| + \\ & \frac{K_y}{m}\left\{V_{wm}^2 + [2V_{wm} + \alpha|\dot{y}-V\varepsilon|]|\dot{y}-V\varepsilon|\right\} + \\ & \alpha\left|\ddot{y}_d + \ddot{\varepsilon}_d - (2\lambda+\ln\gamma)\dot{e} - (\lambda+\ln\gamma)^2 e - \ln\gamma\cdot(\lambda+\ln\gamma)^2 v\right|\end{aligned} \tag{6.83}$$

Besides sliding mode control method, there were several other approaches for lane changing control design. For example, in [65]-[66], O'Brien et. al analyzed the possibility of using a frequency domain H_∞ controller to implement lane changing. They claimed that the proposed controller was able to steer the vehicle though the desired lane change maneuver with no overshoot and shorter settling time than the conventional controller for a wide range of operation conditions. And it was robust to parameter uncertainty, too. In [14], [81], different sliding mode controllers associated with robust controller were applied in lane changing control. It was proved that this method further improves the steering performance.

In some recent researches, i.e. [85], 3D complete vehicle models that consider lateral, longitudinal and vertical dynamics simultaneously are studied. These methods naturally lead to notable system nonlinearity and significant calculation costs. Modestly speaking, all these attempts still need further discussions.

6.5 Reference

1. D. E. Smith, J. M. Starkey, and R. E. Benton, "Nonlinear-gain-optimized controller development and evaluation for automated emergency vehicle steering," Proceedings of American Control Conference, vol. 5, pp: 3586-3591, 1995.
2. J. Forbes, T. Huang, and K. Kanazawa, et. al, "The BATmobile: towards a Bayesian automated taxi," Proceedings of International Joint Conference on Artificial Intelligence, 1995.
3. C. Unsal, P. Kachroo, and J. S. Bay, "Multiple stochastic learning automata for vehicle path control in an automated highway system," IEEE Transactions on Systems, Man and Cybernetics, Part A, vol. 29, no. 1, pp. 120-128, 1999.
4. L. Alvarez and J. Yi, "Adaptive emergency braking control in automated highway systems," Proceedings of IEEE Conference on Decision and Control, vol. 4, pp. 3740-3745, 1999.
5. P. Tsiotras and C. C. de Wit, "On the optimal braking of wheeled vehicles," Proceedings of American Control Conference, vol. 1, pp. 569-573, 2000.
6. U. Lages, "Collision avoidance system for fixed obstacles-fuzzy controller network for robot driving of an autonomous vehicle," Proceedings of IEEE Intelligent Transportation Systems Conference, pp. 489-491, 2001.
7. T. Nishi and T. Takagi, "A proposal of collision avoidance algorithm for driving support system," Proceedings of Annual Conference of the IEEE Industrial Electronics Society, vol. 1, pp. 80-83, 2001.
8. R. Labayrade, C. Royere, and D. Aubert, "A collision mitigation system using laser scanner and stereovision fusion and its assessment," Proceedings of IEEE Intelligent Vehicles Symposium, pp. 441-446, 2005.
9. Z. Shiller and Y.-R. Gwo, "Dynamic motion planning of autonomous vehicles," IEEE Transactions on Robotics and Automation, vol. 7, no. 2, pp. 241-249, 1991.
10. T. Fraichard, "Dynamic trajectory planning with dynamic constraints: A 'state-time space' approach," Proceedings of IEEE/RSJ International

Conference on Intelligent Robots and Systems, vol. 2, pp. 1393-1400, 1993.

11. L. Singh and J. Fuller, "Trajectory generation for a UAV in urban terrain, using nonlinear MPC," Proceedings of the American Control Conference, vol. 3, pp. 2301-2308, 2001.
12. E. J. Bernabeu, J. Tornero, and M. Tomizuka, "Collision prediction and avoidance amidst moving objects for trajectory planning applications," Proceedings of IEEE International Conference on Robotics and Automation, vol. 4, pp. 3801-3806, 2001.
13. T. Acarman, Y. Pan, and U. Ozguner, "A control authority transition system for collision avoidance," Proceedings of IEEE Intelligent Transportation Systems, pp. 466-471, 2001.
14. C.-Y. Chan and H.-S. Tan, "Feasibility analysis of steering control as a driver-assistance function in collision situations," IEEE Transactions on Intelligent Transportation Systems, vol. 2, no. 1, pp. 1-9, 2001.
15. S. Fleury, P. Soueres, and J.-P. Laumond, et. al, "Primitives for smoothing mobile robot trajectories, " IEEE Transactions on Robotics and Automation, vol. 11, no. 3, pp. 441-448, 1995.
16. M. Khatib, H. Jaouni, and R. Chatila, et. al, "Dynamic path modification for car-like nonholonomic mobile robots," Proceedings of IEEE International Conference on Robotics and Automation, vol. 4, pp. 2920-2925, 1997.
17. A. Scheuer and T. Fraichard, "Collision-free and continuous curvature path planning for car-like robots," Proceedings of IEEE International Conference on Robotics and Automation, pp. 867-873, 1997.
18. U. Lages, "Collision avoidance system for fixed obstacles-fuzzy controller network for robot driving of an autonomous vehicle," Proceedings of IEEE Intelligent Transportation Systems, pp. 489-491, 2001.
19. J. Hermosillo, C. Pradalier, and S. Sekhavat, et. al, "Towards motion autonomy of a bi-steerable car: experimental issues from map-building to trajectory execution," Proceedings of IEEE International Conference on Robotics and Automation, vol. 2, pp. 2430-2435, 2003.
20. F. Lamiraux and J.-P. Lammond, "Smooth motion planning for car-like vehicles," IEEE Transactions on Robotics and Automation, vol. 17, no. 4, pp. 498-501, 2001.
21. R. M. Murray and S. S. Sastry, "Nonholonomic motion planning: steering using sinusoids," IEEE Transactions on Automatic Control, vol. 38, no. 5, pp. 700-716, 1993.
22. W. Yossawee, T. Tsubouchi, and S. Sarata, et. al, "Path generation for articulated steering type vehicle using symmetrical clothoid," IEEE International Conference on Industrial Technology, vol. 1, pp. 187-192, 2002.
23. A. Piazzi and C. G. Lo Bianco, "Quintic G^2 -splines for trajectory planning of autonomous vehicles, " Proceedings of the IEEE Intelligent Vehicles Symposium, pp. 198-203, 2000.

24. A. Piazzi, C. G. Lo Bianco, and Bertozzi, M., et. al, "Quintic G^2-splines for the iterative steering of vision-based autonomous vehicles, " IEEE Transactions on Intelligent Transportation Systems, vol. 3, no. 1, pp. 27-36, 2002.
25. J. Hilgert, K. Hirsch, and T. Bertram, et. al, "Emergency path planning for autonomous vehicles using elastic band theory, Proceedings of IEEE/ASME International Conference on Advanced Intelligent Mechatronics, vol. 2, pp. 1390-1395, 2003.
26. A. Simon and J. C. Becker, "Vehicle guidance for an autonomous vehicle," Proceedings of IEEE/IEEJ/JSAI International Conference on Intelligent Transportation Systems, pp. 429-434, 1999.
27. A. Simon, I. Sohnitz, and J. Becker, et. al "Navigation and control of an autonomous vehicle," Proceedings of IFAC Symposium on Control in Transportation Systems, 2000.
28. I. Papadimitriou and M. Tomizuka, "Fast lane changing computations using polynomials," Proceedings of American Control Conference, vol. 1, pp. 48-53, 2003.
29. D. Metz and D. Williams, "Near time-optimal control of racing vehicles," Automatica, vol. 25, no. 6, pp. 841-857, 1989.
30. H. A. Pham and J. K. Hedrick, "A robust optimal lateral vehicle control strategy," Proceedings of IEEE International Conference on Control Applications, pp. 361-366, 1996.
31. J.-S. Choi and B. K. Kim, "Near-time-optimal trajectory planning for wheeled mobile robots with translational and rotational sections," IEEE Transactions on Robotics and Automation, vol. 17, no. 1, pp. 85-90, 2001.
32. L. Li and F.-Y. Wang, "Vehicle trajectory generation for optimal driving guidance," IEEE International Conference on Intelligent Transportation Systems, pp. 231-235, 2002.
33. L. Li and F.-Y. Wang, "A design framework for driver/passenger-oriented trajectory planning," IEEE International Conference on Intelligent Transportation Systems, pp. 1764-1769, 2003.
34. L. Li and F.-Y. Wang, "Trajectory generation for driving guidance of front wheel steering vehicles," IEEE Intelligent Vehicles Symposium, pp. 231-236, 2003.
35. S. A. Nobe and F.-Y. Wang, "An overview of recent developments in automated lateral and longitudinal vehicle controls," Proceedings of IEEE International Conference on Systems, Man, and Cybernetics, vol. 5, pp. 3447-3452, 2001.
36. M. Cherif, "Motion planning for all-terrain vehicles: a physical modeling approach for coping with dynamic and contact interaction constraints," IEEE Transactions on Robotics and Automation, vol. 15, no. 2, pp. 202-218, 1999.
37. D. Wang and F. Qi, "Trajectory planning for a four-wheel-steering vehicle," Proceedings of IEEE International Conference on Robotics and Automation, vol. 4, pp. 3320-3325, 2001.

38. M. Cherif, "Kinodynamic motion planning for all-terrain wheeled vehicles," Proceedings of IEEE International Conference on Robotics and Automation, vol. 1, pp. 317-322, 1999.
39. C. S. Hsu, "Generalized theory of cell-to-cell mapping for nonlinear dynamical systems," ASME Journal of Applied Mechanics, vol. 48, pp. 834-842, 1981.
40. C. S. Hsu, "A discrete method of optimal control based upon the cell state space concept," ASME Journal of Optimization Theory and Applications, vol. 46, pp. 547-569, 1985.
41. C. S. Hsu and R. S. Guttalu, "An unravelling algorithm for global analysis of dynamical systems: an application of cell-to-cell mappings," ASME Journal of Applied Mechanics, vol. 47, pp. 940-948, 1980.
42. W. H. Zhu and M. C. Leu, "Planning optimal robot trajectories by cell mapping," Proceedings of IEEE International Conference on Robotics and Automation, vol. 3, pp. 1730-1735, 1990.
43. L. Guvenc, "Park by wire," IEEE Control Systems Magazine, vol. 25, no. 5, pp. 17-17, 2005.
44. A. Ohata and M. Mio, "Parking control based on nonlinear trajectory control for low speed vehicles," Proceedings of International Conference on Industrial Electronics, Control and Instrumentation, pp. 107-112, 1991.
45. P. Song and A. Goldenberg, "Fundamental principles of design of position architecture and controller design for automated car parking," Proceedings of American Control Conference, pp. 806–810, 1994.
46. I. E. Paromtchik and C. Laugier, "Autonomous parallel parking of a nonholonomic vehicle," Proceedings of IEEE Intelligent Vehicles Symposium, pp. 13–18, 1999.
47. K. Jiang, "A sensor guided parallel parking system for nonholonomic vehicles," Proceedings of IEEE Intelligent Transportation Systems, pp. 270–275, 2000.
48. M. Kochem, R. Neddenriep, and R. Isermann, et. al, "Accurate local vehicle dead-reckoning for a parking assistance system," Proceedings of the 2002 American Control Conference, vol. 5, pp. 4297-4302, 2002.
49. S.-J. Chang, C.-W. Cheng, and S. T.-H. Li, "Design and implementation of fuzzy garage-parking control for a PC-based model car," Proceedings of International Conference on Industrial Electronics, Control and Instrumentation, vol. 3, pp. 1299-1304, 1997.
50. G. Chen and D. Zhang, "Back-driving a truck with suboptimal distance trajectories: a fuzzy logic control approach, " IEEE Transactions on Fuzzy Systems, vol. 5, no. 3, pp. 369-380, 1997.
51. Y. Zhao and E. G. Collins, "Fuzzy parallel parking control of autonomous ground vehicles in tight spaces," Proceedings of IEEE International Symposium on Intelligent Control, pp. 811-816, 2003.
52. D. Nguyen and B. Widrow, "The truck backer-upper: an example of self-learning in neural networks, " Proceedings of International Joint Conference on Neural Networks, pp. 357-363, 1989.

53. D. Gorinevsky, A. Kapitanovsky, and A. Goldenberg, "Neural network architecture for trajectory generation and control of automated car parking," IEEE Transactions on Control Systems Technology, vol. 4, no. 1, pp: 50-56, 1996.
54. L. Li and F.-Y. Wang, "Parking guidance system for front wheel steering vehicles using trajectory generation," Proceedings of IEEE International Conference on Intelligent Transportation Systems, pp. 1770-1775, 2003.
55. M. Wada, K.-S. Yoon, and H. Hashimoto, et. al, "Development of advanced parking assistance system using human guidance," Proceedings of IEEE/ASME International Conference on Advanced Intelligent Mechatronics, pp. 997-1002, 1999.
56. M. Wada, K.-S. Yoon, and H. Hashimoto, "Development of advanced parking assistance system," IEEE Transactions on Industrial Electronics, vol. 50, no. 1, pp. 4-17, 2003.
57. M. Omae, H. Shimize and T. Fujioka, "GPS-based automatic driving control in local area with course of large curvature and parking space," Vehicle System Dynamics, vol. 42, no. 1-2. pp. 59-73, 2004.
58. E. Seignez, A. Lambert, and T. Maurin, "Autonomous parking carrier for intelligent vehicle," Proceedings of IEEE Intelligent Vehicles Symposium, pp. 411-416, 2005.
59. K. Tanaka, T. Kosaki, and H. O. Wang, "Backing control problem of a mobile robot with multiple trailers: fuzzy modeling and LMI-based design," IEEE Transactions on Systems, Man, and Cybernetics, Part C, vol. 28, no. 3, pp. 329-337, 1998.
60. H. Pham, K. Hedrick, and M. Tomizuka, "Combined lateral and longitudinal control of vehicles for IVHS," Proceedings of American Control Conference, vol. 2, pp. 1205-1206, 1994.
61. E. Ono, K. Takanami, and N. Iwama, et. al, "Vehicle integrated control for steering and traction systems by μ -synthesis," Automatica, vol. 30, no. 11, pp. 1639-1647, 1994.
62. W. Chee, M. Tomizuka, and S. Patwardhan, et. al, "Experimental study of lane change maneuver for AHS applications," Proceedings of American Control Conference, vol. 1, pp. 139-143, 1995.
63. T. Pilutti, G. Ulsoy, and D. Hrovat, "Vehicle steering intervention through differential braking, Proceedings of the American Control Conference, vol. 3, pp. 1667-1671, 1995.
64. D. E. Smith, J. M. Starkey, and R. E. Benton, "Nonlinear-gain-optimized controller development and evaluation for automated emergency vehicle steering," Proceedings of American Control Conference, vol. 5, pp. 3586-3591, 1995.
65. R. T. O'Brien, P. A. Iglesias, and T. J. Urban, "Lane change maneuver via H_∞ steering control methods," Proceedings of IEEE Conference on Control Applications, pp. 131-136, 1995.

66. R. T. O'Brien, P. A. Iglesias, and T. J. Urban, "Vehicle lateral control for automated highway systems," IEEE Transactions on Control Systems Technology, vol. 4, no. 3, pp. 266-273, 1996.
67. P. Waltermann, "Modelling and control of the longitudinal and lateral dynamics of a series hybrid vehicle," Proceedings of IEEE International Conference on Control Applications, pp. 191-198, 1996.
68. P. Seiler, B. Song, and J. K. Hedrick, "Development of a collision avoidance system," Proceedings of SAE Conference, pp. 97-103, 1998.
69. E. H. M. Lim and J. K. Hedrick, "Lateral and longitudinal vehicle control coupling for automated vehicle operation," Proceedings of American Control Conference, vol. 5, pp. 3676-3680, 1999.
70. Y. Jia, "Robust control with decoupling performance for steering and traction of 4WS vehicles under velocity-varying motion," IEEE Transactions on Control Systems Technology, vol. 8, 3, pp. 554-569, 2000.
71. R. Outbib and A. Rachid, "Control of vehicle speed: a nonlinear approach," Proceedings of IEEE Conference on Decision and Control, vol. 1, pp. 462-463, 2000.
72. H. Lee and M. Tomizuka, "Coordinated longitudinal and lateral motion control of vehicles for IVHS," ASME Journal of Dynamic Systems, Measurement, and Control, vol. 123, pp. 535-543, 2001.
73. R. White and M. Tomizuka, "Autonomous following lateral control of heavy vehicles using laser scanning radar," Proceedings of American Control Conference, vol. 3, pp. 2333-2338, 2001.
74. R. Saeks, C. J. Cox, and J. Neidhoefer, et. al, "Adaptive control of a hybrid electric vehicle," IEEE Transactions on Intelligent Transportation Systems, vol. 3, no. 4, pp. 213-234, 2002.
75. P. Setlur, D. Dawson, and J. Wagner, et. al, "Nonlinear tracking controller design for steer-by-wire automotive systems," Proceedings of American Control Conference, vol. 1, pp. 280-285, 2002.
76. Ph. Heinzl, P. Luger and M. Plochl, "Stability control of a passenger car by combined additional steering and unilateral braking," Vehicle System Dynamics, Supplement 37, pp. 221-233, 2002.
77. J. R. Zhang, S. J. Xu, and A. Rachid, "Sliding mode controller for automatic path tracking of vehicles," Proceedings of American Control Conference, vol. 5, pp. 3974-3979, 2002.
78. J. Ryu. H.-S. Kim and J.-H. Kim, "An emergency obstacle avoidance control strategy for automated highway vehicles," Vehicle System Dynamics, vol. 38, no. 5, pp. 319–339, 2002.
79. M. Lakehal-Ayat, S. Diop, and E. Fenaux, "An improved active suspension yaw rate control," Proceedings of American Control Conference, vol. 2, pp. 863-868, 2002.
80. C. Hatipoglu, U. Ozguner, and K. A. Redmill, "Automated lane change controller design," IEEE Transactions on Intelligent Transportation Systems, vol. 4, no. 1, pp. 13-22, 2003.

81. F. Tahami, S. Farhangi, and R. Kazemi, "A Fuzzy logic direct yaw-moment control system for all-wheel-drive electric vehicles," Vehicle System Dynamics, vol. 41, no. 3, pp. 203-221, 2004.
82. K. Guo, H. Ding, and J. Zhang, "Development of a longitudinal and lateral driver model for autonomous vehicle control," International Journal of Vehicle Design, vol. 36, no. 1, pp. 50-65, 2004.
83. O. Mokhiamar and M. Abe, "Simultaneous optimal distribution of lateral and longitudinal tire forces for the model following control," ASME Journal of Dynamic Systems, Measurement, and Control, vol. 126, pp. 753-763, 2004.
84. L. Beji and Y. Bestaoui, "Motion generation and adaptive control method of automated guided vehicles in road following," IEEE Transactions on Intelligent Transportation Systems, vol. 6, no. 1, pp. 113-123, 2005.
85. A. R. W. Huang and C. Chen, "A low-cost driving simulator for full vehicle dynamics simulation," IEEE Transactions on Vehicular Technology, vol. 52, no. 1, pp. 162-172, 2003.

Advanced Multiple Vehicles Motion Control

7.1 Introduction

In the last two decades, advanced multiples vehicle motion control becomes a new hot direction in intelligent vehicle research field. Unlike the topics that had been mentioned in the above chapters, the driving processes of more than two vehicles is emphasized instead of a single vehicle's movement at this level. It is proven that proper cooperative driving of multiple vehicles can significantly reduces the probability of traffic accident and make the riding more smoothly.

Different concepts are established to describe intelligent cooperative driving control strategies. The first well discussed idea is the so called vehicle platoon control. A vehicle platoon is a tightly spaced string of vehicles, where the inter-vehicle distances are appropriately maintained as low as three to one meter at highway speeds depending on what sensors and communication devices that are applied. Since each vehicle knows the dynamics of its leading car and might also the platoon's leading car, such short distances are safe enough for the vehicles. Usually, the vehicles need to use radar, laser sensor to directly measure the preceding car's speed and their gap [43]-[57]. In many recent approaches, inter-vehicle communication is also employed to transmit the required messages, i.e. the speed and position of the platoon's leading vehicle [1]-[8].

Urban highways are often crowed or congested, which makes lane merging and changing difficult and raises the possibility of accidents. As results, lane merging and changing driver aid assistance systems were proposed and studied in the last ten years [109]-[111]. Stepping a litter further than platoon control, such systems aim to either find a collision free trajectory without hindering other vehicles, or guarantee collision free movements by notifying other vehicles of the coming lane merging/changing action and guide them to cooperatively speed up or slow down.

More complicated cooperative driving situations are considered recently. For instance, multiple vehicles riding control at intersections received considerable interests in [119]-[133]. Because the vehicles could move in totally different directions, it is much harder to guarantee the ride safety than the longitudinal platoon control in which all the vehicles are moving in the same directions.

Cooperative driving concept was first presented by JSK (Association of Electronic Technology for Automobile Traffic and Driving) in Japan in the early 1990s, [7]. It was originally defined as flexible platooning of automated vehicles with a short inter-vehicle distance over a couple of lanes, which was also named super smart vehicle system (SSVS) at that time [1], [7]. Using appropriate inter-vehicle communication to link the vehicles, cooperative driving may enable each vehicle to perform safe and efficient lane changing and merging, and thus improve the traffic control performance. Then, the feasibility and benefit of cooperative driving was further discussed and examined world-widely in the past ten years, i.e. in California PATH project [120][121], Chauffeur project of EU [122], and Demo 2000 Cooperative Driving System in Japan [116].

Generally, these approaches mainly focus on two problems: how to exchange the information among vehicles and how to guide vehicles using the obtained information. The answer to the former problem is inter-vehicle communication. It enables the vehicles to share information about their driving status and desires, which greatly extends the horizon of each individual drivers or intelligent driving systems. And the latter problem is normally solved by cooperative trajectory planning which will be examined in details in this Chapter, because a thorough discussions of the former problem requires a dedicate book.

7.2 Inter-Vehicle Communication

The inter-vehicle communication plays an important role in cooperative driving, since the necessary driving information of every other vehicle has to be transmitted over the vehicle which needs to make decision. Several different inter-vehicle communication models were designed, implemented and tested in the last decade, i.e. COCAIN (Cooperative Optimized Channel Access for INter-vehicle communication) proposed by Kaltwasser and Kassubek in [9], TELCO (Telecommunication Network for Cooperative Driving) proposed by Verdone in [10], DOLPHIN (Dedicated Omnipurpose inter-vehicle communication Linkage Protocol for HIghway automatioN) proposed by Tokuda, Akiyama and Fujii in [11].

Based on the media for the inter-vehicle communication, different systems can be generally divided into three kinds: infrared [12], optical [13] and radio [14]-[15]. It was shown in [13] that an optical transmitter and receiver using LED and condenser lens can guarantee a bit error rate of 10-8 or less of good communication quality under conditions of ± 20° communication directivity, a communication distance of 30 m, and the number of light emitting diodes to be used as 8. However, optical based inter-vehicle communication systems can only be applied for vehicle platoon control, since the optical transmitter and receiver are mounted at the two ends of the vehicle, which make it difficult for a vehicle to communicate with the vehicle moving beside it.

Most recent inter-vehicle communication systems use VHF waves, micro waves or millimeter waves. Different from those systems with infrared or millimeter waves which are of a type-of-sight, the systems using VHF waves or micro waves work in types of broadcasting. The most important problems that need to solve is vehicle-to-vehicle radio links suffer from multi-path fading and interference from other vehicles. Numerous efforts are reported in the research of wireless communication channel modeling [16]-[23] and performance analysis of the network involving many links [24]-[27]. Different protocols were proposed and tested in the quickly varying traffic environment [28]-[29]. However, there is no standard had been world-widely accepted yet.

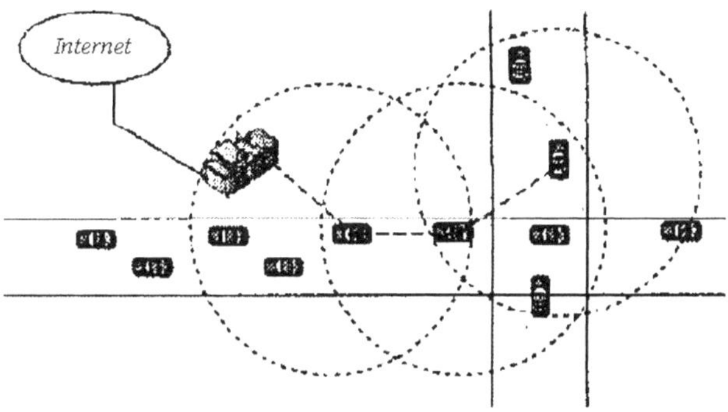

Fig.7.1 Example of message being routed based on digital map information and geographical location of the node, from [30]. (© [2003] IEEE)

The conventional implementation methods of inter-vehicle communication link vehicles through an extra remote service station (i.e. the communication tower mentioned in [30] and [31]). The driving information of a vehicle will be first transformed to the service station and then broadcasted to other relate vehicles, see Fig.7.1. Or, the vehicles need to inquiry service station to locate other vehicles [30]. But such approaches require considerable cost for building and maintaining service station.

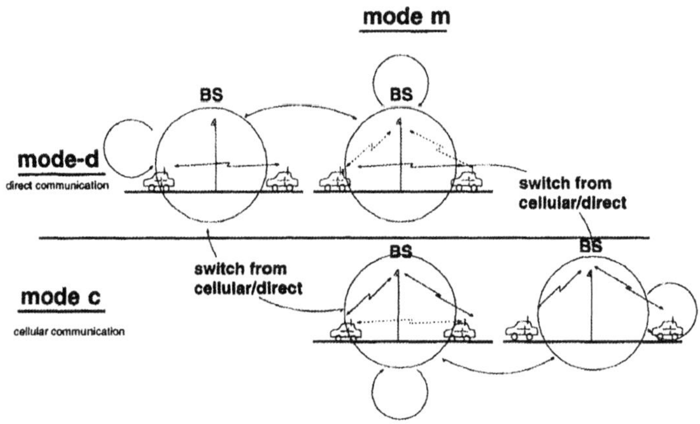

Fig.7.2 Diagram of switching between the two communication modes direct (mode-d) and cellular (mode-c) and the relationship to the monitoring modes (mode-dm, mode-cm), from [34]. (© [2000] IEEE)

Different from above methods, many new designs use the peer-to-peer ad-hoc network to achieve vehicular information sharing. For instance, Fig.7.2 shows a smart switching scheme from two communication modes: direct (mode-d) and cellular (mode-c). As revealed in [32]-[37], peer-to-peer ad-hoc communication networks integrate four valuable features: ad-hoc connectivity, local peer-to-peer networking, short-range and inter-personal communication. However, due to varied driving behavior and high mobility, the performance of inter-vehicle communication network is still under further discussions.

The data format of the inter-vehicle communication also received significant interest [38]-[41]. The resulted communication security problem is another hard topic that need further examine [42]-[43]. Besides, some other related features, i.e. time delay, of the inter-vehicle communication and their effects on advanced multiple vehicles motion control will be discussed in the rest of this Chapter.

7.3 Vehicle Platoon Control

Vehicle platoon control is a newly developing concept about vehicle longitudinal control. Because of its importance, a brief discussion is given here, but dynamics of individual vehicles are simplified to the maximum extent in such a discussion.

A vehicle platoon is a group of two or more closely spaced vehicles traveling with the same velocity in the same lane. A Platoon can have large interplatoon distances (i.e., 40 m) and small intraplatoon distance (i.e., 2 m). The first vehicle of a platoon is called the leader, while all other vehicles in the platoon are called followers; a single vehicle by itself in a platoon is viewed as a free-agent.

These are two main approaches in intelligent cruise control (ICC) to form platoon: autonomous intelligent cruise control (AICC) [44]-[46] and cooperative intelligent cruise control (CICC) [47]-[56]. AICC does not communicate with exteriors sources, the driver need to set the desired speed and headway. Cooperative intelligent cruise control (CICC) allows the vehicle to communicate with the other vehicles in its platoon, roadside monitors, roadway controllers and execute the following platoon maneuvers: joining (two platoons into one) and split (one platoon into two) which allows a vehicle to enter or exit the platoon).

AICC models use the velocity and acceleration of the vehicle and the vehicle immediately preceding it along with such factors as the vehicle's mass, wind drag and acceleration rate, and road traction. For instance, a simple AICC controller is proposed in [44]. As their approach and many other literals including [57]-[58], the dynamic models of vehicle are described by Newton's Second Law approximately in terms of the characteristics of the vehicle's engine dynamics as

$$\begin{cases} m\dot{v} = m\xi - K_d v^2 - d_m \\ \dot{\xi} = -\dfrac{\xi}{\tau(v)} + \dfrac{u}{m\tau(v)} \end{cases} \tag{7.1}$$

where $m\xi$ represents the engine force applied to the vehicle, $K_d v^2$ specifies the force generated by air resistance and d_m denotes the mechanical drag. $\tau(v)$ specifies the vehicle engine time constant at velocity v. m is the vehicle's mass, and u represents the throttle input to the vehicle engine.

Eq.(7.1) can be simplified for the i th vehicle, $i = 1, \ldots, N$, as

$$\dddot{x}_i = b_i(\dot{x}_i, \ddot{x}_i) + a_i(\dot{x}_i)u_i \tag{7.2}$$

where

$$a_i(\dot{x}_i) = \frac{1}{m_i \tau_i(\dot{x}_i)} \tag{7.3}$$

and

$$b_i(\dot{x}_i, \ddot{x}_i) = -\frac{1}{\tau_i(\dot{x}_i)}\left[\ddot{x}_i + \frac{K_{di}}{m_i}\dot{x}_i^2 + \frac{d_{mi}}{m_i}\right] - \frac{2K_{di}}{m_i}\dot{x}_i\ddot{x}_i a_i(\dot{x}_i)u_i \tag{7.4}$$

The following control law was proposed as

$$u_i(\dot{x}_i, \ddot{x}_i) = \frac{1}{a_i(\dot{x}_i)}\left[-b_i(\dot{x}_i, \ddot{x}_i) + c_i\right] \tag{7.5}$$

so that the closed loop dynamics of each vehicle satisfy

$$\dddot{x}_i = c_i \tag{7.6}$$

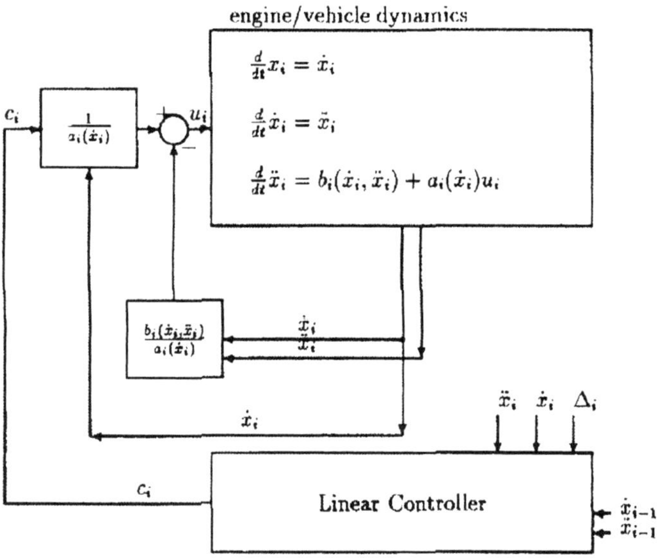

Fig.7.3 Linearized model of the i th vehicle with control input c_i, from [44]. (© [1993] IEEE)

The diagram of the proposed control law and the platoon are shown in Fig.7.3 and Fig.7.4(a) respectively, where the displacement of each gap is denoted as Δ_i. Thus, the platoon system dynamics can be mounted as Fig.7.4(b) if the control law (7.5) is used. In other words, if the system is stable, the transfer functions $h_{\Delta_i w_i}(s)$ or $g(s) = h_{\Delta_i \Delta_{i-1}}(s)$ between each displacement Δ_i should be dissipative.

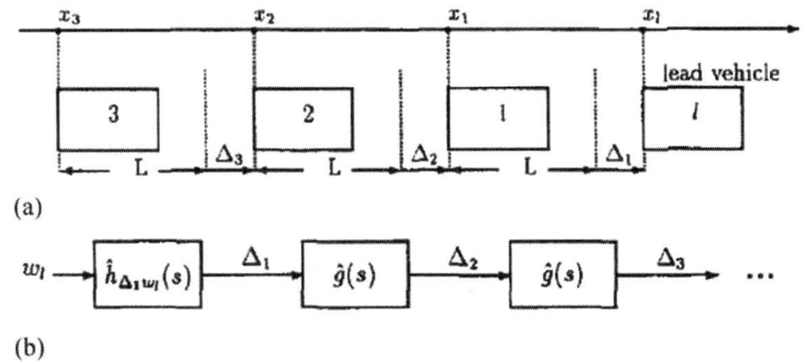

Fig.7.4 (a) diagram of a platoon of vehicles on a straight lane; (b) block diagram for a platoon of linearized vehicle models, from [44]. (© [1993] IEEE)

In [44], a three order linear system is carefully selected for $h_{\Delta_1 w_l}(s)$ and $g(s)$, so that the whole platoon system and each internal gap are dissipative. Fig.7.5 below shows a example of the resulting gap dynamics.

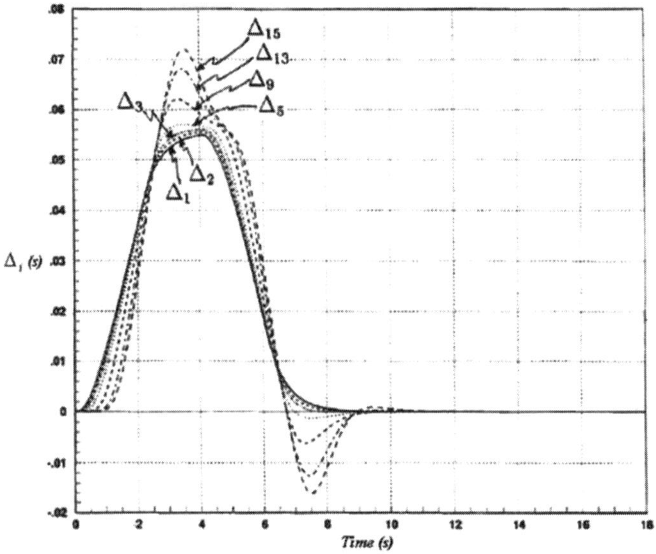

Fig.7.5 Block diagram for a platoon of linearized vehicle models, from [44]. (© [1993] IEEE)

From a different aspect, Swaroop and Rajagopal et. al considered the AICC problems with traffic flow stability as the so called slinky (accordion) effect or string instability problems [48], [54], [59]-[65]. Spacing oscillations propagate through a platoon causing it to stretch or contract length. Their major research method is to use longitudinal control to reducing the slinky effect. Besides, the delay of driver actions is discussed in [66], and the delay of [67].

It should be pointed out that most above approaches assumes that the vehicle longitudinal dynamics can be accurately estimated and well established. However, as discussed in *Chapter 4*, the nonlinearity and parameter uncertainties of the vehicle longitudinal motion control cannot be simply handled sometimes. Thus, fuzzy controller [68]-[70], adaptive controller [71] and some other controllers [72]-[73] were also proposed for vehicle platoon control recently. The steering behaviors of the vehicles and their effects on platoon stability also gained notable attentions [63], [74]-[82].

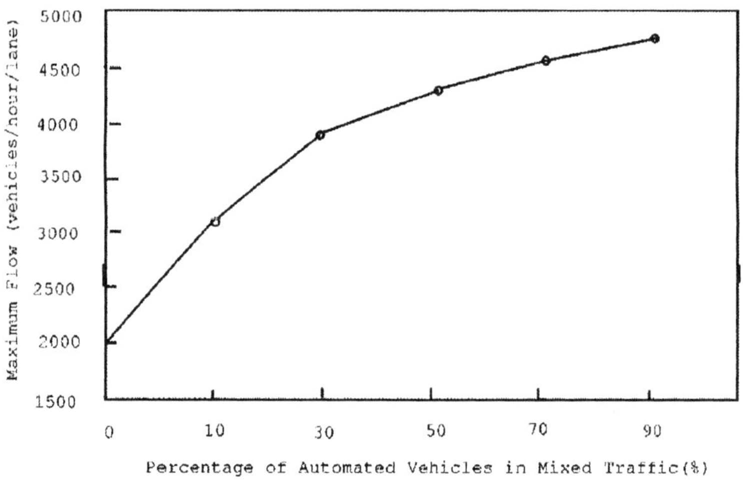

Fig.7.6 Flow grows as the percentage of automated vehicles increases in mixed traffic, from [83]. (© [2000] IEEE)

Different from AICC, CICC links each vehicle via platoon communicating with the vehicle immediately before and after it and with the lead vehicle. For instance, in [64], Rajamani and Shladover tested several controllers for a platoon of eight Buick LeSabres on an eight mile section of I-15 in San Diego. It was reported that "dramatic improvements in the trade-off between ride quality and spacing accuracy can be obtained merely by replacing radar range rate in the autonomous control algorithm with the difference between the measured velocities of the two cars (a rudimentary form of co-operation)." When keeping an inter-vehicle spacing of 30 meters at highway speeds, a theoretical maximum traffic flow was shown to be about 3000 vehicles per hour using AICC and 6400 vehicles per hour using CICC. In [83], Huang, Ren and Chan showed that vehicle flow can be increased from a maximum of 2000 vehicle per hour per lane without autonomous vehicles to a maximum of 5000 vehicle per hour per lane with 90% of the vehicle with CICC. Fig.7.6 shows the simulation results based on averages of both lanes. The first thing to note is the increase in the maximum sustainable flow as the proportion of AVDS-equipped vehicles increases. In addition, platoon with higher proportion of AVDS-equipped vehicles tends to be more stable. Besides, one manifestation of the increased traffic stability is that the traffic speed varies less with vehicle density when the proportion of automated vehicles is high.

Besides, the delay in communication and its effect on CICC were studied in [84]. The fault toleration CICC becomes a hot topic in the new century [85]-[88]. But all these approaches still need further examinations.

7.4 Lane Changing and Lane Merging Control

7.4.1 Vehicle Lane Changing Control

Observation reveals that experienced human drivers normally divide the lane changing process into three steps subconsciously. First, he/she is engaged in information gathering and decision making to determine whether and when the conditions are permutable for a lane change. When he/she thinks that lane changing can be successfully completed, the driver will notify other vehicles of their intent through certain signals, i.e., turning lights, thumb gestures. Finally, he/she changes to another lane.

However, if the driver failed to collect critical information or failed to provide a signal, or the other drivers failed to notice or take wrong action, accidence may occur due to too short spacing left. Therefore, numerous lane changing systems were proposed to increase the safety level of lane changing [89]-[97].

The first problem that needs to be addressed is how to calculate the spacing required for safe lane changing. Several different disciplines to estimate the required space were studied nowadays. Basically, if the driver has no information about the other vehicles' decisions and braking capabilities, it has to make the worst-case assumptions to allow for a safety margin that is long enough [96]. Otherwise, inter-vehicle communication will significantly reduced the required safe space since each vehicle involved can be informed of the other vehicles' dynamic capabilities, [93].

In [109], Kanaris, Kosmatopoulos and Ioannou thought that a spacing has to be kept at least equal to the sum of the required safety distance between itself and the lane changing vehicle, associated with the length of the merging vehicle. Since the relative speed of the vehicle that intends to change lane may greatly vary during the lane changing, it is difficult to accurately estimate the required safety distance.

In [98], Bascunna discussed the conditions for safe and unsafe lane changing. The conditions were obtained by assuming only two vehicles, A and B, involved in the lane changing maneuver. Initially the vehicles are traveling in two neighboring lanes and vehicle A changes lane to another lane. To make the problem more tractable, He just considered the following four cases:

(1) the initial longitudinal velocity of vehicle A is less than that of vehicle B, and vehicle A intends to complete the lane changing with constant longitudinal velocity, and then follows vehicle B;

(2) the initial longitudinal velocity of vehicle A is less than that of vehicle B, and vehicle A intends to complete the lane changing by applying constant longitudinal acceleration and leads vehicle B;

(3) the initial longitudinal velocity of vehicle A is greater than that of vehicle B, and vehicle A intends to complete the lane changing with constant longitudinal velocity and leads vehicle B;

(4) the initial longitudinal velocity of vehicle A is greater than vehicle B, and vehicle A intends to complete the lane changing with constant longitudinal deceleration and then follows vehicle B.

In [99], Jula, Kosmatopoulos and Ioannou extended these cases as follows. They assumed the lane changing vehicle M moves from its current position between the vehicles L_o and F_o to a new position between the vehicles L_d and F_d in the neighboring lane, see Fig.7.7.

Without losing too much generality, Jula, Kosmatopoulos and Ioannou 2000 assumed that vehicle M starts the lane-changing maneuver at time $t = 0$. This maneuver consists of two parts. Initially, it adjusts the longitudinal velocity and spacing for a time-interval t_{adj}, and then applies lateral acceleration to merge to the destination lane. In other words, t_{adj} is the required time for the merging vehicle to adjust its longitudinal position and velocity before it starts merging to the destination lane.

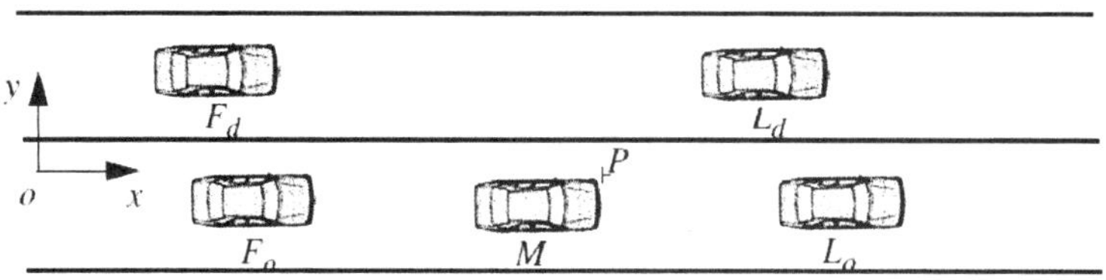

Fig.7.7 Pre-lane changing configuration showing position of merging vehicle M, from [99]. (© [2000] IEEE)

To measure the lateral and longitudinal positions of the vehicles involved in the maneuver, an arbitrary origin O is selected. The axis is directed toward the destination lane and the axis is aligned with the lateral side of the merging vehicle which is closer to the destination lane. The origin O and the axes $x-y$, in Fig.7.7 were assumed to be fixed till the end of maneuver.

Moreover, the longitudinal acceleration/deceleration, the longitudinal velocity, the longitudinal position, and the lateral position of vehicle were denoted by $a_i(t)$, $v_i(t)$, $x_i(t)$ and $y_i(t)$ respectively, where $i \in \{L_d, F_d, F_o, L_o, M\}$. $x_i(t)$ and $y_i(t)$ denotes the longitudinal and lateral distances between the front-left corner of the vehicle and O.

A "lane change crash" occurs when the vehicle attempts to change its lane and strikes or is struck by a vehicle in the adjacent lane. The following model (7.7) was considered as an accurate model for many simple lane change/merge maneuvers, i.e. [47].

$$a_{lat}(t) = \begin{cases} \dfrac{2\pi H}{t_{lat}^2} \cdot \sin\left(\dfrac{2\pi}{t_{lat}}(t - t_{adj})\right), & t_{adj} \le t \le t_{adj} + t_{lat} \\ 0 \end{cases} \tag{7.7}$$

where $a_{lat}(t)$ is the lateral acceleration of the vehicle that is used to complete the lane change maneuver. H is the total lateral displacement for the merging vehicle, t_{adj} here is the time elapse before lateral acceleration applies, and t_{lat} is the total time for the lane change.

Noting that the lateral acceleration $a_{lat}(t)$ is positive within the first half of the lateral displacement, and negative in the second half, it is easy to compute the lateral velocity $v_{lat}(t)$, and lateral position $y_{lat}(t)$ of the front-left corner of the merging vehicle M given $a_{lat}(t)$.

Using the lane change model described above, Jula, Kosmatopoulos and Ioannou analyzed the longitudinal acceleration profiles of the five vehicles in Fig.7.7 to find the initial minimum longitudinal spacing between M and each other vehicle such that during a specified time-interval $[0,T]$, no collision of any type occurs. The length of the time interval T is the time under consideration. In all cases, vehicle M is assumed to start its lane changing maneuver at $t=0$ by adjusting its longitudinal position, and applying lateral movement at $t=t_{adj}$ according to Eq.(7.7).

For example, the minimum longitudinal safety spacing between M and L_d can be determined as what is shown in Fig.7.8. Here S is the initial lateral distance between the upper side of the vehicle M and the lower side of vehicle L_d. Since the leading vehicle remains in the destination lane, an angle and/or a side-wipe collision may occur as passes the line LS in Fig.7.8, which is the tangent to the lower side of the leading vehicle L_d. The front-left corner of M is the first point of vehicle M which passes the line LS at the point C. Since the lateral acceleration of the leading vehicle L_d is zero, the lateral position of vehicle L_d is constant.

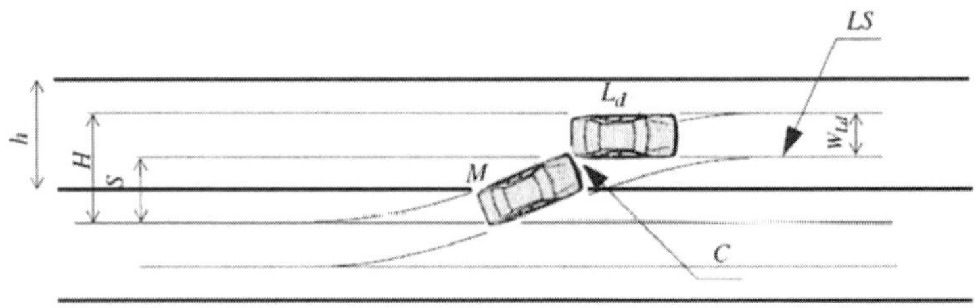

Fig.7.8 The marginal collision point between the vehicle M and the leading vehicle L_d, from [99]. (© [2000] IEEE)

Let $t_{adj}+t_C$ be the time-instant at which the front-left corner of the merging vehicle is at the point C in Fig.7.8. Its value can be found by solving

$$y_{lat}(t) = S = y_{Ld} - w_{Ld} \tag{7.8}$$

where w_{Ld} is the width of the leading vehicle L_d. By taking all types of collisions mentioned above into account, the condition of collision free between M and L_d can be given as

$$x_M(t) < x_M(t) - l_{Ld} - w_M \cdot \sin(\theta(t)),\ \forall t \in [t_{adj}+t_C, T] \tag{7.9}$$

where l_{Ld} is the length of leading vehicle L_d, w_M is the width of vehicle M, and $\theta(t)$ is the angle between the tangent of the lane changing trajectory at the point $y_{lat}(t)$ and the horizontal axis.

Here, $w_M \cdot \sin(\theta(t))$ is to prevent any angle collision between any point on the front bumper of the vehicle M and the rear-right corner of leading vehicle L_d in the time interval $[t_{adj} + t_C, t_{lat} + t_C]$. More precisely

$$\tan(\theta(t)) = \frac{\partial y_{lat}(t)}{\partial x_M(t)} = \frac{\partial y_{lat}(t)/t}{\partial x_M(t)/t} = \frac{v_{lat}(t)}{v_M(t)} \tag{7.10}$$

Define $l_{L1} = l_{Ld} + w_m \cdot \sin(\theta(t_C + t_{adj}))$, Eq.(7.10) can be further simplified for $\forall t \in [t_{adj} + t_C, T]$ as

$$x_M(t) < x_{Ld}(t) - l_{L1} \tag{7.11}$$

Let $S_r(t)$ be the longitudinal spacing between point P of vehicle M and the rear end (bumper) of vehicle L_d. Since $y_{Ld}(t) = 0$, for $\forall t \in [t_{adj} + t_C, T]$, it has

$$S_r(t) = x_{Ld}(t) - l_{L1} - x_M(t) \tag{7.12}$$

Obviously, if the longitudinal spacing is greater than zero, i.e., $S_r(t) > 0$ for $t \ge t_{adj} + t_C$, no collision will occur during the lane-changing maneuver. Combine (7.11) and (7.12), for $\forall t \in [t_{adj} + t_C, T]$, it has

$$S_r(t) = \left(S_r(0) + \int_0^t \int_0^\lambda (a_{Ld}(\tau) - a_M(\tau)) d\tau d\lambda + (v_{Ld}(0) - v_M(0))t \right) > 0 \tag{7.13}$$

where $S_r(0) = x_{Ld}(t) - l_{L1} - x_M(0)$.

The objective is to find the initial minimum value of $S_r(0)$ that guarantee collision free between the leading vehicle L_{Ld} and the vehicle M. Therefore the minimum value of $S_r(0)$ should be the minimum initial longitudinal relative space $MSS(L_d, M)$, which is calculated using (7.12) for $\forall t \in [t_{adj} + t_C, T]$ as:

$$MSS(L_d, M) = \underset{t}{Max}\left(\int_0^t \int_0^\lambda (a_M(\tau) - a_{Ld}(\tau)) d\tau d\lambda + (v_M(0) - v_{Ld}(0))t \right) \tag{7.14}$$

Similarly, the minimum longitudinal safety spacing between M and F_d should be

$$MSS(F_d, M) = \underset{t}{Max}\left(\int_0^t\int_0^\lambda (a_{Fd}(\tau) - a_M(\tau))d\tau d\lambda + (v_{Fd}(0) - v_{Md}(0))t\right) \tag{7.15}$$

For $\forall t \in [0, t_{adj} + t_C]$, the minimum longitudinal safety spacing between M and L_o as

$$MSS(L_o, M) = \underset{t}{Max}\left(\int_0^t\int_0^\lambda (a_M(\tau) - a_{Lo}(\tau))d\tau d\lambda + (v_M(0) - v_{Lo}(0))t\right) \tag{7.16}$$

and for $\forall t \in [0, t_{adj} + t_C]$, the minimum longitudinal safety spacing between M and F_o as

$$MSS(F_o, M) = \underset{t}{Max}\left(\int_0^t\int_0^\lambda (a_{Fo}(\tau) - a_M(\tau))d\tau d\lambda + (v_{Fo}(0) - v_{Md}(0))t\right) \tag{7.17}$$

In [99], Jula, Kosmatopoulos and Ioannou examined two scenarios based on the obtained solution above. In the first scenario, vehicle M performs lane changing with constant longitudinal velocity.

The above solutions are then revalued as

$$MSS(L_d, M) = \begin{cases} (v_M - v_{Ld}) \cdot T, & \text{if } v_M - v_{Ld} \geq 0 \\ (v_M - v_{Ld}) \cdot (t_{adj} + t_C), & \text{otherwise} \end{cases} \tag{7.18}$$

and

$$MSS(F_d, M) = \begin{cases} (v_{Fd} - v_M) \cdot T, & \text{if } v_{Fd} - v_M \geq 0 \\ (v_{Fd} - v_M) \cdot (t_{adj} + t_C), & \text{otherwise} \end{cases} \tag{7.19}$$

and

$$MSS(L_0, M) = \begin{cases} (v_M - v_{Lo}) \cdot (t_{adj} + t_C), & \text{if } v_M - v_{Lo} \geq 0 \\ 0, & \text{otherwise} \end{cases} \tag{7.20}$$

and

$$MSS(F_o, M) = \begin{cases} (v_{Fo} - v_M) \cdot (t_{adj} + t_C), & \text{if } v_{Fo} - v_M \geq 0 \\ 0, & \text{otherwise} \end{cases} \tag{7.21}$$

Simulations were used to further demonstrate the conclusion of (7.14)-(7.15) in reference of Jula etc. 2000. In these simulations, the time was set to $T = 50$ s, the adjustment time was $t_{adj} = 0$ s, the lateral time was $t_{lat} = 5$ s, and the lateral displacement was $H = 12$ feet.

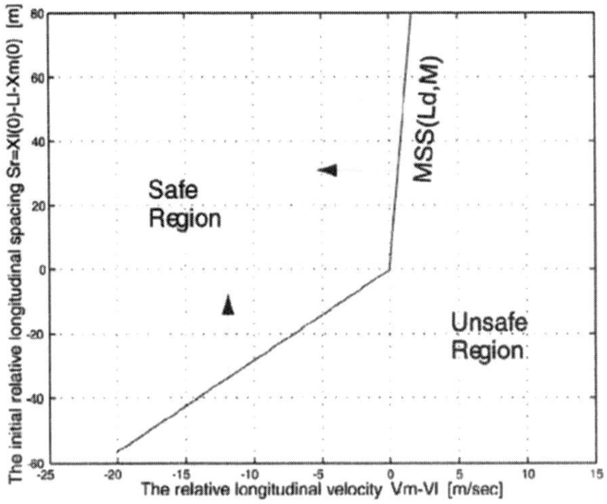

Fig.7.9 The collision region between the vehicles L_d and M where M is assumed to have constant velocity, from [99]. (© [2000] IEEE)

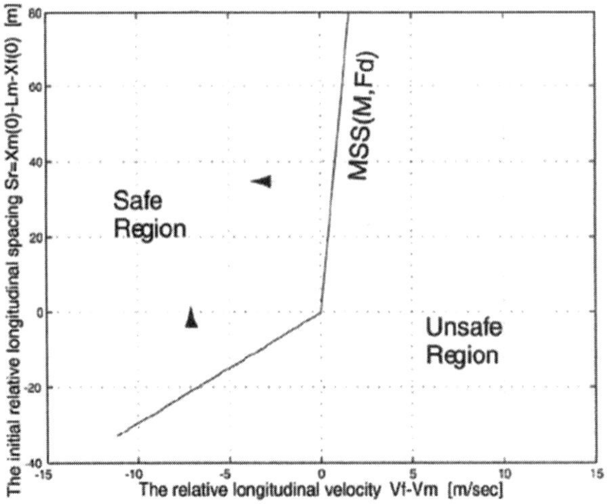

Fig.7.10 The collision region between the vehicles F_d and M where M is assumed to have constant velocity, from [99]. (© [2000] IEEE)

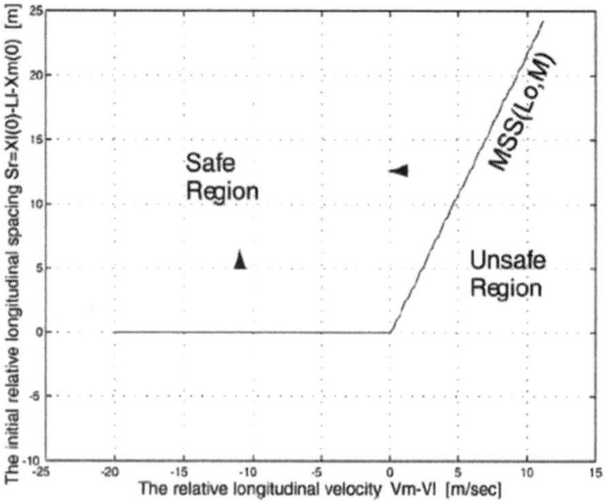

Fig.7.11 The collision region between the vehicles L_o and M where M is assumed to have constant velocity, from [99]. (© [2000] IEEE)

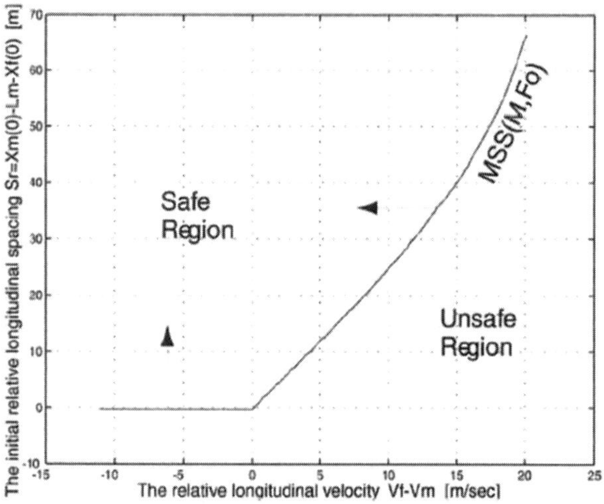

Fig.7.12 The collision region between the vehicles F_o and M where M is assumed to have constant velocity, from [99]. (© [2000] IEEE)

Fig.7.9-Fig.7.12 show the initial relative longitudinal spaces versus the relative longitudinal velocity between M and the other four vehicles involved in the lane-changing maneuver. The solid lines (which will be called thereafter safety margins) in these figures represents the margins between safe and unsafe lane changing regions.

In the second scenario, the lane changing strategy was proposed by Kanaris, Kosmatopoulos and Ioannou in [109], which divides the lane change process into two separated process: the longitudinal acceleration process and the lateral acceleration process.

Let V_d and V_o denote the average velocity of destination and originating lanes, respectively. Two longitudinal acceleration profiles were designed in [109] as follows. When $V_d > V_o$, i.e. in the case where the vehicle speed in the destination lane is higher than that in the originating one, the longitudinal acceleration profile is constructed as Fig.7.13. The lane changing vehicle's acceleration initially decreases linearly with respect to time until it reaches a limit $-a_{comf}$, where is appropriately chosen to maintain safety and comfort of the passengers in the vehicle. Then, the acceleration remains constant $-a_{comf}$ and equal to until a sufficient spacing has been created in the originating lane and then it switches from decelerating to accelerating. The acceleration starts increasing linearly until it reaches the positive acceleration limit a_{comf}. This acceleration rate remains constant until the lane changing vehicle's velocity equals to V_d.

The constant t_{ch} in Fig.7.13 denotes the time-instant at which the merging vehicle switches from decelerating to accelerating, while the constant t_{long} is such that the longitudinal velocity of the merging vehicle equals to the vehicle speed V_d in the destination lane.

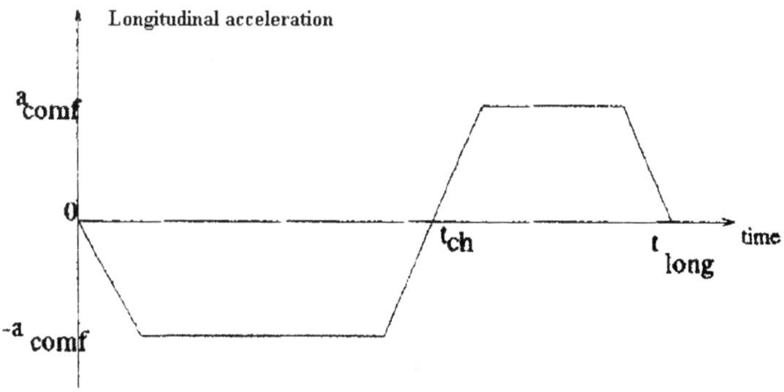

Fig.7.13 Longitudinal acceleration profile in the case where the vehicle speed in destination lane is higher than that in the originating one, from [109]. (© [2001] IEEE)

When $V_d \le V_o$, i.e. in the case where the vehicle speed in the destination lane is lower than that in the originating one, the longitudinal acceleration profile is constructed as what is shown in Fig.7.14. Similar to the previous case, the vehicle's acceleration initially decreases linearly with respect to time until it reaches a limit $-a_{comf}$, where a_{comf} is appropriately chosen to maintain safety and comfort of the passengers in the vehicle. Then, the acceleration remains constant and equal to $-a_{comf}$ until both a sufficient spacing has been created in the originating lane and the merging vehicle's velocity equals to V_d. When both the spacing in the originating lane guarantees a safe and collision-free lane changing maneuver and the velocity of the merging vehicle is equal to V_d, the acceleration is linearly increased to zero.

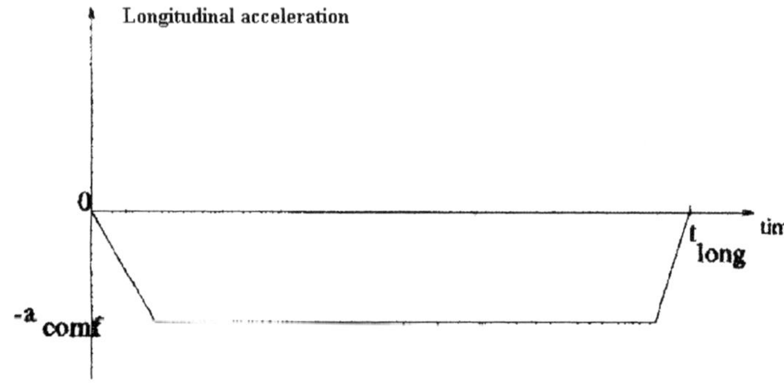

Fig.7.14 Longitudinal acceleration profile in the case where vehicles in the destination lane move slower than in the originating one, from [109]. (© [2001] IEEE)

For the case $t_{adj} = 0$, the proposed lateral acceleration model for the vehicle was still (7.7). Therefore, the value of the longitudinal acceleration of the vehicle M can be written as follows:

$$a_M = \begin{cases} \dfrac{v_{Ld} - v_M(0)}{t_{long}}, & t \le t_{long} \\ 0, & otherwise \end{cases} \tag{7.22}$$

According to (7.13), the condition for collision free is

$$S_r(t) = \left(S_r(0) + (v_{Ld}(0) - v_M(0)) \cdot (t - \frac{t^2}{2t_{long}}) \right) > 0, \ \forall t \in [t_{long}, T] \tag{7.23}$$

Thus, the minimum initial longitudinal relative space $MSS(L_d, M)$ should be

$$MSS(L_d, M) = \begin{cases} (v_M(0) - v_{Ld}) \cdot t_{long} / 2, & if \quad v_M(0) - v_{Ld} \geq 0 \\ (v_M(0) - v_{Ld}) \cdot t_C, & otherwise \end{cases} \quad (7.24)$$

Assume that $t_{long} = 10$ s, the safe and unsafe regions between vehicles L_d and M are shown in Fig.7.15. For positive relative velocity $v_M(0) - v_{Ld}$, the safety margin corresponds to a line with slope equal to $t_{long} / 2$; while for negative relative velocities, it is a line with tangent equal to 2.8, which is indeed the value of $t_{adj} + t_C$ in Eq.(7.8).

Comparison between Fig.7.9 and Fig.7.15 indicates that the safe region has been expanded. Therefore the switching longitudinal acceleration policy with $t_{adj} = 0$ is more reliable than the constant longitudinal velocity policy for the case of vehicles and L_d.

For the case $t_{adj} > 0$, let's define the following state space variables:

$$\begin{cases} x_1 = x_{Ld} - x_M - l_{Ld} \\ x_2 = v_M - v_{Ld} \end{cases} \quad (7.25)$$

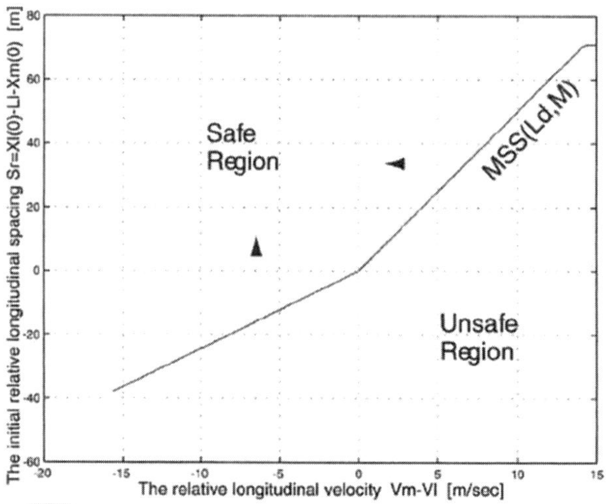

Fig.7.15 The collision region between the vehicles L_d and M where M is assumed to have switching longitudinal acceleration, from [99]. (© [2000] IEEE)

By differentiating (7.25), it has

$$\begin{cases} \dot{x}_1 = \dot{x}_{Ld} - \dot{x}_M = v_{Ld} - v_M = -x_2 \\ \dot{x}_2 = \dot{v}_M = a_{adj} \end{cases} \tag{7.26}$$

Solving (7.26) yields the isoclines as

$$x_1 = -\frac{x_2^2}{2a_{adj}} + c \tag{7.27}$$

where the constant c is the integration constant which depends on the initial values $x_1(0)$ and $x_2(0)$.

Fig.7.16 shows the isoclines that correspond to different values of a_{adj} for the simulation in Fig.7.15. The initial state (initial relative spacing and velocity) has been chosen to be in the unsafe region. Applying negative a_{adj}, it is possible to move into the safe region in order to start the merging maneuver. The larger the absolute value of a_{adj} is, the faster it moves into the safe region. The minimum value of t_{adj} for each is determined by the point of intersection between the corresponding isoclines curves of a_{adj} and the safety margin in Fig.7.16.

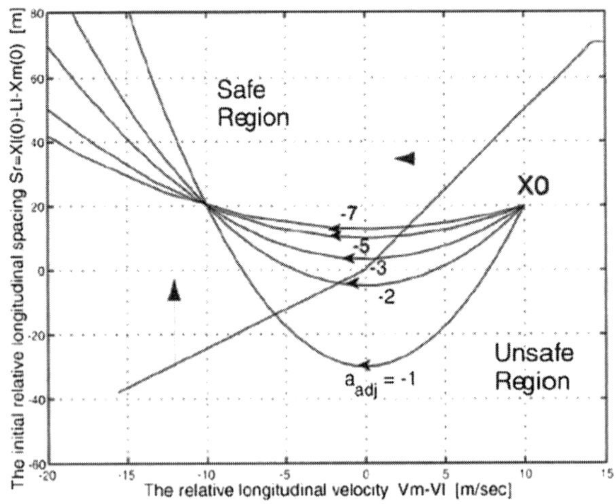

Fig.7.16 The collision region between the vehicles L_d and M where a_{adj} is applied a to move from unsafe area into safe area, from [99]. (© [2000] IEEE)

There were some other approaches that focused on the acceleration profiles design. For example in [95], Hatipoglu, Ozguner and Unyelioglu considered time optimal open loop lane changing profile design with nonlinear constraints on the design parameters. Recently, researchers [80] also tried to incorporate both acceleration profile design and vehicle lateral dynamic control models together. Constrained by the book length, they will not be discussed in details here.

7.4.2 Vehicle Lane Merging Control

Vehicle merging is another important multiple vehicles cooperating situation [100]-[111]. For simplicity, Uno, Sakaguchi and Tsugawa assumed in [100] that all vehicles are moving at the same speed and every vehicle has the same dynamics, see Fig.7.17-Fig.7.18.

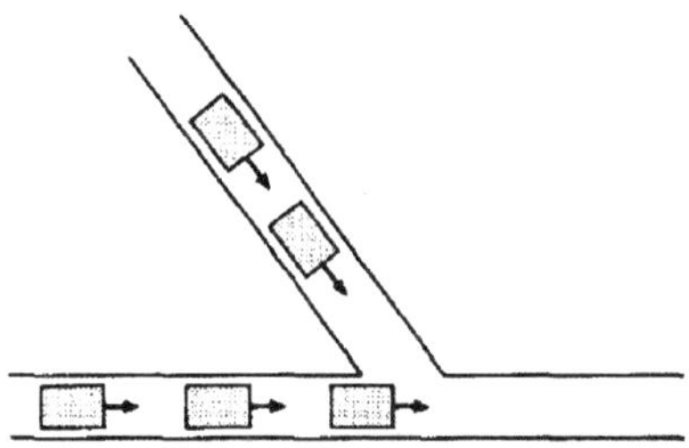

Fig.7.17 Lane merging at a ramp, from [100]. (© [1999] IEEE)

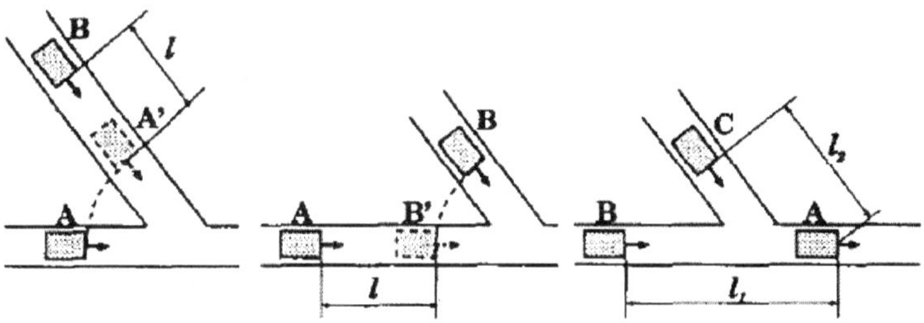

Fig.7.18 Classification of merging and generation of a virtual vehicle at a ramp, from [100]. (© [1999] IEEE)

As shown in Fig.7.18, the lane merging was classified in [100] to three cases as:

(1) a vehicle on a sub lane merges after the last vehicle in a platoon on a main lane;

(2) a vehicle on a sub lane merges before the first vehicle in a platoon on a main lane;

(3) a vehicle on a sub lane merges between two vehicles in a platoon on a main lane.

Here, merging of platoons is regarded as a series of merging a single vehicle. To clearly describe the merging control algorithm, Uno, Sakaguchi and Tsugawa proposed the concept of virtual vehicle and also a data transmission algorithm for inter-vehicle communication. The details of the proposed communication data are not mentioned here, since they have little to do with the main topic of this section.

From a longitudinal control point of view, there is no difference between the merging at a ramp and that on lane changing. Generation process of the virtual vehicles can be formulated as follows:

(1) in this case, when vehicle B is to be positioned at the back the vehicle A after merging, it must make a headway L at present. And vehicle B need to generate the virtual vehicle A' by mapping vehicle A into its own lane using the data transmitted to it over inter-vehicle communication. the longitudinal control procedure will be performed between virtual vehicle A' and vehicle B to simulate a preceding vehicle and a following one;

(2) in this case, when vehicle B is to be positioned in front of vehicle A after merging, it is mapped onto a main lane or the next lane as virtual vehicle B' which will be taken as the preceding vehicle for vehicle a on longitudinal control;

(3) in this case, vehicle A and C will both set up inter-vehicle communication links between vehicle B. In terms of the interaction between vehicle A and C, the merging of vehicle C after A is the same as in Case (a). In terms of the interaction between vehicle A and B, the longitudinal control will eventually make the headway $2L$.

Under such formulations, the lane merge problem will be equivalent to the lane change problem shown in Fig.7.19, in which the ramp joins the highways at 30 degrees. Acceleration of each vehicle is assumed to be bounded. Results indicate that it is possible to merge the two platoons smoothly using the proposed algorithms.

In [101], Kachroo and Li addressed the vehicle merging dynamics and provided three control guidance laws, namely, linear, optimal and parabolic speed profiles to describe the desired behaviors of the merging vehicle based on the merging quality and the merging vehicle. On the basis of it, they build both longitudinal and lateral sliding mode controllers. They claimed that the applied controllers were capable to handle the nonlinear and model uncertainties of the vehicle dynamics. Different from them, Park and Ryu designed in [104] a special Neuro-Fuzzy controller to guide lane merging process. They showed some encouraging simulation results which indicate the feasibility of this method. Besides the control algorithms, the side view radar and corresponding visual sensing problems, i.e. vehicle and lane detection, platoon velocity estimation, also achieved careful discussions recently. However, constrained by the length of this book, they are skipped here.

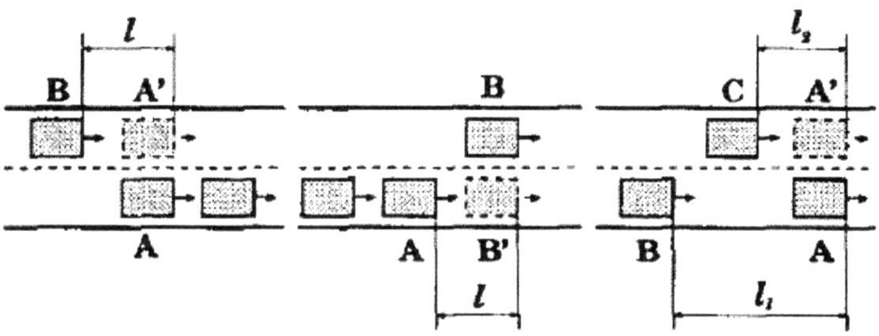

Fig.7.19 Classification of merging and generation of a virtual vehicle at a ramp, from [100]. (© [1999] IEEE)

To carefully compare adaptive cruise control without communication (ACCWC) and cooperative adaptive cruise control (CACC), simulation tests were carried out in [110]. Fig.7.20(a) and Fig.7.20(b) are the trajectories of vehicles in the main lane between 900 second and 1000 second for ACCWC and CACC highway simulation. Each curve corresponds to the trajectory of one vehicle. The x-axis is time in seconds and the y-axis is the position in meters. The merge-in point is at 510 meters, and the horizontal line at 510 meter represents queued vehicles. Merging vehicles can be identified by curves which lie entirely above this line. A shock wave propagating upstream can be clearly found in Fig.7.20(a), i.e. in the opposite direction of the traffic. However, this shock wave is smoothed in Fig.7.20(b).

Fig.7.21 shows the range and range rate in braking scenario using different lane merging control strategy. It evidently shows: the more ACC vehicles in the highway, the smaller the average braking effort is. With the same percentage of ACC vehicles, CACC performs better than ACC in saving the braking effort.

Fig.7.22 shows the length of the queue of the waiting to merge vehicles in the merge-in lane from 600 to 1000 second. Results for four cases are plotted here. Clearly, the larger the percentage of ACC (CACC) vehicles, the shorter the waiting queue. For the same percentage, the queue in CACC highway is up to 5 vehicles shorter than in ACC highway. Due to the communication, the relevant main lane vehicles have longer time to make a good gap in front.

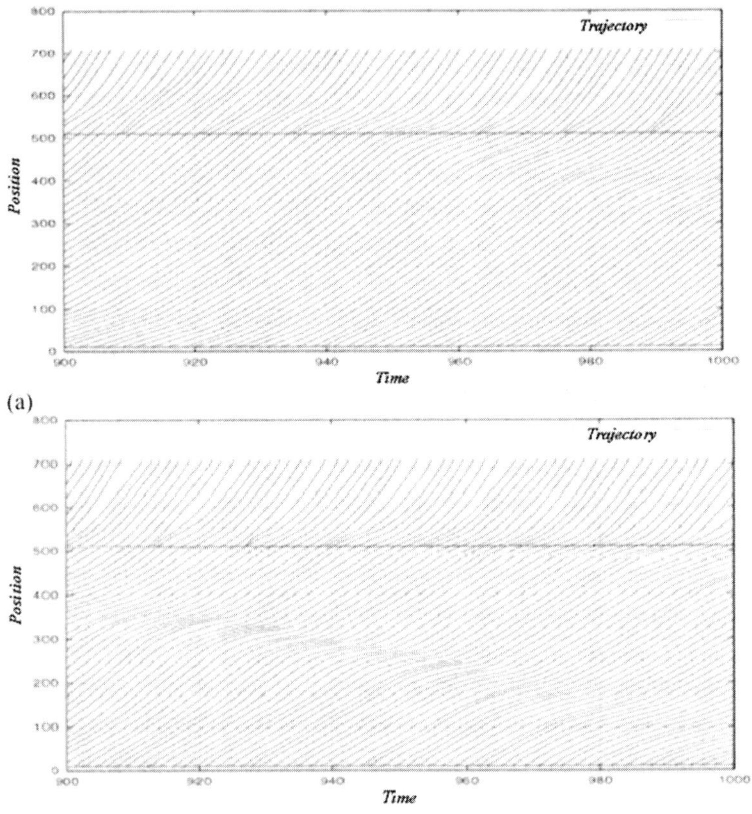

Fig.7.20 Trajectories of main lane vehicles in: (a) ACC highway merging and (b) CACC highway merging, from [110]. (© [2002] IEEE)

Fig.7.22 shows the length of the queue of the waiting to merge vehicles in the merge-in lane from 600 to 1000 second. Results for four cases are plotted here. Clearly, the larger the percentage of ACC (CACC) vehicles, the shorter the waiting queue. For the same percentage, the queue in CACC highway is up to 5 vehicles shorter than in ACC highway. Due to the communication, the relevant main lane vehicles have longer time to make a good gap in front.

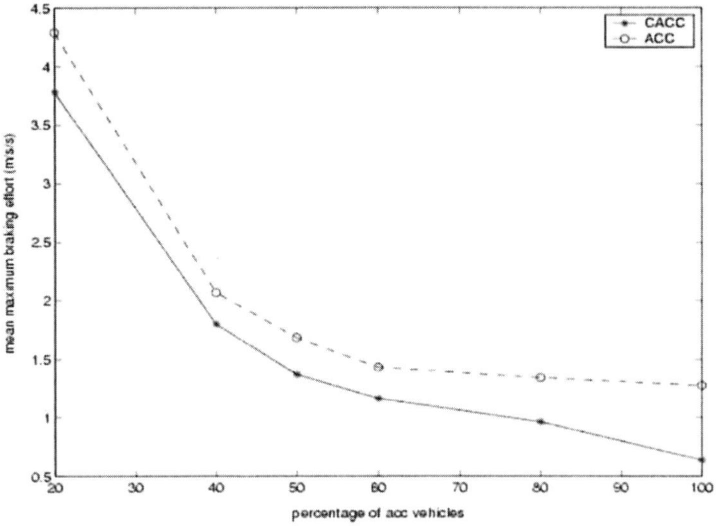

Fig.7.21 Average braking effort for different market penetration of ACC/CACC in Highway merging simulation, from [110]. (© [2002] IEEE)

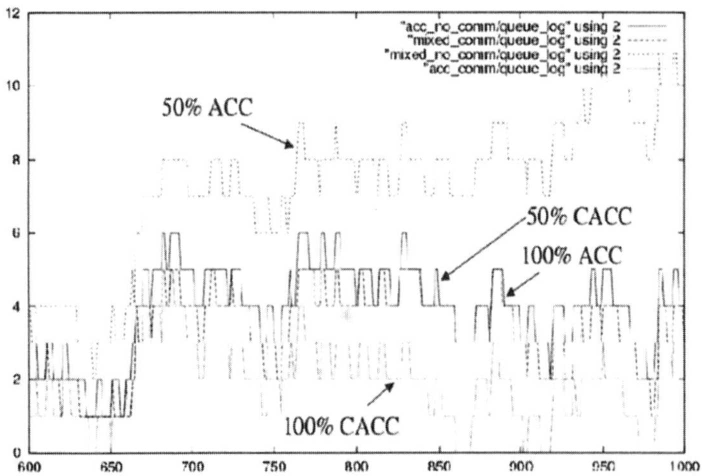

Fig.7.22 Queue length of waiting-for-merge in Highway merging simulation, from [110]. (© [2002] IEEE)

As pointed out in [110], the reason for the shown differences lies in two facts. "First is that the relevant main lane vehicle receives the warning message well in advance, thus it can brake gently to increase the gap to preceding vehicle. Second is when the merging vehicle enters the main lane, the gap it is in is already large enough for safety, and the main lane vehicles behind it do not need to brake hard, as they have to do in ACC scheme." "Therefore in average the merging cars wait shorter in CACC highway, and the queue behind it is shorter if exists at all."

7.5 Cooperative Driving at Intersections

7.5.1 Uncooperative Driving at Intersections

In the above driving scenarios, all the vehicles are assumed to move in the same or completely opposite direction. However, collision risk may also be raised by other vehicles which are moving in explicit different but not opposite directions [112]-[118]. Among those situations, collision avoidance in road crossing and lane merging were specially discussed.

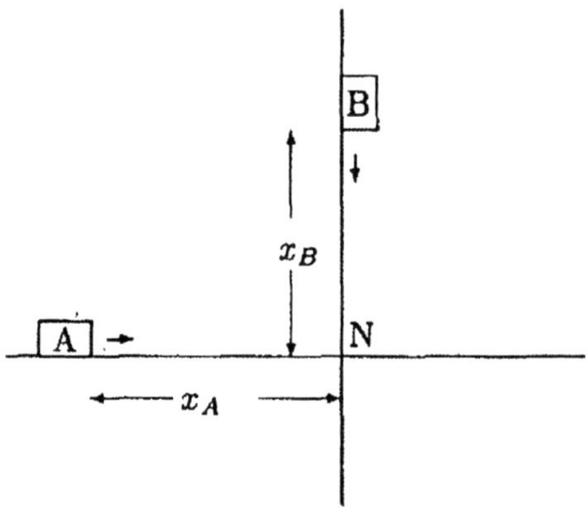

Fig.7.23 Diagram of road junction and vehicle movement, from [112]. (© [2000] IEEE)

In [112], Arora, Raina and Mitta investigated the collision free control strategy for a typical road junction without traffic lights; see Fig.7.23. Such road junctions are very popular in the urban areas. In Fig.7.23, only two vehicles are considered. It is assumed that the two roads are perpendicular to each other. Each vehicle is modeled by state space model at the continuous level. It was assumed that no communication between the vehicles and thus the action of one vehicle is "disturbance" to the neighboring vehicle. And the design objective is to generate an acceptable control law for the control inputs, mainly refer to the acceleration/ deceleration rate here, which will give the required system performance despite the possible "disturbance".

Suppose only vehicle A is controllable here. In other words, the movement of vehicle B is unknown. The control law of A should guarantee the safety for the worst possible actions of vehicle B. It may undergo varied velocity changes. But its velocity can only change continuously within the permissible range.

Arora, Raina and Mittal denoted the vehicle dynamics through x_A and x_B which represents their positions in the pre-selected world coordinates. Their velocities were constrained by

$$\dot{x}_A \in [\dot{x}_{A\min}, \dot{x}_{A\max}],\ \dot{x}_B \in [\dot{x}_{B\min}, \dot{x}_{B\max}] \tag{7.28}$$

and

$$\ddot{x}_A \in [\ddot{x}_{A\min}, \ddot{x}_{A\max}],\ \ddot{x}_B \in [\ddot{x}_{B\min}, \ddot{x}_{B\max}] \tag{7.29}$$

where $\dot{x}_A$ and $\dot{x}_B$ represents the speed of vehicle A and B respectively. $\ddot{x}_A$ and $\ddot{x}_B$ are the acceleration rates respectively.

To make the problem more simply, $\ddot{x}_A$ was written as the control input $u(t)$ and $\dot{x}_B$ was treated as disturbance $d(t)$ directly in [112]. Then, it leads to the general sate space model as

$$\dot{x} = \begin{bmatrix} 0 & 1 & 0 \\ 0 & 0 & 0 \\ 0 & -1 & 0 \end{bmatrix} x + \begin{bmatrix} x_A \\ \dot{x}_A \\ r \end{bmatrix} \ddot{x}_A + \begin{bmatrix} x_A \\ \dot{x}_A \\ r \end{bmatrix} \dot{x}_B = Ax + Bu + Dd \tag{7.30}$$

where $x = \begin{bmatrix} x_1 \\ x_2 \\ x_3 \end{bmatrix} = \begin{bmatrix} x_A \\ \dot{x}_A \\ r \end{bmatrix} \in R^3$ and $r = x_B - x_A$.

For the collision avoidance, it requires that the relative distance r between the two vehicles is never less than a safe value. Notice that

$$x(t) = e^{At} x^0 + \int_0^t e^{A(t-\tau)} Bu(\tau) d\tau + \int_0^t e^{A(t-\tau)} Dd(\tau) d\tau \tag{7.31}$$

where x^0 denotes the initial state of $x(t)$.

Simplifying (7.31) leads to

$$x(t)=\begin{bmatrix} x_1^0+tx_2^0 \\ x_2^0 \\ -tx_2^0+x_3^0 \end{bmatrix}+\int_0^t\left\{\begin{bmatrix} t-\tau \\ 1 \\ -(t-\tau) \end{bmatrix}u(\tau)+\begin{bmatrix} 0 \\ 0 \\ 1 \end{bmatrix}d(\tau)\right\}d\tau \tag{7.32}$$

Based on (7.32), the problem can be tackled as two player zero sum game problem, where one player i.e. the control action of vehicle A competes with the other player i.e. velocity disturbance in vehicle B. The safety criterion can be expressed as the cost function:

$$J(x^0,u,d)=|x_3(T)| \tag{7.33}$$

where T is the time when vehicle A reaches the junction area. Thus, the safety requirement is encoded by:

$$J(x^0,u,d)>C=|x_3(T)| \tag{7.34}$$

The safe gap requirement and the length of the vehicles is taken into account by selecting an appropriate positive value for C. Using Game Theory, Arora, Raina and Mittal built two saddle strategies for vehicle A. The first one is:

$$u_1^*=\begin{cases} a_{\min}, & if \quad (t\le T_{11})\wedge(x_A\ne 0) \\ 0, & if \quad T_{11}<t\le T_{31} \end{cases} \tag{7.35}$$

and

$$d_1^*=d_{\min}, \quad if \quad t\le T_{21} \tag{7.36}$$

where T_{11} is the time taken by vehicle to reach its minimum velocity under $a_{\max}$, and T_{31} is the total time in which vehicle A reaches the junction area. T_{21} is the time taken by vehicle B to reach junction area under $d_{\max}$. It has

$$T_{11}=\frac{v_{\min}^A-x_2^0}{a_{\min}},\quad T_{21}=\frac{x_B^0}{\dot{x}_{\min}^B}=\frac{x_B^0}{d_{\min}},\quad T_{31}=T_{11}-\frac{x_A(T_{11})}{v_{\min}^A} \tag{7.37}$$

The second strategy is

$$u_2^*=\begin{cases} a_{\max}, & if \quad (t\le T_{12})\wedge(x_A\ne 0) \\ 0, & if \quad T_{12}<t\le T_{32} \end{cases} \tag{7.38}$$

$$d_1^*=d_{\max}, \quad if \quad t\le T_{22} \tag{7.39}$$

T_{12} is the time taken by vehicle A to reach its maximum velocity under a_{max}, and T_{32} is the total time in which it reaches to the junction. T_{32} is the time taken by vehicle B to reach junction under d_{max}. Or

$$T_{12} = \frac{v_{max}^A - x_2^0}{a_{max}},\ T_{22} = \frac{x_B^0}{\dot{x}_{max}^B} = \frac{x_B^0}{d_{max}},\ T_{32} = T_{12} - \frac{x_A(T_{12})}{v_{max}^A} \tag{7.40}$$

The saddle property of the obtained solution (u_1^*, d_1^*) was proven by Arora, Raina and Mittal through the following way: change the strategies of one player and then show that this player will become the loser.

The sate variable of interest is x_3. From (7.40), it yields

$$x_3^*(t) = \begin{bmatrix}0 & -t & 1\end{bmatrix} x^0 - \int_0^t \left[(t-\tau)u_1^*(\tau) + d_1^*(\tau)\right] d\tau \tag{7.41}$$

Suppose the state $x_3(t)$ is under varied strategy (u_1, d_1^*), then

$$x_3(t) = \begin{bmatrix}0 & -t & 1\end{bmatrix} x^0 - \int_0^t \left[(t-\tau)u_1(\tau) + d_1^*(\tau)\right] d\tau \tag{7.42}$$

Thus,

$$x_3^*(t) - x_3(t) = \int_0^t \left\{(t-\tau)\left[u_1^*(\tau) - u_1(\tau)\right]\right\} d\tau \tag{7.43}$$

Notice that the only possible change in the control due to physical constraints leads to $u_1(\tau) \geq u_1^*(\tau)$. It has

$$x_3^*(t) - x_3(t) \leq 0 \tag{7.44}$$

and

$$J(x^0, u_1, d_1^*) \leq J(x^0, u_1^*, d_1^*) \tag{7.45}$$

Meanwhile, suppose the disturbance is varied and the new strategy (u_1^*, d_1), it has

$$x_3(t) = \begin{bmatrix}0 & -t & 1\end{bmatrix} x^0 - \int_0^t \left[(t-\tau)u_1^*(\tau) + d_1(\tau)\right] d\tau \tag{7.46}$$

Thus

$$x_3^*(t) - x_3(t) = \int_0^t \left\{(t-\tau)\left[d_1^*(\tau) - d_1(\tau)\right]\right\} d\tau \tag{7.47}$$

Similarly, since $d_1^*(\tau) - d_1(\tau) = \dot{x}_B(\tau) - \dot{x}_B^*(\tau) \leq 0$, it has

$$J(x^0, u_1, d_1^*) \leq J(x^0, u_1^*, d_1^*) \leq J(x^0, u_1^*, d_1) \tag{7.48}$$

Ineq.(7.48) indicates that the obtained solutions are saddle solutions. Further analysis shows that the cost function for this game is

$$J(x^0, u, d) = |x_3(T)| > C \tag{7.49}$$

The safe/unsafe initial condition set $\{x^0\}$ can be determined by substituting different position and velocity values for vehicle B in the previous equation sets.

The above method assumed there is no communication between vehicles, which is quite common in current traffic situations. However, such Game Theory based approaches often lead to slow traffic flow or unexpected traffic congestion. Recently, more and more researchers discussed the possibility to improve the driving performance using inter-vehicle communication.

7.5.2 Cooperative Driving at Intersections

There were numerous different cooperative driving strategies proposed for blind crossing situations. In this section, the control strategy that was proposed in [119] is addressed.

In [119], it is assumed that vehicle flows arrive continuously at a junction area. At a particular time, only a few vehicles which are moving in the vicinity of the crossing need to be considered. Under this consideration, the continuous traffic flow can be truncated into small segments, which simplifies the problem greatly.

The simple grouping algorithm used here is to label the vehicles by the times they enter the virtual circle centered at the junction area. As shown in Fig.7.24(left), the four shadow vehicles inside the circle will be considered as a group to take part in cooperative driving; while the other three vehicles will not be considered temporarily. The radius of this virtual circle should be determined appropriately by inter-vehicle communication protocol that is selected for this application.

Moreover, it is also assumed in this Chapter that all vehicles have relatively low speeds when approaching the junction and they will not change lanes after entering the virtual circle since it is too close to safely change lanes within the virtual circle. Slowing down makes the vehicles have more time to negotiate with each others and prepare for emergence.

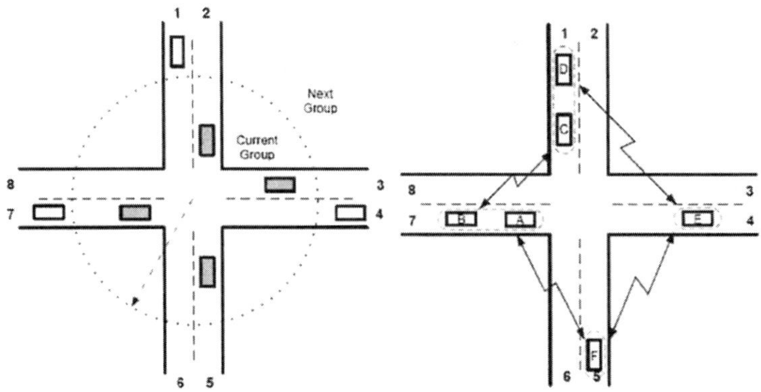

Fig.7.24 (left) vehicle grouping at crossings, and (right) vehicle sub-grouping and subgroup inter-vehicle communication, from [119]. (© [2005] IEEE)

As discussed in [21]-[23], in blind crossing driving scenario, the encountered vehicles need to share the following information among each other:

(1) vehicle ID; This ID can be generated by the unique communication hardware that is equipped on the vehicle;
(2) vehicle classification (length and width). Especially, vehicle size is an important parameter that affects intersection collisions;
(3) the current moving lane, the desired moving lane number and current speed variation profile of each vehicle before the cooperative driving plan is made;
(4) the driving plan of each vehicle after the cooperative driving plan is made;
(5) the real-time position (represented as lane number and distance from the crossings) and speed of each vehicle;
(6) the emergency signal if needed.

When approaching the crossing areas, the original vehicle platoon splits into several independent groups. The maximum allowable size of such a group is set as three in this Chapter. The initial groups are constructed by randomly assigning several temporary dominant vehicles and letting them pick up their neighbor vehicles into its group. The groups contain more than three vehicles will reject the tail vehicles to meet the limits. The rejected vehicles will try to merge into other groups or formulate a new group. In this way, the vehicles are supposed to be self-organized into several small groups as shown in Fig.7.24(right), before they enter the virtual circle.

A vehicle is assumed to store the real-time positions, speeds and desired driving lane information of all vehicles in the same group. It communicates with each other periodically to update position and speed information. Simulations show that three is an appropriate size for an independent group for blind crossing scenarios.

When two vehicles, for instance vehicle ***A*** and ***C*** shown in Fig.7.24(b), in two different groups communicate with each other, vehicle ***A*** will transfer the driving information of all vehicles in its same group to ***C***. The received driving information from ***C*** will be soon delivered to other vehicles in the group of vehicle ***A***. So does vehicle ***C*** to ***A***. And a vehicle will store all the information it received.

It is also assumed that all vehicles will exchange their "temporal" velocity variation profile before the cooperative driving plan being set. Here, "temporal" means the velocity change plan applied before the cooperative driving plan begins. Thus, one vehicle can predict the movements of other vehicles based on received driving information before the cooperative driving plan is set.

Although the data transmission rate is constrained by several factors including media and communication protocol, it is assumed that the vehicles within the virtual circle can properly and timely acquire the necessary driving information of each other. Based on the experiment results, such assumptions are valid for most blind crossings, because the radius of the virtual circle and the number of encountered vehicles are both limited in those driving scenarios.

The inter-vehicle communication network will use tags to label the groups/vehicles encountered. If all the vehicles' moving information has been collected by one vehicle, it will proceed to deal with the trajectory planning. It will simultaneously send messages to block off other vehicles for taking this job.

After the cooperative driving plan is made, the schedule will be properly delivered. Constrained by the communication rate, the driving plan should be represented concisely.

When one vehicle leaves the virtual circle, it should soon leave the current communication group and join a new platoon. And emergency signals will be broadcasted to all vehicles with the highest priority.

In order to guarantee the safety of driving, the concept of safe driving patterns is introduced first. Generally, the collision may occur when

(1) two vehicles move along the same lane, and the lagging one runs into the leading one;

(2) two vehicles moves on different lanes but pass the same junction area simultaneously.

The most frequently used control strategy for blind crossings is the zone blocking strategy. It divides the crossing zone into server sub zones and only one vehicle is allowed to enter a zone at any time to avoid collisions.

Based on a similar idea, a simplified strategy is proposed here, which only allows safe driving patterns (vehicle pairs) to pass the junctions simultaneously. It is similar to the concept of phase of traffic flows in traffic lights control. For instance, consider the two-lane road junction. Suppose one vehicle moves from lane 1 and another vehicle moves from lane 5. Then there only exist four driving patterns that allow two vehicles cross the junction area safely and simultaneously, see Fig.7.25.

In terms of safe driving patterns, a driving schedule can be represented as an ordered series of safe patterns. To illustrate this idea, consider a typical driving scenario shown in Fig.7.25, in which vehicle A needs to move from lane 7 to lane 4, vehicle $\boldsymbol{B}$ from lane 7 to lane 6, vehicle $\boldsymbol{C}$ from lane 1 to lane 8, vehicle $\boldsymbol{D}$ from lane 1 to lane 4.

One possible driving schedule for this scenario is to let vehicle A pass the junction first; then let vehicle $\boldsymbol{B}$ and $\boldsymbol{C}$ pass the junction at the same time; and finally let vehicle $\boldsymbol{D}$ pass. Apparently, it represents this schedule as the following sequence

$$A_4^7 \mid \boldsymbol{B}_4^7, \boldsymbol{C}_8^1 \mid \boldsymbol{D}_4^1 \tag{7.49}$$

where the superscript denotes the original lane of a vehicle and the subscript denotes the destination lane.

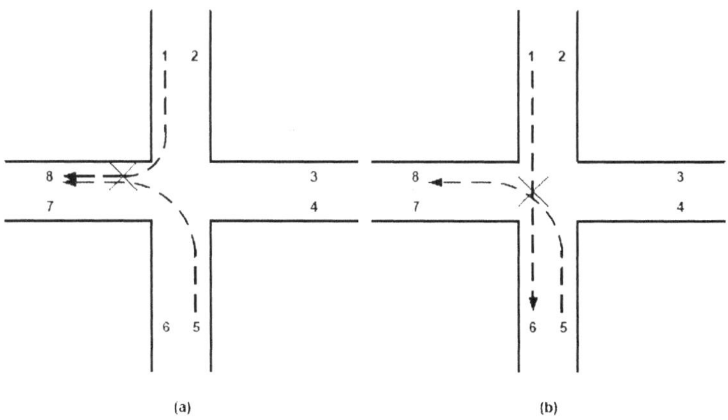

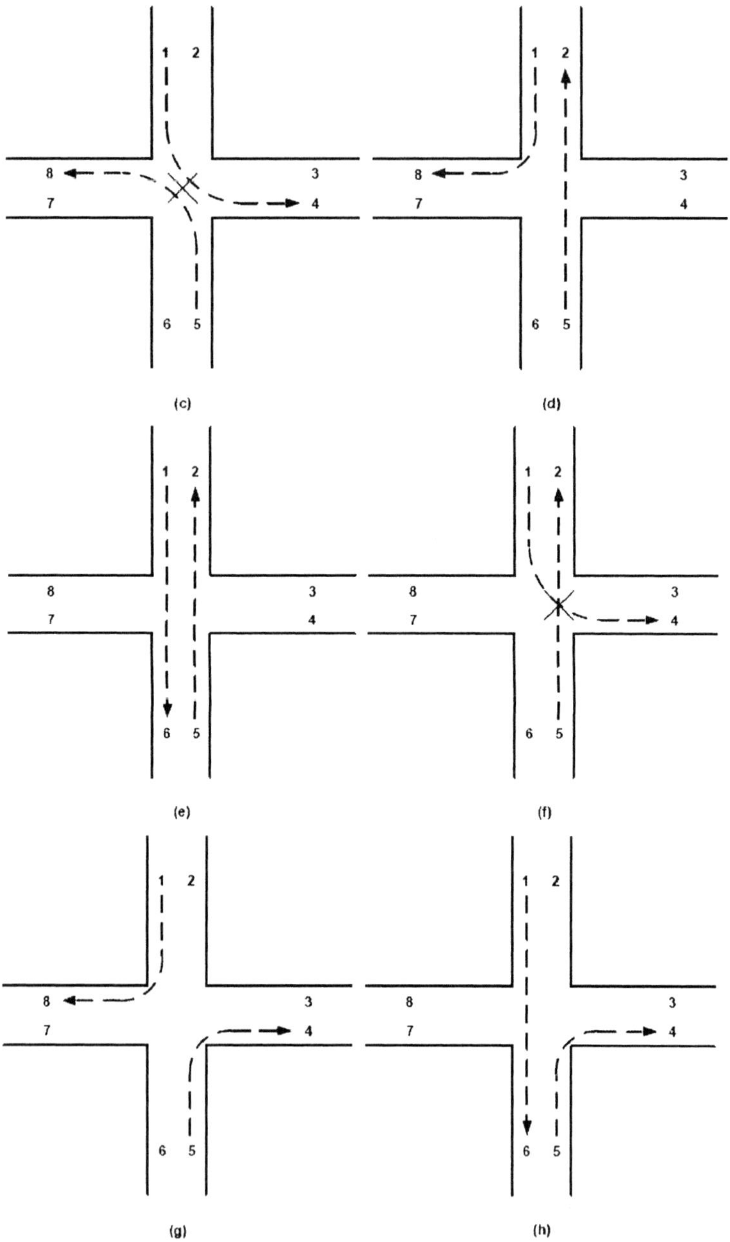
(c)
(d)
(e)
(f)
(g)
(h)

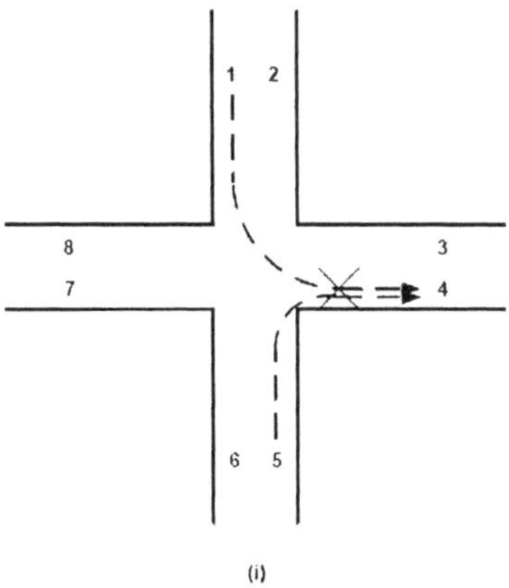

Fig.7.25 All the driving patterns of two vehicles that simultaneously enter junction areas from lane 1 and lane 5 respectively.

For simplicity, $\boldsymbol{A}_4^7 \mid \boldsymbol{B}_4^7, \boldsymbol{C}_8^1 \mid \boldsymbol{D}_4^1$ may also be written as $\boldsymbol{A} \mid \boldsymbol{B}, \boldsymbol{C} \mid \boldsymbol{D}$ in the rest of this Chapter, if the absence of superscripts and subscripts does not make the driving plan ambiguous.

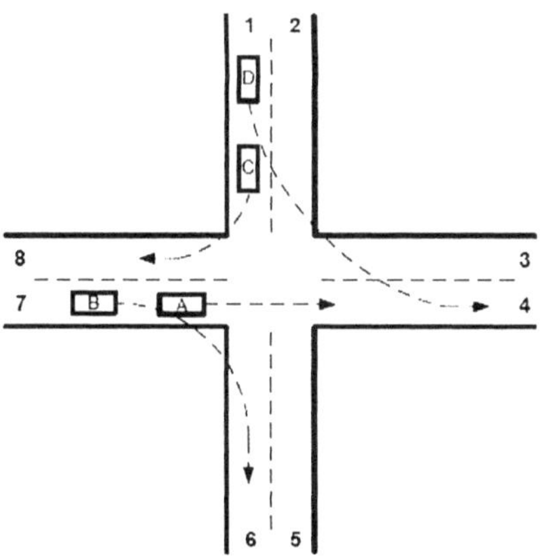

Fig.7.26 A four-vehicle driving scenario for a two-lane junction, from [119]. (© [2005] IEEE)

Here "|" is a separator symbol that divides the sequence into three subsets: ***A***, ***B*** and ***C***, ***D***. Specially, ***B*** and ***C*** are in one subset indicating they will pass the junctions at the same time. Generally, consider N vehicles

i_1, i_2, ... , i_N moving towards the junction. Their order of passing the junction area, i.e. the cooperative driving schedule is specified as

$$i_1, i_5, \ldots, | \; i_2, i_4, \ldots, | \; i_3, \ldots, i_N \tag{7.50}$$

When a spefcial driving schedule is determined, the corresponding trajectory planning process for each vehicle will then be carried out. Notice that different driving schedules lead to different passing times. An optimal (suboptimal) cooperative driving plan needs to find the schedule that completes the total driving process with the (nearly) least time.

In general, searching the allowable driving schedule will yield a tree in which each node represents a particular driving plan (sequence) except for the root node. One basic algorithm to generate such a tree is as follows:

Basic Solution Tree Generation Algorithm

Suppose there are vehicles under consideration.

1) Generate the root node of the tree;

2) Generate children for the root node which represent all the possible permutation orders of the vehicle sequence without any separator symbols. It is apparent that all these orders can be enumerated through basic permutation algorithm.

3) For each node in the second level of the tree, generate $N-1$ children by inserting only one separator into the driving sequence that is represented by it, since there are only $N-1$ positions that a separator can be inserted.

4) For each node in the third level of the tree, generate $N-2$ children by inserting only one separator into the driving sequence that is represented by it, since there are only $N-2$ positions that a separator can be inserted.

……

N) For each node in the $N-1$ level of the tree, generate one child by inserting only one separator into the driving sequence that is represented by it, since there is only one position that a separator can be inserted.

However, a great number of nodes in this tree can be discarded since they represent invalid driving schedules, for which no trajectory planning is needed. One apparent fact is that leading vehicles will always pass the junction areas earlier than lagging vehicles in the platoon, because it is assumed that the lagging vehicle will not change lanes. For instance, in the driving scenario shown in Fig.7.26, vehicle $\boldsymbol{A}$ should always pass the crossing earlier than vehicle $\boldsymbol{B}$. Thus the node $\boldsymbol{B}$ $\boldsymbol{A}$ $\boldsymbol{C}$ $\boldsymbol{D}$ and all its children should be invalid.

Based on this fact, the above algorithm can be modified as

Modified Solution Tree Generation Algorithm

Suppose there are vehicles under consideration.

1) Generate the root node of the tree;

2) Generate children for the root node which represent all the possible permutation orders of the vehicle sequence without considering separator symbols. Then, prune all the obtained nodes that represent invalid order of driving.

3)-N) Same as Basic Solution Tree Generation Algorithm.

To guarantee a collision free movement, the safety of the simultaneously moving subnets in a driving sequence has to be checked. It is obvious that each subset in a valid driving schedule should constitute of a safe driving pattern.

Note that M vehicles can safely cross the junctions at the same time if and only if every pair of these vehicles is a safety pattern. Therefore, a labeling algorithm is given by:

Safety Pair Labeling Algorithm

If there are M ($M \geq 2$) vehicles j_1, j_2, ..., j_M pass the crossing the junctions simultaneously, check the $M(M-1)/2$ distinct vehicle combinations (j_1, j_2), (j_1, j_3), ..., (j_{M-1}, j_M) to see whether they are all safety patterns. If not, label the corresponding node as an unsafe node.

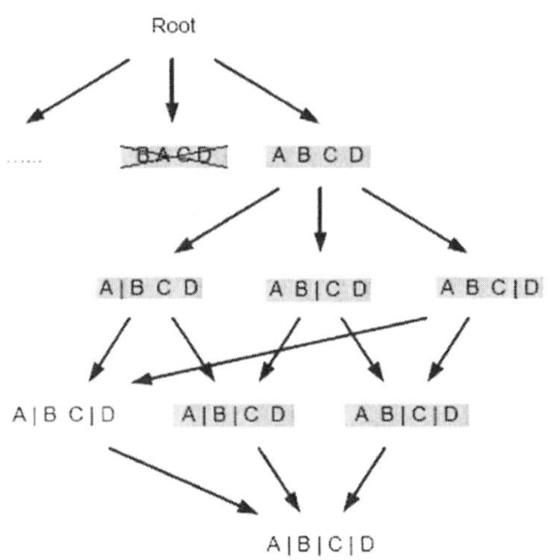

Fig.7.27 A schedule tree stemmed from the driving scenario shown in Fig.7.26. The shadow nodes represent invalid driving schedule, from [119]. (© [2005] IEEE)

Using these two considerations, most unsafe driving schedules can be directly pruned from the solution tree. For example, the driving plan tree stemmed from scenarios in Fig.7.26 only has a few valid nodes after pruning and labeling, i.e. there are only two valid driving schedules for the ordered sequence ***A***, ***B***, ***C***, ***D***, see Fig.7.27. The first schedule is specified by Eq.(7.49) and the second is to simply let vehicle ***A***, ***B***, ***C***, ***D*** pass the junction area sequentially.

After labeling, trajectory planning will be executed for every valid node with respect to its driving schedule. Some driving plans will be discarded since they are implicitly forbidden by the vehicle dynamics constraints.

In general, a vehicle's trajectory crossing the junctions can be divided into three sequential stages:

(1) approach the junction. The vehicle should avoid collision between the leading vehicle and itself. Moreover, it should also avoid collision between itself and the last vehicle that had passed the junction before it but does not form a safety pair with it;

(2) cross the junction, which is considered as a decelerate- accelerate process;

(3) leave the junction. It should avoid collision between the new leading vehicle and itself.

In order to solve the lane merging trajectory planning problems, the virtual vehicle mapping technique was proposed and employed by Uno, Sakaguchi and Tsugawa in [31] and [32]. The concept of a virtual vehicle is to map a vehicle on a lane onto an object lane; then the interested vehicle can be controlled with respect to the virtual vehicle to guarantee safety.

Considering the risk of failure, an algorithm similar to what is proposed in [31] and [32] is applied here. The only difference is that the leading vehicle is not mapped onto the symmetry position of the desired lane. Actually, it will be mapped into a position that lags off the mirror point to compensate the communication delay and vehicle deceleration at steering.

For example, as shown in Fig.7.28, vehicle ***A*** moves form lane 7 to lane 6 first; then vehicle ***B*** moves from lane 1 to lane 6. Since vehicle ***B*** passes the junction area right after vehicle ***A***, it must make enough headway to avoid collision. Therefore, it generates a virtual vehicle ***A'*** by mapping vehicle A into its own lane using the data transmitted via the inter-vehicle communication. Classical longitudinal control will then be performed between virtual vehicle ***A'*** and vehicle ***B***.

The lengths of the two headways should be determined by the dynamic properties and length/size of encountered vehicles as well as the applied inter-communication protocols.

The trajectory profile of each vehicle in the driving schedule will be generated one by one with respect to the corresponding driving order passing the junction. Every vehicle can "know" the trajectories of the other related vehicles that need to map onto its lane.

Since it requires to improve traffic efficiency, the driving plan should keep the headway between the potential leading/virtual vehicle and itself to the minimum safe distance.

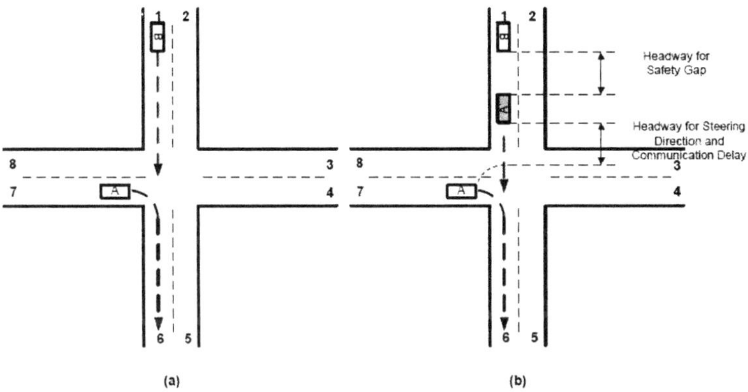

Fig.7.28 Trajectory generation considering one virtual vehicle: (a) with actual vehicles only, (b) with virtual vehicle A' introduced., from [119]. (© [2005] IEEE)

Generally, various driving scenarios can be classified into the following four cases:

(1) there is neither a leading vehicle moving in the same lane nor a virtual vehicle that needs to map before the planning vehicle enters the junction area;

(2) there is a leading vehicle moving in the same lane but no virtual vehicles that need to be mapped before the planning vehicle enters the junction areas. Indeed, it equivalently means that the planning vehicle will pass the junction area right after the leading vehicle;

(3) there is not a leading vehicle moving in the same lane but a virtual vehicle that needs to be mapped before the planning vehicle enters the junction area;

(4) there is a leading vehicle moving in the same lane, and there is also a virtual vehicle that needs to be mapped before the planning vehicle enters the junction area. Actually, it means that the planning vehicle will first follow the leading vehicle and then follow the virtual vehicle before it enters the junction.

Suppose the position of the planning vehicle is denoted by x_p, and the accelerate/decelerate rate of the vehicle is denoted by $\ddot{x}_p$. Their dynamics are constrained by

$$\ddot{x}_{p\min} \le \ddot{x}_p \le \ddot{x}_{p\max}, \; 0 \le \dot{x}_p \le v_{\max} \tag{7.51}$$

where $\ddot{x}_{p\min}$ and $\ddot{x}_{p\max}$ are the accelerate/decelerate rate bounds; and $v_{\max}$ is the maximum allowable safe speed.

If the planning vehicle needs to steer directions, there is an additional constraint

$$\dot{x}_p(t_3) \le v_{s\max} \tag{7.52}$$

where $v_{s\max}$ is the maximum allowable speed for steering, and t_3 is the time when the planning vehicle enters the junction area.

For example, the corresponding trajectory planning algorithms for Scenario 1 above can be formulated as follows.

Trajectory Planning Algorithm for Case A.

1) Solve the time optimal trajectory planning problem before the planning vehicle enters the junction areas;

Obviously, the objective is

$$\min_u t_3 \tag{7.53}$$

constrained by the given boundary conditions

$$x_p(t_0) = x_{p0}, \; x_p(t_3) = x_{p3} \tag{7.54}$$

where x_{p0} is the start position and x_{p3} is the final position at the boundary of junction areas; t_0 is the time that planning starts.

2) There is no planning at this stage. It simply assumes that the time consumed in crossing junction areas only depends on the speed of the vehicle when it enters the junction areas;

The time that the planning vehicle leaves the junction areas should be

$$t_4 = t_s + f_1(\dot{x}_p(t_3)) \tag{7.55}$$

and the speed should be

$$\dot{x}_p(t_4) = f_2(\dot{x}_p(t_3)) \tag{7.56}$$

where t_4 is the time that the planning vehicle leaves junction areas.

3) Solve the tracking problem for the new leading vehicle before the virtual vehicle leaves the junction area.

Suppose the trajectory of the new leading vehicle is given as

$$\ddot{x}_n = f_3(t) \tag{7.57}$$

where x_n is the position of the new leading vehicle and $f_3(t)$ is its movement descriptive function in terms of time.

Thus, the error e_{np} in headway from the pre-selected safe distance L_3 can be written as

$$e_{np} = x_n - x_p - L_3 \tag{7.58}$$

Obviously, the objective is

$$\min_u \int_{t=t_4}^{t_4+T} |e_{lp}| dt \tag{7.59}$$

constrained by (6), (7) and boundary condition (10). Here T is a predetermined time span that is long enough to appropriately describe the planning vehicle's movement after it leaves the junction areas.

It should be pointed out that the choice of vehicle dynamic models and/or longitudinal driving controllers will not vary the feasibility of the proposed planning framework. Furthermore, some other driving performance index such as ride comfortableness can be formulated and added here. Some related discussion on objective choice of vehicle trajectory planning can be found in our previous works [33]-[36].

However, there is still one important issue need to be clarified. Limited by the communication rate, the cooperative driving plan should not be too complicated. Otherwise, the cooperative driving plan cannot be correctly delivered to all the encountered vehicles.

Here, Li and Wang further constrain the trajectory of one vehicle constitutes of at most five steady acceleration/deceleration process. Thus, one vehicle's trajectory will be determined by at most seven dataset as follows:

<*star time* t_0, *velocity* $\dot{x}_p(t_0)$>,..., <*end time* t_5, *velocity* $\dot{x}_p(t_5)$> (7.60)

The velocity between time t_i and t_{i+1} can be obtained as

$$\dot{x}_p(t) = \dot{x}_p(t_i) + [\dot{x}_p(t_{i+1}) - \dot{x}_p(t_i)] \cdot \frac{t - t_i}{t_{i+1} - t_i} \tag{7.61}$$

This method greatly reduces the computation costs of the above trajectory planning problems without losing too much generality. Moreover, this technique also relieves the burden of the inter-vehicle communication network, since there are only twelve variables in (23) need to be encoded and delivered to the involved vehicle. The vehicle will resolve the control inputs from this simplified velocity profile based on its dynamic equation.

The total time cost of trajectory is counted from the time when the cooperative driving process begins to the time when the last vehicle leaves the junction area. All the un-discarded driving plans will be compared, and the one with least time cost will be chosen as the actual driving plan. The diagram of this planning framework is shown in Fig.7.29.

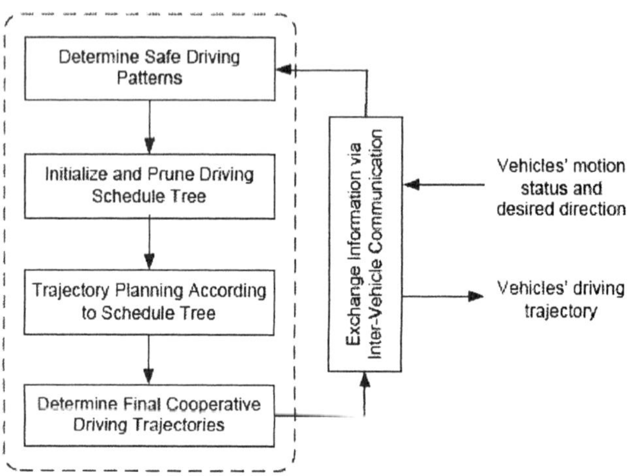

Fig.7.29 The cooperative driving planning framework.

In the searching process of optimal driving plan, the upper level nodes in the schedule tree will be examined earlier with respect to the lower level plans. If one plan is proven to be valid, then all its children nodes will be omitted, because a valid node always represents the plan that uses less time than that of its children plans. For example, if solution $A \mid B \;\; C \mid D$ in Fig.7.27 is a valid plan, then its child node $A \mid B \mid C \mid D$ needs no analysis.

7.6 Summary

Cooperative multiple vehicles driving with inter-vehicle communication attracts increasing attentions recently. It aims to improve driving safety and efficiency using appropriate motion scheduling of all the encountered vehicles. Under cooperative driving control, the motion of individual vehicles can be conducted in a safe, deterministic and smooth manner. This is particularly useful to heavy vehicles, since their acceleration/deceleration capacity is relatively low.

Constrained by the book length, some interesting topics are left unmentioned in this Chapter. For instance:

(1) the integrated high-level traffic control and low-level cooperative driving is expected to further increase riding efficiency [123]-[124];

(2) The geometry constraints of vehicles and junction areas have not been well discussed in this Chapter, but they should not be neglected in real conditions;

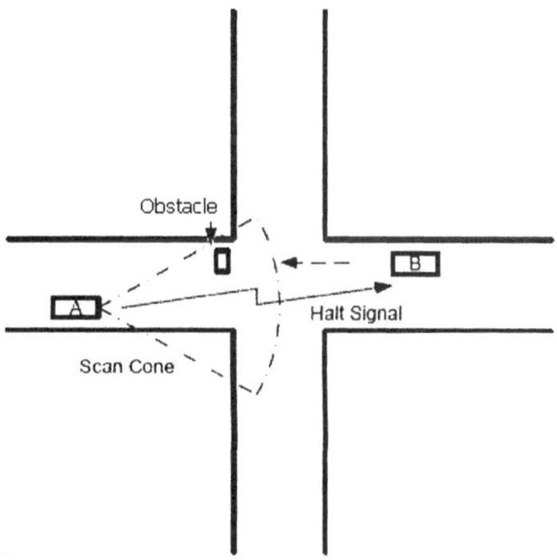

Fig.7.30 Handling an emergency case by broadcasting message, from [119]. (© [2005] IEEE)

(3) The actual driving scenario at junctions might be more complex than what is discussed here. One important case is how to deal with emergencies. One potential answer is to share each vehicle's sensor information through the inter-vehicle communication. If an emergence happens, the first 'knowing' vehicle will soon stop all the related vehicles synchronically. For example, as shown in Fig.7.30, an emergent halt signal sent out from vehicle A quickly stops vehicle B that intends to move to the blocked lane, when it suddenly 'sees' an obstacle on that lane. The cooperative driving plan will be regenerated when the detected obstacles moved away. Furthermore, sensor network implemented by multiple vehicles sensor fusion is believed to another hot topic within the near future [29], [119], [125]-[133]; see Fig.7.31.

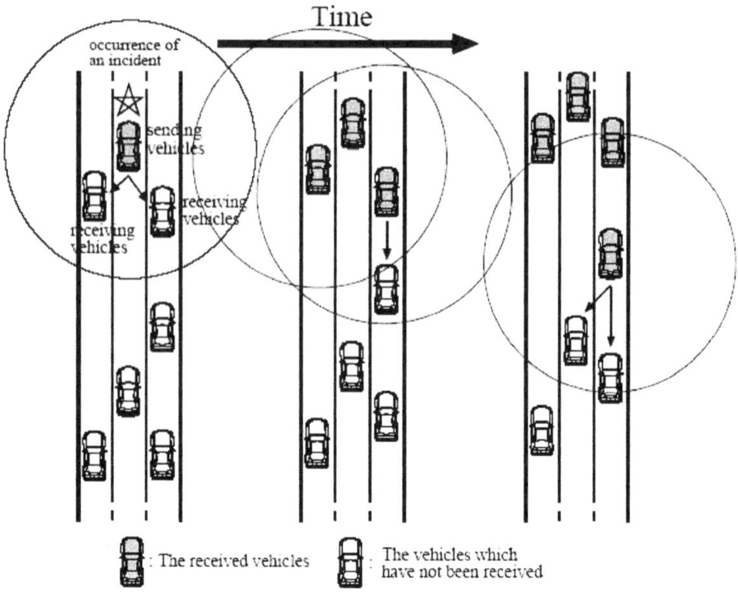

Fig.7.31 Handling an emergency case by broadcasting message, from [29]. (© [2005] IEEE)

7.7 Reference

1. S. Tsugawa, T. Saito, and A. Hosaka, "Super smart vehicle system: AVCS related systems for the future," Proceedings of IEEE Intelligent Vehicles Symposium, pp: 132-137, 1992.
2. S. Yamada, "The strategy and deployment plan for VICS," IEEE Communications Magazine, vol. 34, no. 10, pp. 94-97, 1996.
3. M. Aoki and H. Fujii, "Inter-vehicle communication: technical issues on vehicle control application," IEEE Communications Magazine, vol. 34, no. 10, pp. 90-93, 1996.
4. T. Tank and J.-P.M.G. Linnartz, "Vehicle-to-vehicle communications for AVCS platooning," IEEE Transactions on Vehicular Technology, vol. 46, no. 2, pp. 528-536, 1997.
5. A. Cochran, "AHS communications overview," Proceedings of IEEE Intelligent transportation Systems Conference, 1997.

6. R. Kohno, "Evolution and demands of SoC for ITS and Mobile communication systems," Proceedings of International Symposium on Semiconductor Manufacturing, pp. 12-17, 2000.
7. S. Tsugawa, "Inter-vehicle communications and their applications to intelligent vehicles: an overview," Proceedings of IEEE Intelligent Vehicle Symposium, vol. 2, pp. 564-569, 2002.
8. N. Navet, Y. Song, and F. Simonot-Lion, "Trends in automotive communication systems," Proceedings of IEEE, vol. 93, no. 6, pp. 1204-1223, 2005.
9. J. Kaltwasser and J. Kassubek, "A new cooperative optimized channel access for inter-vehicle communication," Proceedings of Vehicle Navigation and Information Systems Conference, pp. 145-148, 1994.
10. R. Verdone, "Communication systems at millimeter waves for ITS applications," Proceedings of Vehicular Technology Conference, vol. 2, pp. 914-918, 1997.
11. K. Tokuda, M. Akiyama, and H. Fujii, "DOLPHIN for inter-vehicle communication system," Proceedings of the IEEE Intelligent Vehicles Symposium, pp. 504-509, 2000.
12. J. S. Kwak and J. H. Lee, "Infrared transmission for intervehicle ranging and vehicle-to-roadside communication systems using spread-spectrum technique," IEEE Transactions on Intelligent Transportation Systems, vol. 5, no. 1, pp. 12-19, 2004.
13. T. Tomimoto and H. Ogawa, "Optical transmitter and receiver for inter-vehicle communication," Oki Technical Review, vol. 63, no. 158, pp. 7-10, 1997.
14. L. B. Michael and M. Nakagawa, "Interference characteristics in intervehicle communication from oncoming vehicles," Electronics and Communications in Japan, Part 2, vol. 84, no. 2, pp. 2034-2042, 2001.
15. S. Lindenmeier, K. Boehm, and J. F. Luy, "A wireless data link for mobile applications," IEEE Microwave and Wireless Components Letters, vol. 13, no. 8, pp. 326-328, 2003.
16. L. B. Michael and M. Nakagawa, "Interference characteristics in intervehicle communication from oncoming vehicles," Institute of Electronics, Information and Communication Engineers Transactions on Communications, vol. J82-B, no.11, pp. 2034-2042, 1999.
17. L. B. Michael and M. Nakagawa, "Non-platoon inter-vehicle communication using multiple hops," Institute of Electronics, Information and Communication Engineers Transactions on Communications, vol. E82-B, no. 10, pp. 1651-1658, 1999.
18. T. Wada, M. Maeda, and M. Okada, et. al, "Theoretical analysis of propagation characteristics in millimeter-wave intervehicle communication system," Electronics and Communications in Japan, Part II, vol. 83, no. 11, pp. 33-43, 2000.
19. A. Kato, K. Sato, and M. Fujise, "Wave propagation characteristics of inter-vehicle communication on an expressway," Proceedings of World Congress on Intelligent Transportation Systems, 2001.

20. T. Ikegami and E. Yamashita, "A study of inter-vehicle communication protocol based on frequency hopping technique," Proceedings of International Symposium on Wireless Personal Multimedia Communications, vol. 1, pp. 320-323, 2002.
21. Y. Karasawa, "Multipath fading due to road surface reflection and fading reduction by means of space diversity in ITS vehicle-to-vehicle communications at 60 Ghz," Electronics and Communications in Japan, Part I, vol. 85, no. 1, pp. 35-42, 2002.
22. S. K. Shamnugam and H. Leung, "A robust chaotic spread spectrum inter-vehicle communication scheme for ITS, " Proceedings of IEEE Intelligent Transportation Systems, vol. 2, pp. 1540-1545, 2003.
23. A. Ueda, K. Mizui, and T. Ihara, "Intervehicle communication and ranging system using code-hopping spread spectrum technique," Electronics and Communications in Japan, Part III, vol. 88, no. 5, pp. 50-60, 2005.
24. S. Y. Wang, "On the intermittence of routing paths in vehicle-formed mobile ad hoc networks on highways," Proceedings of International IEEE Conference on Intelligent Transportation Systems, pp. 803-809, 2004.
25. H. Sawant, J. Tan, and Q. Yang, et. al, "Using Bluetooth and sensor networks for intelligent transportation systems," Proceedings of International IEEE Conference on Intelligent Transportation Systems, pp. 767-772, 2004.
26. J. Blum, A. Eskandarian, and L. J. Hoffman, "Performance characteristics of inter-vehicle ad hoc networks," Proceedings of IEEE Intelligent Transportation Systems, vol. 1, pp. 114-119, 2003.
27. W. Chen and S. Cai, "Ad hoc peer-to-peer network architecture for vehicle safety communications," IEEE Communications Magazine, vol. 43, no. 4, pp. 100-107, 2005.
28. J. Zhu and S. Roy, "MAC for dedicated short range communications in intelligent transport system," IEEE Communications Magazine, vol. 41, no. 12, pp. 60-67, 2003.
29. F. Watanabe, M. Fujii, and M. Itami, et. al, "An analysis of incident information transmission performance using MCS/CDMA scheme," Proceedings of IEEE Intelligent Vehicles Symposium, pp. 249-254, 2005.
30. A. M. K. Cheng and K. Rajan, "A digital map/GPS based routing and addressing scheme for wireless ad-hoc networks," Proceedings of IEEE Intelligent Vehicles Symposium, pp. 17-20, 2003.
31. J.-P. Hubaux, S. Capkun, and J. Luo, "The security and privacy of smart vehicles," IEEE Security & Privacy Magazine, vol. 2, no. 3, pp. 49-55, 2004.
32. L. Chisalita and N. Shahmehri, "A peer-to-peer approach to vehicular communication for the support of traffic safety applications," Proceedings of IEEE International Conference on Intelligent Transportation Systems, pp. 336-341, 2002.

33. J. Blum, A. Eskandarian, and L. J. Hoffman, "Performance characteristics of inter-vehicle ad hoc networks," Proceedings of IEEE Intelligent Transportation Systems, vol. 1, pp: 114-119, 2003.
34. L. B. Michael, S. Kikuchi, and T. Adachi, et. al, "Combined cellular/direct method of inter-vehicle communication," Proceedings of IEEE Intelligent Vehicles Symposium, pp. 534-539, 2000.
35. H.-J. Reumerman, M. Roggero, and M. Ruffini, "The application-based clustering concept and requirements for intervehicle networks," IEEE Communications Magazine, vol. 43, no. 4, pp. 108-113, 2005.
36. L. B. Michael, "Adaptive layered data structure for inter-vehicle communication in ad-hoc networks," Proceedings of International World Congress on Intelligent Transport Systems, 2001.
37. L. B. Michael and R. Kohno, "Adaptive data structure for efficient inter-vehicle communication," IEICE Transactions on Information and Systems, vol. E85-D, no. 11, pp. 1830-1838, 2002.
38. H. Fujii, O. Hayashi, and Y. Hirao, "Experimental research on protocol of inter-vehicle communication for multiple vehicles," Proceedings of Annual World Congress on ITS, 1996.
39. L. B. Michael and R. Kohno, "Adaptive data class strutuctre for efficient inter-vehicle communication," IEICE Transactions on Information and Systems, vol. E85-D, no. 11, pp.1830-1838, 2002
40. T. Hasegawa, K. Mizui, and H. Fujii, et. al, "A concept reference model for inter-vehicle communications (Report 2)," Proceedings of International IEEE Conference on Intelligent Transportation Systems, pp. 810-815, 2004.
41. S. Y. Wang, "On the effectiveness of distributing information among vehicles using inter-vehicle communication," Proceedings of IEEE Intelligent Transportation Systems, vol. 2, pp. 1521-1526, 2003.
42. J. P. Hubaux, S. Capkun, and J. Luo, "The security and privacy of smart vehicles," IEEE Security & Privacy Magazine, vol. 2, no. 3, pp. 49-55, 2004.
43. J. Blum and A. Eskandarian, "The threat of intelligent collisions," IEEE IT Professional, vol. 6, no. 1, pp. 24-29, 2004.
44. S. Sheikholeslam and C. A. Desoer, "Longitudinal control of a platoon of vehicles with no communication of lead vehicle information: a system level study," IEEE Transactions on Vehicular technology, vol. 42, no. 4, pp. 546-554, 1993.
45. C. E. Smith, C. A. Richards, and S. A. Brandt, et. al, "Visual tracking for intelligent vehicle-highway systems," IEEE Transactions on Vehicular Technology, vol. 45, no. 4, pp. 744-759, 1996.
46. M. J. Woo and J. W. Choi, "A relative navigation system for vehicle platooning," Proceedings of SICE Annual Conference, pp. 28-31, 2001.
47. S. E. Shladover, C. A. Desoer, and J. K. Hedrick, et. al, "Automated vehicle control development in the PATH program," IEEE Transactions on Vehicular Technology, vol. 40, no. 1, pp. 114-130, 1991.

48. D. Swaroop and J. K. Hedrick, "String stability of interconnected systems," IEEE Transactions on Automatic Control, vol. 41, no. 3, pp. 349-357, 1996.
49. P. Li, L. Alvarez, and R. Horowitz, et. al, "A safe velocity tracking controller for AHS platoon leaders, " Proceedings of Conference on Decision and Control, vol. 2, pp. 2283-2288, 1996.
50. P. Li, L. Alvarez and R. Horowitz, "AVHS Safe control laws for platoon leaders," IEEE Transactions on Control Systems and Technology, Vol. 5, No. 6, pp. 614-628, 1997.
51. L. Alvarez and R. Horowitz, "Safe platooning in automated highway systems. Part I: safety regions design," Vehicle System Dynamics Special Issue: IVHS, vol. 32, no. 1, pp. 23-56, 1999.
52. L. Alvarez and R. Horowitz "Safe platooning in automated highway systems. Part II: velocity tracking controller," Vehicle System Dynamics Special Issue: IVHS, vol. 32, no. 1, pp. 57-84, 1999.
53. R. Rajamani, S. B. Choi, and B. K. Law, et. al, "Design and experimental implementation of longitudinal control for a platoon of automated vehicles," ASME Journal of Dynamic Systems, Measurement, and Control, vol. 122, pp. 470-476, 2000.
54. D. Swaroop, J. K. Hedrick, and S. B. Choi, "Direct adaptive longitudinal control of vehicle platoons," IEEE Transactions on Vehicular Technology, vol. 50, no. 1, pp. 150-161, 2001.
55. T. S. No, K.-T. Chong, and D.-H. Roh, "A Lyapunov function approach to longitudinal control of vehicles in a platoon," IEEE Transactions on Vehicular Technology, vol. 50, no. 1, pp. 116-124, 2001.
56. K. Li and P. Ioannou, "Modeling of traffic flow of automated vehicles," IEEE Transactions on Intelligent Transportation Systems, vol. 5, no. 2, pp. 99-113, 2004.
57. S. Sheikholeslam and C. A. Desoer, "Longitudinal control of a platoon of vehicles," Proceedings of American Controrl Conference, vol. 1, pp. 291-297, 1990.
58. S. Sheikholeslam and C. A. Desoer, "A system level study of the longitudinal control of a platoon of vehicles," ASME Journal of Dynamic Systems, Measurement and Control, vol. 114, no. 2, pp. 286-292, 1992.
59. D. V. A. H. G. Swaroop and K. R. Rajagopal, "Intelligent cruise control system design based on a traffic flow specification," Vehicle System Dynamics, vol. 30, no. 5, pp. 319-344, 1998.
60. D. V. A. H. G. Swaroop and J. K. Hedrick, "Constant spacing strategies for platooning in automated highway systems," ASME Journal of Dynamic Systems, Measurement and control, vol. 121, no. 3, pp. 462-470, 1999.
61. D. V. A. H. G. Swaroop and K. R. Rajagopal, "Intelligent cruise control systems and traffic flow stability," Transportation Research, Part C, vol. 7, no. 6, pp. 329-352, 1999.
62. R. Rajamani, S. B. Choi, and J. K. Hedrick, "Design and experimental implementation of control for a platoon of automated vehicles," Pro-

ceedings of the ASME Dynamic Systems and Control Division, pp. 681-689, 1998.

63. R. Rajamani, H. Tan, and B. K. Law, "Demonstration of integrated longitudinal and lateral control for the operation of automated vehicles in platoons," IEEE Transactions on Control Systems Technology, vol. 8, no. 4, pp. 695-708, 2000.
64. R. Rajamani and S. E. Shladover, "An experimental comparative study of autonomous and cooperative vehicle follower control system," Transportation Research, Part C, vol. 9, no. 1, pp. 15-31, 2001.
65. K. Santhanakrishnan and R. Rajamani, "On spacing policies for highway vehicle automation," IEEE Transactions on Intelligent Transportation Systems, vol. 4, no. 4, pp. 198-204, 2003.
66. J. Bengtsson, R. Johansson, and A. Sjogren, "Modeling of drivers' longitudinal behavior," Proceedings of IEEE/ASME International Conference on Advanced Intelligent Mechatronics, vol. 2, pp. 1076-1081, 2001.
67. D. Yanakiev and I. Kanellakopoulos, "Longitudinal control of automated CHVs with significant actuator delays," IEEE Transactions on Vehicular Technology, vol. 50, no. 5, pp. 1289-1297, 2001.
68. H. M. Kim, J. Dickerson, and B. Kosko, "Fuzzy throttle and brake control for platoon of smart cars," Fuzzy Sets and Systems, vol. 84, no. 3, pp. 209-234, 1996.
69. L. Wu, Y. Xu, and F.-Y. Wang, , et. al, "Supervised learning of longitudinal driving behavior for intelligent vehicles using neuro-fuzzy networks: initial experimental results," International Journal of Intelligent Control and Systems, vol. 3, no. 4, pp. 443-463, 1999.
70. J. E. Naranjo, C. Gonzalez, and J. Reviejo, et. al, "Adaptive fuzzy control for inter-vehicle gap keeping," IEEE Transactions on Intelligent Transportation Systems, vol. 4, no. 3, pp. 132-142, 2003.
71. Z. Bareket, P. S. Fancher, and H. Peng, et. al, "Methodology for assessing adaptive cruise control behavior," IEEE Transactions on Intelligent Transportation Systems, vol. 4, no. 3, pp. 123-131, 2003.
72. S. C. Warnick and A. A. Rodriguez, "A systematic antiwindup strategy and the longitudinal control of a platoon of vehicles with control saturations," IEEE Transactions on Vehicular Technology, vol. 49, no. 3, pp. 1006-1016, 2000.
73. Y. Liu and U. Ozguner, "Effect of inter-vehicle communication on rear-end collision avoidance," Proceedings of IEEE Intelligent Vehicles Symposium, pp. 168-173, 2003.
74. H. Pham, K. Hedrick, and M. Tomizuka, "Combined lateral and longitudinal control of vehicles For IVHS," Proceedings of American Control Conference, vol. 2, pp. 1205-1206, 1994.
75. R. T. O'Brien, P. A. Iglesias, and T. J. Urban, "Lane change maneuver via H_∞ steering control methods," Proceedings of IEEE Conference on Control Applications, pp. 131-136, 1995.

76. D.B. Maciuca, J.K. Hedrick, "Advanced nonlinear brake system control for vehicle platooning," Proceedings of European Control Conference, 1995.
77. R. T. O'Brien, P. A. Iglesias, and T. J. Urban, "Vehicle lateral control for automated highway systems," IEEE Transactions on Control Systems Technology, vol. 4, no. 3, pp. 266-273, 1996.
78. C. Canudas de Wit and B. Brogliato, "Stability issues for vehicle platooning in automated highway systems," Proceedings of IEEE International Conference on Control Applications, vol. 2, pp. 1377-1382, 1999.
79. S. S. Stankovic, M. J. Stanojevic, and D. D. Siljak, "Decentralized overlapping control of a platoon of vehicles," IEEE Transactions on Control Systems Technology, vol. 8, no. 5, pp. 816-832, 2000.
80. R. White and M. Tomizuka, "Autonomous following lateral control of heavy vehicles using laser scanning radar," Proceedings of American Control Conference, vol. 3, pp. 2333-2338, 2001.
81. S. A. Nobe and F.-Y. Wang, "An overview of recent developments in automated lateral and longitudinal vehicle controls," Proceedings of IEEE International Conference on Systems, Man, and Cybernetics, vol. 5, pp. 3447-3452, 2001.
82. I. Papadimitriou and M. Tomizuka, "Lateral control of platoons of vehicles on highways: the autonomous following based approach," International Journal of Vehicle Design, vol. 36, no. 1, pp. 24-37, 2004.
83. S. Huang, W. Ren, and S. C. Chan, "Design and performance evaluation of mixed manual and automated control traffic," IEEE Transactions on Systems, Man and Cybernetics, Part A, vol. 30, no. 6, pp. 661-673, 2000.
84. X. Liu, A. Goldsmith, and S. S. Mahal, et. al, "Effects of communication delay on string stability in vehicle platoons," Proceedings of IEEE Intelligent Transportation Systems Conference, pp. 625-630 , 2001.
85. R. Rajamani, A. S. Howell, and C. Chieh Chen, et. al, "A complete fault diagnostic system for automated vehicles operating in a platoon," IEEE Transactions on Control Systems Technology, vol. 9, no. 4, pp. 553-564, 2001.
86. J. Yi, L. Alvarez, and A. Howell, et. al, "A fault management system for longitudinal vehicle control in AHS," Proceedings of American Control Conference, pp. 1514-1518, 2000.
87. An Examination of Fault, Unsafe Driving Acts, and Total Harm in Car-Truck Collisions, U.S. Federal Highway Administration, Summary Report 04-085, 2004.
88. J. Miura, M. Itoh, and Y. Shirai, "Toward vision-based intelligent navigator: its concept and prototype," IEEE Transactions on Intelligent Transportation Systems, vol. 3, no. 2, pp. 136-146, 2002.
89. W.-S. Chee and M. Tomizuka, "Lane change maneuver for AHS applications," Proceedings of International Symposium on Advanced Vehicle Control, pp. 420-425, 1994.

90. W. Chee and M. Tomizuka, "Lane change maneuver of automobiles for the intelligent vehicle and highway system (IVHS)," Proceedings of American Control Conference, vol. 3, pp. 3586-3587, 1994.
91. W. Chee, M. Tomizuka, and S. Patwardhan, et. al, "Experimental study of lane change maneuver for AHS applications," Proceedings of American Control Conference, vol. 1, pp. 139-143, 1995.
92. J. L. Bascunana, "Analysis of lane change crash avoidance," SAE#951895, 1995.
93. J. Kaneko and A. Shimamura, "A design of lane change maneuver for automated vehicles," Proceedings of IEEE Conference on Decision and Control, vol. 1, pp. 1031-1033, 1998.
94. C. Hatipoglu, U. Ozguner, and K. A. Redmill, "On optimal design of a lane change controller," Proceedings of Intelligent Vehicles Symposium, pp.436-441, 1995
95. C. Hatipoglu, U. Ozguner, and K. A. Redmill, "Automated lane change controller design," IEEE Transactions on Intelligent Transportation Systems, vol. 4, no. 1, pp. 13-22, 2003.
96. T. Shamir, "How should an autonomous vehicle overtake a slower moving vehicle: design and analysis of an optimal trajectory," IEEE Transactions on Automatic Control, vol. 49, no. 4, pp. 607-610, 2004.
97. J. Mar and H.-T. Lin, "The car-following and lane-changing collision prevention system based on the cascaded fuzzy inference system," IEEE Transactions on Vehicular Technology, vol. 54, no. 3, pp. 910-924, 2005.
98. J. L. Bascunana, "Analysis of lane change collision avoidance," System Issues of ITS, 1997.
99. H. Jula, E. B. Kosmatopoulos, and P. A. Ioannou, "Collision avoidance analysis for lane changing and merging," IEEE Transactions on Vehicular Technology, vol. 49, no. 6, pp. 2295-2308, 2000.
100. A. Uno, T. Sakaguchi, and S. Tsugawa, "A merging control algorithm based on inter-vehicle communication," Proceedings of IEEE/IEEJ/JSAI International Conference on Intelligent Transportation Systems, pp. 783-787, 1999.
101. P. Kachroo and Z. Li, "Vehicle merging control design for an automated highway system," Proceedings of IEEE Conference on Intelligent Transportation System, pp. 224-229, 1997.
102. M. Antoniotti, A. Desphande, and A. Girault, "Microsimulation analysis of multiple merge junctions under autonomous AHS operation," Proceedings of IEEE Conference on Intelligent Transportation System, pp. 147-152, 1997.
103. R. W. Hall and H.-S. J. Tsao, "Capacity of automated highway systems: merging efficiency," Proceedings of American Control Conference, vol. 3, pp. 2046-2050, 1997.
104. J.-H. Park and S.-H. Ryu, "Neuro-fuzzy control of converging vehicles for automated transportation systems," Proceedings of American Control Conference, vol. 6, pp. 4193-4197, 1999.

105. T. Sakaguchi, A. Uno, and S. Tsugawa, "Inter-vehicle communications for merging control," Proceedings of IEEE International Vehicle Electronics Conference, pp. 365-370, 1999.
106. X.-Y. Lu and K. J. Hedrick, "Longitudinal control algorithm for automated vehicle merging," Proceedings of IEEE Conference on Decision and Control, vol. 1, pp. 450-455, 2000.
107. X.-Y. Lu, and H.-S. Tan and S. E. Shladover, et. al, "Automated vehicle merging maneuver implementation for AHS," Vehicle System Dynamics, vol. 41, no. 2, pp. 85-107, 2004.
108. S. Halle and B. Chaib-draa, "Collaborative driving system using teamwork for platoon formations," Applications of Agent Technology in Traffic and Transportation, Whitestein Series in Software Agent Technologies, F. Klggl et. al, eds., Birkhauser Verlag, 2005.
109. A. Kanaris, E. B. Kosmatopoulos, and P. A. Ioannou, "Strategies and spacing requirements for lane changing and merging in automated highway systems," IEEE Transactions on Vehicular Technology, vol. 50, no. 6, pp. 1568-1581, 2001.
110. Q. Xu, K. Hedrick, and R. Sengupta, et. al, "Effects of vehicle-vehicle/roadside-vehicle communication on adaptive cruise controlled highway systems," Proceedings of IEEE Vehicular Technology Conference, vol. 2, pp. 1249-1253, 2002.
111. P. A. Ioannou and M. Stefanovic, "Evaluation of ACC vehicles in mixed traffic: lane change effects and sensitivity analysis," IEEE Transactions on Intelligent Transportation Systems, vol. 6, no. 1, pp. 79-89, 2005.
112. S. Arora, A. K. Raina, and A. K. Mittal, "Collision avoidance among AGVs at junctions," Proceedings of the IEEE Intelligent Vehicles Symposium, pp. 585-589, 2000.
113. S. Tsugawa, S. Kato, and T. Matsui, et. al, "An architecture for cooperative driving of automated vehicles," Proceedings of IEEE Intelligent Transportation Systems, pp. 422-427, 2000.
114. T. Nishi and T. Takagi, "A proposal of collision avoidance algorithm for driving support system," Proceedings of Annual Conference of IEEE Industrial Electronics Society, vol. 1, pp. 80-83, 2001.
115. S. Tsugawa, S. Kato, and K. Tokuda, et. al, " A cooperative driving system with automated vehicles and inter-vehicle communications in Demo 2000," Proceedings of IEEE Intelligent Transportation Systems Conference, pp. 918-923, 2001.
116. S. Kato, S. Tsugawa, and K. Tokuda, "Vehicle control algorithms for cooperative driving with automated vehicles and inter-vehicle communications," IEEE Transactions on Intelligent Transportation Systems, vol. 3, no. 3, pp. 155-161, 2002.
117. W. Enkelmann, "FleetNet - applications for inter-vehicle communication," Proceedings of IEEE Intelligent Vehicles Symposium, pp. 162-167, 2003.
118. A. Dogan, G. Korkmaz, and Y. Liu, et. al, "Evaluation of intersection collision warning system using an inter-vehicle communication simula-

tor," Proceedings of International IEEE Conference on Intelligent Transportation Systems, pp. 1103-1108, 2004.
119. L. Li and F.-Y. Wang, "Cooperative driving and lane changing at blind Crossings," Proceedings of IEEE Intelligent Vehicles Symposium, pp. 435–439, 2005.
120. J. K. Hedrick, M. Tomizuka, and P. Varaiya, "Control issues in automated highway systems," IEEE Control Systems Magazine, vol. 14, no. 6, pp: 21-32, 1994.
121. A. R. Girard, J. B. de Sousa, and J. A. Misener, et. al, "A control architecture for integrated cooperative cruise control and collision warning systems," Proceedings of IEEE Conference on Decision and Control, vol. 2, pp: 1491-1496, 2001.
122. O. Gehring and H. Fritz, "Practical results of a longitudinal control concept for truck platooning with vehicle to vehicle communications," IEEE Conference on Intelligent Transportation System, pp: 117-122, 1997.
123. X. Y. Lu, "Coordination layer control and decision making for automated ground vehicles," Proceedings of American Control Conference, vol. 4, pp. 3034-3039, 2002.
124. A. Vahidi and A. Eskandarian, "Research advances in intelligent collision avoidance and adaptive cruise control," IEEE Transactions on Intelligent Transportation Systems, vol. 4, no. 3, pp. 143-153, 2003.
125. S. Tsugawa and S. Kato, "Evaluation of incident information transmission on highways over inter-vehicle communications," Proceedings of IEEE Intelligent Vehicle Symposium, pp. 12-16, 2003.
126. S. Muller, M. Uchanski and J. K. Hedrick, "Cooperative estimation using vehicle communications," Vehicle System Dynamics, vol. 39, no. 2, pp. 121-133, 2003.
127. W. Li and H. Leung, "Simultaneous registration and fusion of multiple dissimilar sensors for cooperative driving," IEEE Transactions on Intelligent Transportation Systems, vol. 5, no. 2, pp. 84-98, 2004.
128. D. Huang and H. Leung, "An expectation-maximization-based interacting multiple model approach for cooperative driving systems," IEEE Transactions on Intelligent Transportation Systems, vol. 6, no. 2, pp. 206-228, 2005.
129. C. Chan and B. Bougler, "Evaluation of cooperative roadside and vehicle-based data collection for assessing intersection conflicts," Proceedings of IEEE Intelligent Vehicles Symposium, pp. 165–170, 2005.
130. Y. Liu, U. Ozguner, and E. Ekici, "Performance evaluation of intersection warning system using a vehicle traffic and wireless simulator," Proceedings of IEEE Intelligent Vehicles Symposium, pp. 171–176, 2005.
131. K. Tischler, and B. Hummel, "Enhanced environmental perception by inter-vehicle data exchange," Proceedings of IEEE Intelligent Vehicles Symposium, pp. 313-318, 2005.
132. B.-W. Chuang, J.-H. Tarng, and J. Lin, et. al, "System development and performance investigation of mobile ad-hoc networks in vehicular envi-

ronments," Proceedings of IEEE Intelligent Vehicles Symposium, pp. 302-307, 2005.

133. A. Benmimoun, J. Chen, and D. Neunzig, et. al, "Communication-based intersection assistance," Proceedings of IEEE Intelligent Vehicles Symposium, pp. 308-312, 2005.

Intelligent Vehicle Vision Systems

8.1 Introduction

Vision sensors become increasingly important in smart transportation systems in the last twenty years. From the viewpoint of traffic infrastructure, the application of vision based traffic flow monitoring and associated control techniques offers considerable improvements over those existing methods of traffic data collection and road traffic monitoring [23]-[25]. On the other side, with the help of visual sensors, modern driver assistance system can alert the driver for dangerous situations or wrong actions, i.e. swerving off lane/road or disregarding traffic signs/lights; or even autonomously take the control of vehicle. Now, vision based driving safety system is world- widely viewed as an essential component of the next generation vehicles [1]-[22].

During the last three decades, vision based driving safety research has attracted great interests and resulted in a wealth of promising results. From their functions, varied vision sensors can be divided into two types: "looking-in" sensors that detect out-vehicle environment and "looking-out" sensors that monitor in-vehicle environment; Fig.8.1.

CMOS/CCD cameras are the most frequently used vision sensors in intelligent vehicle. As pointed out in many literals, CMOS/CCD cameras are superior to many other sensors due to their fast response, easy installation/operation/maintenance, and ability to monitor wide areas. But CMOS/CCD cameras perform perfect only when they work under good environment conditions. In real scenarios, shadows of trees/clouds and dim lighting etc. prevent us from obtaining clear contrasting digital images occasionally.

Recently, applications of mm-wave radar, laser and infrared sensors as alternatives to cameras got significant considerations. One important merit of these new sensors is that they are not sensitive to weather/lighting conditions. Although CMOS/CCD cameras cannot be thoroughly alternated, they are outperformed by these new sensors in lots of applications. In this

Chapter, these new sensors are also viewed as general vision sensors and discussed. And the sensor fusion between different vision sensors is addressed, too.

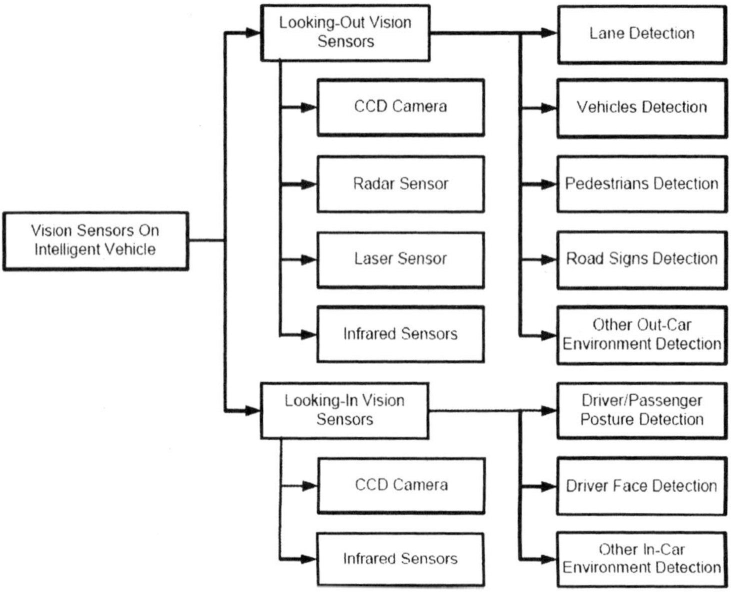

Fig.8.1 Divisions and functions of "looking-in" and "looking out" intelligent vision sensors.

As mentioned in [9], [22], the basic visual sensing tasks of an intelligent vehicle can be outlined as follows:

out-vehicle vision detection:

(1) extracting lane boundaries, especially when they are not clearly marked and/or in bad weather conditions;

(2) detecting vehicles nearby, estimating their positions, speeds, and accelerations;

(3) recognizing the relevant traffic signs and traffic lights;

(4) detecting the unexpected traffic participants/obstacles, such as pedestrians and big stones;

in-vehicle detection:

(5) monitor the driver's eyes movement for fatigue detection;

(6) monitor the driver's hands movement for driving behavior learning;
(7) monitor the driver/passenger's posture and head for applying smart-airbag.

To offer a snapshot of current activities around the world for researchers who are involved in this field, this Chapter looks into these research areas sequentially. Some related research and remained challenges are also discussed, in which the following two topics are emphasized: vision and non-vision sensor fusion; vision information sharing via infrastructure-vehicle/inter-vehicle communications. The focus of this Chapter is on recent literals, since excellent reviews already existed in the previous associated studies, i.e. [1]-[25], [93]-[94], [162]-[164].

8.2 Advances in Vision Based Lane/Road Detection

Lane detection is a key function of any driving assistance systems, and one of the privileges of computer vision. Its main task is to determine the relevant road/lane boundaries without prior knowledge of the road geometry. To achieve this goal, varied methods had been proposed in the last thirty years [26]-[70].

From the aspect of detection sensors, related research can be divided into two classes: conventional CMOS/CCD cameras and radars based approaches, and novel LiDAR based ones. From the viewpoint of detection algorithms, related research can be distinguished into two classes, namely lane-region detection and lane-border detection (lane markings and road edges). On the other side, different approaches can also be categorized by the number of sensors applied, etc.

8.2.1 Lane/Road Detection Using CMOS/CCD Camera/Radars

In ideal scenarios, lane/road detection is easy to solve using standard edge-line detection methods, especially in highway scenarios when the road markings are clear and the lane has a well-defined geometry.

Lane/road detection via lane-border detection using camera/radar image/video is the most frequently used approach in this filed. Its main procedure can be described as follows:

Algorithm 8.1: Hough transform based lane-border detection

1) set up the projection relationship between the world coordinate and the image coordinate; map the obtained image into the corresponding bird view image;
2) assume the lane-borders are marked as (nearly) straight lines; find first the edges and then the lines in the bird view image, i.e. using Hough transformation [26];
3) determine the potential lane-border from these possible lines.

The detection accuracy for above algorithm mainly depends on quality of images, edge/line detection accuracy and speed. Because how to find edges in an image had been systematically studied in [27] and the literals follows, now the main trend is to design fast, adaptive and robust Hough transformation for road detection [28]-[31]. Especially, multi-resolution and parallel Hough transform received notable attentions, since detection speed is quite important to intelligent vehicle applications.

The biggest constraint for Algorithm I is that it assumes the lane borders are straight lines, which usually does not fit actual situations; see Fig.2(a). Therefore, template based lane-border detection methods received constant efforts from early 90s, which can be roughly represented as follows:

Algorithm 8.II: Template based lane-border detection
1) same as Algorithm 8.I.1);
2) assume the lane-borders can be described by certain curves in the bird view image coordinates; find the edges and then the lines in the bird view image;
3) determine the potential lane-borders by matching the template lane-border curves and those edge points obtained in 2).

The core part of a template based lane detection algorithm is to find an appropriate function set to describe the lane-border. First, such a curve function set should be able to depict varying road curvatures. Second, it needs to be easy to identify. Third, it must be robust to the noises that are generated by image capture and transformation, shadows of trees, dim light and unclear road markings. For instance, Likelihood of Image Shape (LOIS) lane detection algorithm was presented as a successful template based method in [32]-[35]. Some heuristic likelihood functions were used to provide a relative measure of how well a given set of shape parameters match the data in a particular image of the road scene. Finally, different optimization algorithms had been applied to determine the best set of road shape parameters for a given image by maximizing the likelihood function.

More precisely, the descriptive function in [32] is chosen as the parabolic curve written as

$$x = k_{y1}y + k_{y2}y^2 \pm b \tag{8.1}$$

where the parameters k_{y1}, k_{y2} denote the road curvature, m is the tangential orientation of the road, and b is the offsets of left and right lane border. x and y are the horizontal and vertical coordinates of the edge points on the road curves.

For instance, in [35], the road boundaries were described as parabolic curves on a flat earth, which were written as

$$x = \frac{1}{2}ky^2 + my + b \tag{8.2}$$

where the parameter k is the road curvature, m is the tangential orientation of the road, and b is the offsets of left and right road boundaries, with respect to the host vehicle. Here, x and y denotes the world coordinates of points on the road curves.

where the parameter k is the road curvature, m is the tangential orientation of the road, and b is the offsets of left and right road boundaries, with respect to the host vehicle. Here, x and y denotes the world coordinates of points on the road curves.

$$x_L = \frac{1}{2}ky_L^2 + my_L + b_L, \quad x_R = \frac{1}{2}ky_R^2 + my_R + b_R \tag{8.3}$$

In the perspective 2D projection image for the road ahead, the general parabolic curves (8.1) would be transform into hyperbolic curves in the image plane and be rewritten as

$$c = \frac{k'}{r - hz} + b' \cdot (r - hz) + vp' \tag{8.4}$$

where

$$k' = \eta_k k, \quad vp' = \eta_m m + \eta_{m,k} k + \eta, \quad b' = \eta_b b + \eta_{b,m} m + \eta_{b,k} k \tag{8.5}$$

In (8.3)-(8.5), hz is the row in the image plane corresponding to the horizon of the ground plane. k' is linearly proportional to the curvature of the arc on the ground plane. vp' is a function of the tangential orientation of the curve on the ground plane, with some coupling to the arc curvature as well. b' is a function of the offset of the curve from the camera on the ground plane, with couplings to arc curvature and tangential orientation. r and c denotes the row and column position of those curve points on the image.

The penalty function proposed is written as

$$f(\alpha, x) = 1/(1+\alpha x^2) \tag{8.6}$$

This penalty function had been proven to be fairly robust on marked roads, since it encoded the knowledge that edges of the lane should be near intensity gradients whose orientation should be perpendicular to the lane edge.

Let $gm(r,c)$ be the gradient magnitude at pixel (r,c), with $gd(r,c)$ being the corresponding gradient direction, and the prior likelihood function is

$$P(s) = \frac{2}{\pi} \cdot \operatorname{atan}(\alpha_{width} \cdot (b'_R - b'_L)) \tag{8.7}$$

where b'_R and b'_L represent the function of the offset of the left and right curve on the ground plane. And s denotes the given parameter set.

Eq.(8.7) embodies two prior constraints on the lame geometry. First is that left boundary is to the left of the right boundary. Thus, Eq.(8.7) should be positive for right road parameters. Second is that the road segment is not too narrow. The parameter α_{width} in Eq.(8.7) is used to handle different road width. In [35], α_{width} is set to be 10.

In general, the likelihood function for a given road image and a given parameter set s is defined as

$$\begin{aligned} L(s) = P(s) \cdot \sum_{r,c} [&gm(r,c) \cdot f(\alpha_{prox}, c - E_L(s)) \cdot \\ &f(\alpha_{orient}, \cos(grad(r,c) - \operatorname{atan}(\frac{d}{dr} E_L(s)))) + \\ &gm(r,c) \cdot f(\alpha_{prox}, c - E_R(s)) \cdot \\ &f(\alpha_{orient}, \cos(grad(r,c) - \operatorname{atan}(\frac{d}{dr} E_R(s))))] \end{aligned} \tag{8.8}$$

where

$$E_L(s) = \frac{s_1}{r - hz} + s_2 \cdot (r - hz) + s_3, \; E_L(s) = \frac{s_1}{r - hz} + s_4 \cdot (r - hz) + s_3 \tag{8.9}$$

In (8.8), the contribution of a pixel to the likelihood is the gradient magnitude at that pixel multiplied by a function whose value decrease as the pixel column gets further form the curve and a function whose value decreases as the gradient direction at the pixel moves further away from perpendicular to the tangent to the curve. Here the weight coefficients α_{width} and α_{width} were set at 0.01 and 1.13.

Finally, the global optimal parameters set for $L(s)$ will lead to the best estimation of road curves. However, usually only local optimal solutions will be obtained, which indicate acceptable estimation of the road segments. In [35], a metropolis algorithm was applied to search the optimal matched parameter set, which is indeed a geometric annealing schedule. Besides, GA was also used to solve such kind of optimization problems in [43].

Besides Eq.(8.8), there were also some other likelihood evaluation functions. For example, in [33], the following likelihood evaluation function was employed:

$$
\begin{aligned}
L(s) = {} & I_{b'_L > b_L}(b'_L, b_L) \cdot I_{b'_R > b_R}(b'_R, b_R) \cdot \delta(k' - \eta_k k) \\
& \cdot \delta(vp' - [\eta_m m + \eta_{m.k} k + \eta]) \\
f(\alpha_{orient} & , \cos(grad(r,c) - a\tan(\frac{d}{dr} E_L(s)))) + \\
& \cdot \frac{2}{\pi} \operatorname{atan}[s'_1 \cdot (b'_R - b'_L)] \cdot \frac{2}{\pi} \operatorname{atan}[s'_2 \cdot (b'_R - b'_L)] \cdot \\
& \cdot \frac{2}{\pi} \operatorname{atan}[s'_3 \cdot (b_R - b_L)] \cdot \frac{2}{\pi} \operatorname{atan}[s'_4 \cdot (b_R - b_L)]
\end{aligned}
\tag{8.10}
$$

where $I_A(x, y)$ is an indicator function that satisfies

$$
I_A(x, y) = \begin{cases} 1, & \text{if } (x, y) \text{ satisfy relation } A \\ 0, & \text{otherwise} \end{cases}
\tag{8.11}
$$

Obviously, the first two terms in Eq.(8.10) impose the constraint that the lane markers should be within the road region, while the last two terms impose the constraint that the lane boundaries' curvature and orientation should be precisely related to the pavement boundaries' curvature and offset in Eq.(8.8).

The likelihood function in [32] is chosen to embody several prior constraints on the lane geometry:

(1) left boundary is to the left of the right boundary;

(2) the lane segment is not too narrow and the lane width should be relatively constant;

(3) the curvature of the road should not vary too intensively.

It had been proven that these three heuristic rules capture the most important features of lane border [32]-[35]. There are also some other heuristic rules proposed recently, i.e. road direction (or orientation) mentioned in [36]. Since all of these objective functions are too complex, they are not expressed here.

Besides polynomial functions, some different functions have also been used to describe lane-border curves. For instant, The road curves was modeled as a piecewise function Eq.(8.12). in [31]. B-spline snake in [37], cubical function set in [38], and spline model in [39] were applied to represent road boundaries, also.

$$\begin{cases} x = k_{y3}y, & \text{if} \quad y \le y_m \\ x = k_{y1}y + k_{y2}y^2 \pm b, & \text{if} \quad y > y_m \end{cases} \tag{8.12}$$

Different from likelihood functions, there are also several other functions that are chosen to be the evaluation objectives for the template algorithms [40]-[41]. For example in [40], a lane-curve function (LCF) is used by directly transforming the defined parabolic function in the world coordinates.

Another important problem of template based lane border detection is how to quickly find the best set of road shape parameters. The evaluation objectives are so complex that they normally contain lots of local optimal solutions. Thus, different approaches i.e. simulation annealing algorithm [42], genetic algorithm [43], dynamic programming [44], Cellular Neural networks (CNN) [45] had been tested and reported.

To improve lane detection accuracy, different cameras and radars including wide angle camera [46] and omni-directional camera [47] had been tested. Notice that to project the 3-D environments into a single 2-D image will unavoidably lose the depth information, lane-border detection based on stereo vision techniques had been addressed in the last ten years [48]-[54]. One important benefit of using stereo visions is that the road slope can be further described by extending the curve functions in terms of road surface latitude variations; see Fig.8.2(middle).

Fig.8.2 Three frequently encountered problems: (left) lane detection on roads with high horizontal curvature and shaded lane markers; (middle) lane detection on non-flat roads with vertical curvature; and (right) lane detection with the presence of obstacles on current lane; from [53]. (© [2004] IEEE)

For example, in [51]-[53], the 3-D reconstruction algorithms are used to capture the distortion of road curves in bird view coordinate caused by road slopes, which may not be correctly explained by 3-D model like Eq.(1). Out of the many possible lane boundary candidates, the best one is then chosen as the one at a minimum distance from the previous lane vector according to a weighted distance metric in which each feature is assigned a different weight. But the road curve matching/identification cost increase simultaneously, since a third dimensional variable z is added.

To use image compression and feature extraction method to sharpen the lane borders is also an interesting topic in this filed [34], [41]. For example in [34], discrete cosine transform (DCT) was carried out. The obtained road image was divided into 8×8 blocks of pixels. Each pixel block is then orthogonally decomposed in terms of a set of 64 DCT basis elements. Each of these elements, with respect to spatial domain edges of a certain strength and orientation. Out of these 64 elements, diagonally dominant edges are best represented by a set of 12. The corresponding feature image is finally gotten by summing the squares of its 12 special DCT decompositions. Despite the original image having features/edges of various strengths and orientations, the corresponding DCT feature images contain only information about those edges which are diagonally dominant. The authors claimed that DCT-based features are capable to appropriately de-noise and help to precisely locate the lane markings, see Fig.8.3.

Fig.8.3 DCT features of typical roadway scenery, (left) original image and (right) edge map build by using the corresponding DCT features, from [34]. (© [1999] IEEE)

Different from the above "hard" computation algorithms, some "soft" decision/computation algorithms were developed during the past decade, too. Generally, its idea can be presented as follows:

Algorithm 8.III: Soft decision based lane-border detection

1) same as Algorithm 8.I.1);

2) assume the lane-borders can be described by certain curves in the bird view image coordinates; find the edges and then the lines in the bird view image;

3) directly eliminate the false lane-border lines using some special image filters or "human-like" decision rules.

For example, the so called "steerable filter" was used in [54], which is indeed a space direction filter based on road starting position and direction/orientation. It can provide robustness to lighting changes and shadows and is claimed to perform well in picking out circular reflector road markings and painted line road markings. In [55], Bayesian judgment based algorithms were used to directly localize the potential lane-border lines. In [56], the characteristics of the gray level histogram of the road were exploited to detect lane markers. The common extracted features include:

(1) Maximum angle spanned by the object with respect to the vanishing (parallel lane-border converge) point in the image;

(2) Absolute value of the difference between pre-estimated lane-border angle and the real line angle in the image.

For example, in [56], each lane marker is analyzed using a special decision tree, and finally the relations between the lane markers are analyzed to create the structures defining the lane boundaries.

Lane detection task is sometimes quite complex, i.e. in the following scenarios:

(1) the road and surrounding areas are hard to discriminate. One typical example for those northern countries is the road that is covered by ices and snows; see Fig.8.4(left). Another case is the unstructured roads, such as those rural roads consist of mud, clay or sand;

(2) the lane markings are shaded by vehicles or other obstacles on the road; see Fig.8.2(right).

In these scenarios, the road/non-road borders are spatially fuzzy and have low intensity contrast, the overall road shape is uncertain and may even change drastically. Thus, lane-region detection is normally applied, in which road detection is viewed as a process of segmentation and amalgamation. The final two segments obtained will denote road and non-road respectively.

As pointed out in [57]-[64], the frequently used features for road segmentation include:

(1) color/gray-level/texture similarity. It is normally assumed that road surface would be more-or-less consistent in their mix of colors/gray-levels;

(2) road width and directions. Usually, the manual constructed road has relatively regular contours and width in the obtained image.

Generally, the associated approaches can be divided into two kinds, whose general ideas can be written as follows:

Algorithm 8.IV: Direct binary segmentation based lane-region detection

1) same as Algorithm 8.I.1);

2) directly divide the image into two parts: road and non-road, with coherent color/grey-levels.

Algorithm 8/V: Segmentation and amalgamation based lane-region detection

1) same as Algorithm 8.I.1);

2) divide the image into different small segments with coherent color/grey-levels;

3) merge the obtained segments into two groups: road and non-road, according to the road boundaries, width and direction hints.

In [60], the histogram of the whole image's gray-level is frequently used to directly distinguish the road and non-road parts. In this way, the process is adjusted to the different illumination conditions and variable contrast between tracks and ground: a constant density of edges is obtained by lowering the threshold when the luminance difference is low and raising the threshold when the contrast is relatively high. Here, the threshold value is determined from a cumulative histogram of the gray-level intensity values. In [61], the chromatic and luminance distances among every image pixel are analyzed to distinguish the road and non-road parts. Besides, a brightness attenuation technique had been devised analogously to handle the shadows etc.

Morphological technique is a representative segmentation and amalgamation lane-region method. For instance, Beucher et. al proposed the so called watershed segmentation algorithm in [62]-[63].

Let's consider a graph f as a topographic surface in terms of brightness. The surface presents the "minima", which are connected regions.

From those "minima" areas, it is impossible to reach a point that locates at lower altitude by an always descending path, see Fig.8.4. Suppose that "minima" are pierced and the topographic surface is immersed in water. Apparently, the water will pour down to the holes, especially to those deepest ones. During the flooding process, dams are built at any point where waters coming form two different minima may merge. Therefore, at the end of the flooding, divide lines will appear, which is called watersheds of graph f . Those different connected components separated by the watershed lines are called catchments basins, each on being associated to a single "minima", see Fig.8.5.

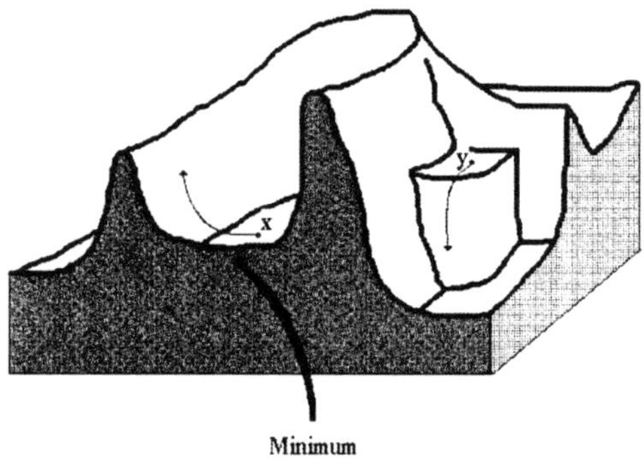

Fig.8.4 "Minima" of a graph surface, from [62].

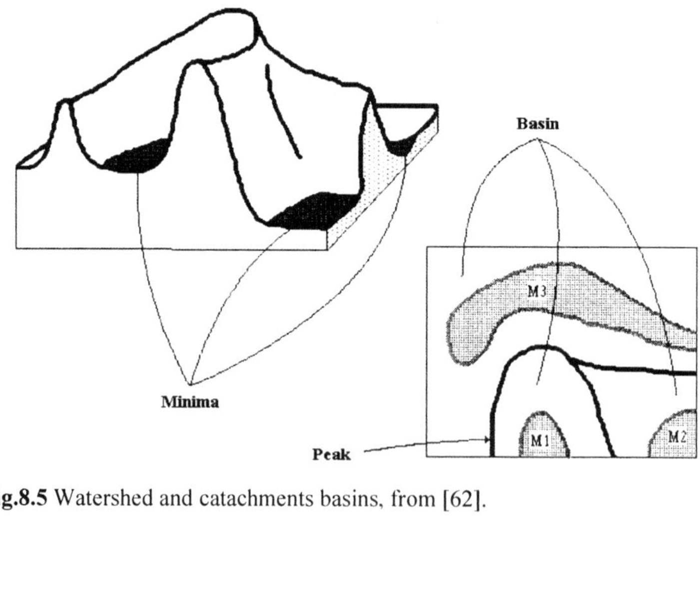

Fig.8.5 Watershed and catachments basins, from [62].

In the segmentation process by watersheds, it is assumed that any object in an image is characterized by a homogeneous texture and hence a weak gradient. The objects in an image will lead to the "minima" in the morphological gradients, and their contours can be determined by the watersheds too. Finally, eliminate those trivial "minima", the reconstructed graph surface can be achieved, which has been appropriately segmented, see Fig.8.6. Fig.8.7 demonstrates its detection process for a road covered by snows. It sequentially (left-top to right-bottom) displays the original image, the corresponding morphological gradients, the initial "minima" that are found, the merge process which discard the trivial edge information, the final road segmentation results and the final detection result represented in normal image.

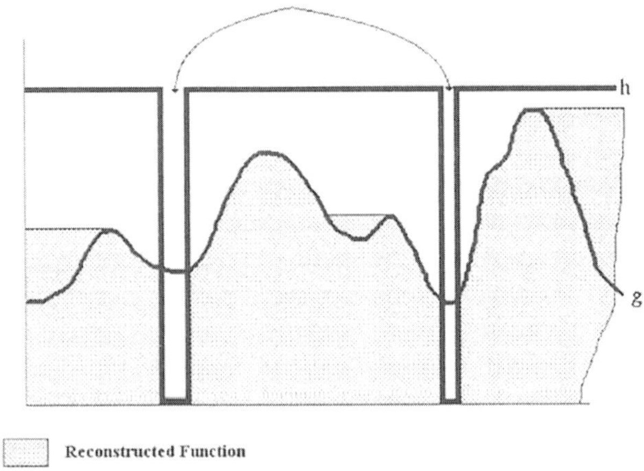

Fig.8.6 Surface reconstruction, from [62].

Fig.8.7 A typical watershed road segment process; from [62].

8.2.2 Lane/Road Detection Using LiDAR and Laser Sensors

Recently, LiDAR based road surface detection approaches gained significant attentions [64]-[70]. One basic approach can be roughly explained as follows. The scanner is mounted in front of the test vehicle in such a way that the measuring plane cuts the ground plane at a certain distance, see Fig.8.8(left-top). Because of different reflect coefficients of ground, the road surface will show as a straight line in the cutting curves, see Fig.8.8(right-top). Therefore, the detection algorithm can be written as:

Algorithm 8.VI: road detection using front LiDAR

1) same as Algorithm 8.I.1);

2) detect the road surface road by detecting horizontal line segments in the transformed image;

3) By recording all the found road edge segments, the road boundaries can be sequentially detected precisely as the vehicle moves ahead. An illustrative detection result is shown in Fig.8.8(bottom).

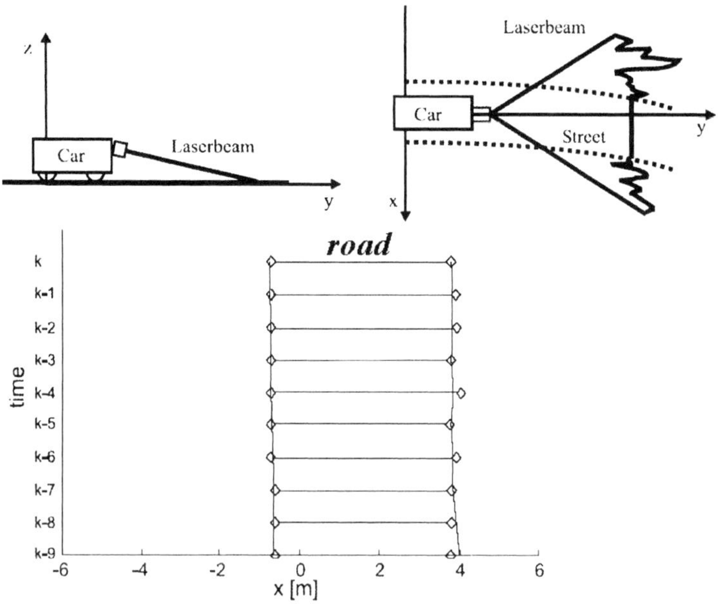

Fig.8.8 Embedding of the laser scanner and road tracking result; from [65].

It is pointed out in [70] that the lane detection performance can be influenced by mounting position and quantity of the sensors. An additional aspect for the quality of object and lane detection is the horizontal resolution between the single laser beams. As shown in Fig.8.9, the distance measurement d should follow

$$y = h(x) = d \sin \gamma \sin \beta \tag{8.13}$$

where the angle γ represents the vertical inclination angle and β denotes the horizontal angle of the outgoing beam. The measurement Eq.(3) is expressed in terms of world Cartesian coordinates for simplicity.

Based on Eq.(8.13), it is obvious that the uncertainty introduced into y result from the measurements of distance and angle. Because a change of the curvature c has low impact in near range where the sensor is mainly expected to detect the lane, to improve detection accuracy, the following circular model (8.14) is chosen to create the EKFs in [68]-[70],

$$y = h(x) = (1/c + \Delta y_{off})\cos \Delta\psi \pm \sqrt{(1/c - ab) - (x - (1/c + \Delta y_{off})\sin \Delta\psi)^2} \tag{8.14}$$

where a is used to distinguish the right ($a = 0$) and the left ($a = 1$) lane. Δy_{off} describes the lateral distance of the vehicle's center to the right lane, b the lane's width, and $\Delta \psi$ is the angle between the vehicle's longitudinal axis and the lane.

As pointed out in [67], a very interesting point is that the resulting error of y is very small and the system is very robust towards pitch movement. To use laser sensor for road boundary detection is another hot topic recently. Its main goal is to accurately determine the distance from the vehicle to the curb at one location. By linking these points that indicate the curb, the road boundary will be determined [71]-[73]. More precisely, the light of a laser is fanned out to form a plane of light, see Fig.8.10.

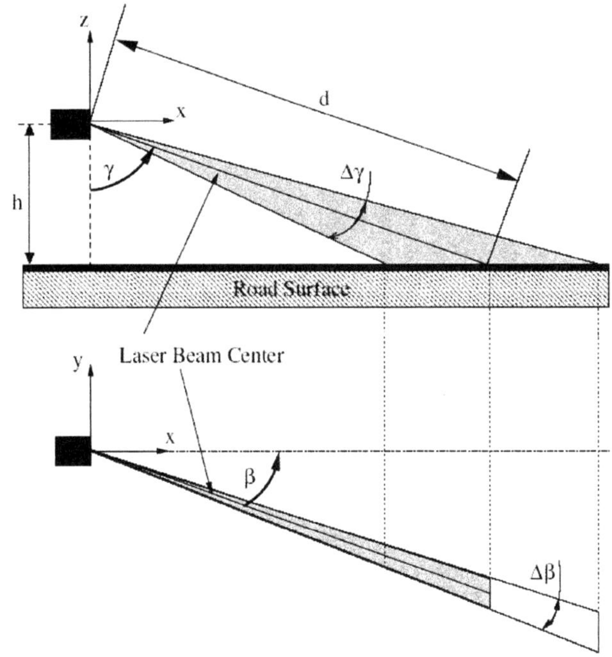

Fig.8.9 LiDAR sensor and its detection ranges, from [67]. (© [2003] IEEE)

If an object intersects the plane, the reflected light is observed by a camera placed at a distance from the laser. The distance between laser and object can be determined by triangulation:

$$d = s \cdot \cos \alpha \tag{8.15}$$

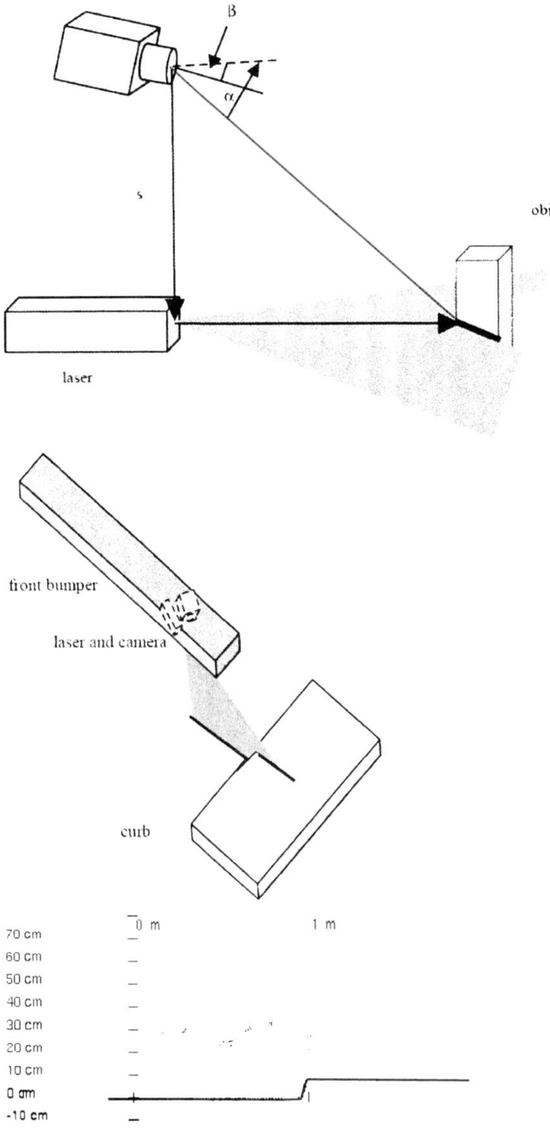

Fig.8.10 (top) Geometric configuration of a laser line striper; (middle) Schematic of the curb detector; (bottom) Profile of the road/curb observed by the laser line striper, where some erroneous readings can be seen, from [72]. (© [2003] IEEE)

where s is the camera-laser distance and α the angle at which the camera sees the object, see Fig.8.10(top).

To make a system work outside, where the sun produces a strong background of light, two techniques were employed in [72] to reduce the background. In the one hand, the camera is equipped with a narrow filter and in the other hand the laser is pulsed and in synchronization with the fast shutter of the camera. For the purpose of curb detection the laser line striper is placed beneath the front bumper of the vehicle looking to the side. A schematic of the arrangement is shown in Fig.8.10(middle) and Fig.8.10 (bottom). Besides the above tests, laser based 3D environment, including road, reconstruction were also described in [74]-[76].

8.2.3 Integration of Lane/Departure Detection and Localization

During the last twenty years, people gradually realized that lane detection and lane departure monitor problems should be considered simultaneously, especially for the following two reasons. First, the vehicle's heading angle may affect the lane detection results; second the lane curvature information is necessary for vehicle departure measurement [77]-[86].

The position and vision sensor fusion received considerable interests now. For example in [83], a LiDAR is used to measure the lane and vehicle's heading angle in an indoor scenario. Since LiDAR exhibits less measurement accuracy than Inertial Navigation Systems, the authors combined the two sensor systems for better results. However, it should be pointed out that for the crowed urban areas where the obtained view is too narrow, magnetic position/navigation system seems to be a better alternative than vision systems [87]-[89].

Localization is also an important functionality for navigating intelligent vehicles. However, the data obtained from GPS and cameras is sometimes uncertain and or even momentarily unavailable (in urban areas, for example). The problem of GPS and vision sensor fusion based localization was studied in [90], which combined GPS absolute localization data with the data computed by a vision system to provide accurate position and orientation measurements. The position and orientation data was transformed into a global reference using a map of the environment and then estimate localization parameters using a particle filter. This lets them manage multimodal estimations, because the vehicle can be in the left or right lane. The best precision can supposedly reach 48 cm along the road axis and 8 cm along the axis normal to the road.

In [91], a framework is presented for robust estimation of vehicle roll, pitch, yaw and obstacles detection using stereo vision. On the other side, analysis of the effect form vehicle movement on vision accuracy was also discussed in some literals including [92]. Besides, how to handle lane detection error in vehicle lateral control is another important problem. Some discussions can be found in [93]-[94].

8.3 Advances in Vision Based Vehicle Detection

To avoid collision, an intelligent vehicle system should be able to detect static (normally obstacles) or moving objects (normally vehicles or pedestrians) nearby. Due to the difficulty of this function, numerous efforts had been put into this field in the last thirty years.

8.3.1 Vehicle Detection Using CMOS/CCD Cameras

Generally, the majority of reported approaches follow three basic steps: Hypothesis Generation (HG) in which the Area of Interests (AOI) are located; Hypothesis Verification (HV) in which tests are performed to verify the presences of a vehicle, and vehicle tracking.

1) Hypothesis Generation

Most HG methods can be categorized into three kinds [19]: knowledge-based methods using 2D image, stereo vision based methods, and motion based methods using video. Knowledge-based methods employ a-priori knowledge to find potential vehicles in an image. Usually, the following cues are taken:

(1) vehicle geometry structures: it is frequently assumed that road vehicles, especially cars and lesser extent lorries, consists of a large number of horizontal structures, particularly when viewed approximately from the rear, e.g. rear-window, boot, bumper; Fig.8.11(a). This leads to vehicle detection based on geometry structures [95]-[96]. For instance in [95], the horizontal overlap assumption was applied to each image column may result in several Areas of Interest. The horizontal edge response in each image column is summed and smoothed with a triangular filter. And each locally maximal peak which is extracted from the smoothed column responses will indicate a potential vehicle, see Fig.8.11(b).

(2) shadow beneath the vehicle: in daytime, the vehicle's vertical location need to be further examined by checking the position of the shadow under a vehicle, see Fig.8.11(c). The major advantage of shadow method is that all potential "vehicles" will be detected, [95], [97]-[99].

(a)

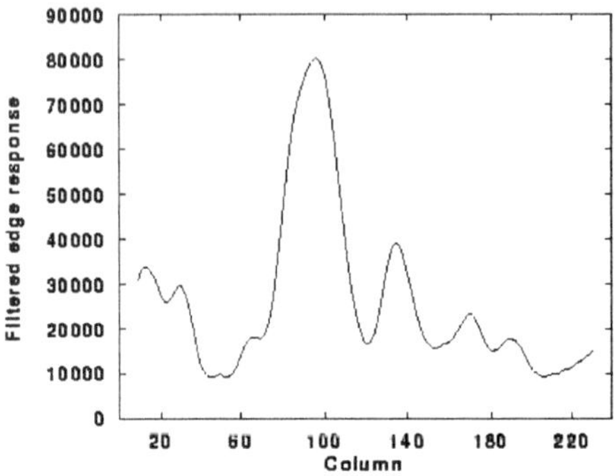

(b)

(c)

Fig.8.11 Detected candidate vehicle from column information of image (a), filtered edge response column sums (b) and under-vehicle shadows (b), from [95].

(3) texture: usually vehicle will have smoother textures than the background. Several different methods are developed using this cue for vehicle detection, i.e. [99]-[101];

(4) symmetry: the figure of almost any vehicle is symmetry on the center axis of level or vertical direction. This feature represents the presence of vehicles of distinctively different nature, and it is normally assumed that their corruption by noise is independent. How to find all the interested symmetry areas in an image is still a hard problem to be thoroughly resolved, [98]- [99], [101]-[104];

(5) color: since color image contains much more information than grey image, some newly developed methods investigated vehicle color cues recently. For instance in [17], it was shown that the vehicles may have distinct colors with respect to its surroundings, which is a useful cue for candidate vehicle location algorithm. And the information of color consistency is also used to classify possible obstacle candidates, since usually the color of a vehicle surface should be consistent, [105]-[106];

(6) rear-lights: this method searches for bright spots in image regions that are most likely to contain rear lights. To reduce the search time, sometimes only the red component of each image frame is analyzed, which is sufficient for rear light detection [126]. This method is especially useful for vehicle detection in nighttime. Daytime vehicle detection is more complicated than nighttime vehicle detection, since it is much more difficult to distinguish the vehicle from the background scene in daytime.

Stereo vision based vehicle detection methods gained more attentions with respect to knowledge-based methods during the last ten years. Normally, the approaches can be divided into two types: disparity mapping and inverse perspective mapping.

The difference in the left and right images between the corresponding pixels is called disparity. Re-mapping the right image onto the left one, or both images onto the road coordinate system, based on the given model of the road in front of the vehicle (e.g. flat straight road) will form the so called disparity- map, which can be further converted into a 3D scene around the vehicle. Obviously, this reconstructed 3D scene can be used to nearby vehicle detection [107]-[110]. One important problem of this method is how to reduce the reconstruction computation cost. Another problem is how to deal with the distortions, i.e. caused by road slope. Some discussion on this topic can be found in [111]-[113].

Another promising approach called as inverse perspective mapping (IPM) is to remove the inherent perspective effects from acquired single or stereo images. The IPM can be applied to stereovision [49], [114]-[116], by re-mapping both right and left images into a common (road) domain.

Suppose the world and image Euclidean spaces are defined as follows, and the relationship between the two coordinate systems is shown in Fig.8.12.

(1) $W = \{(x, y, z)\} \in E^3$, representing the 3D world space (world coordinate), where the real world is defined;

(2) $I = \{(u, v)\} \in E^2$, representing the 2D image space (screen coordinate), where the real world is projected.

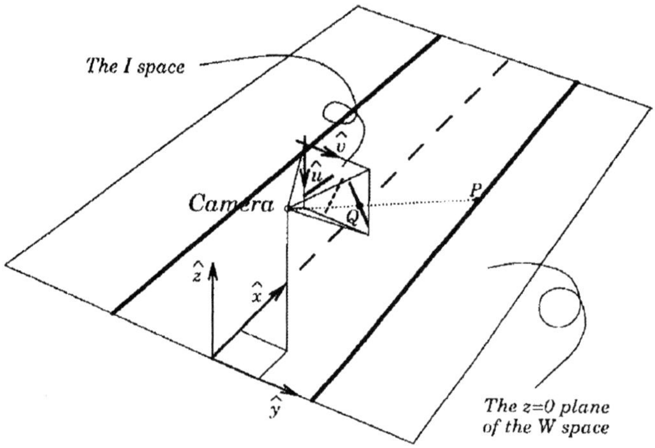

Fig.8.12 Relationship between the two coordinate systems, from [49].

As shown in Fig.8.12, suppose the camera position (viewpoint) in the world coordinate is defined as $C = (l, d, h) \in W$. The optical axis $\hat{o}$ is caculted by the following two angles $\bar{\gamma}$ and $\bar{\theta}$, which detnoes the angle formed by the projection of the optical axis $\hat{o}$ on the plane $z = 0$ and the x axis, and the angle formed by the optical axis $\hat{o}$ and $\hat{\eta}$, respectively.

Thus, the inverse transform $g : W \rightarrow I$ can be written as

$$u(x, y, 0) = \frac{arctg\left[\dfrac{h \sin \gamma(x, y, 0)}{y - d}\right] - (\bar{\theta} - \alpha)}{\dfrac{2\alpha}{n - 1}} \tag{8.16}$$

and

$$v(x, y, 0) = \frac{arctg\left[\dfrac{y-d}{x-l}\right] - (\gamma - \alpha)}{\dfrac{2\alpha}{n-1}} \tag{8.17}$$

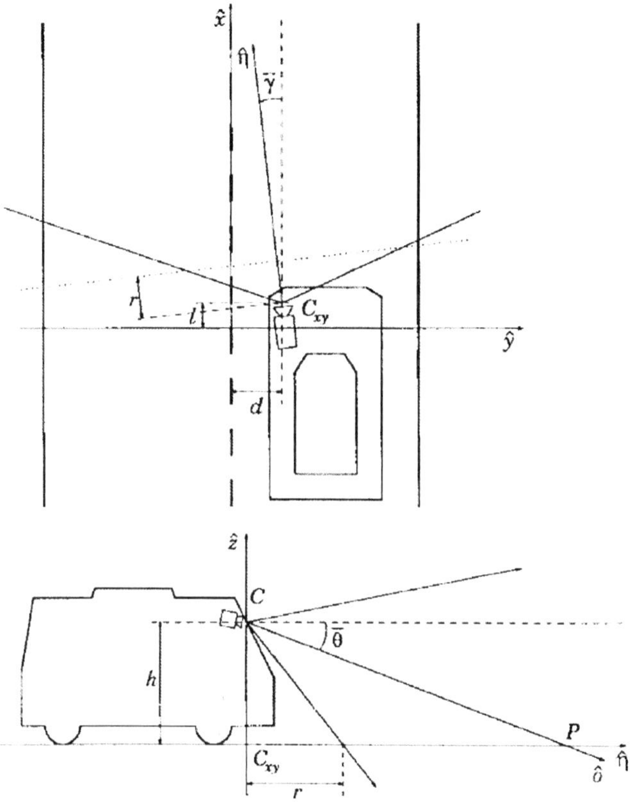

Fig.8.13 (top) The xy plane in the W space and (bottom) the z plane, from [49].

Using this approach, the localization of the lane and detection of generic obstacles on the road can be performed without 3D-world reconstruction. Its prime theory is briefly explained as shown in Fig.8.14. In an ideal case, the difference between the two remapped images presents two triangles due to different coordinates of the projection of the two cameras. Fig.8.14(a) and Fig.8.14(b) depict the left and right views of an ideal white square obstacle on a grid-marked dark background, where the distance between the two cameras are just one grid. Fig.8.14(c) and Fig.8.14(d) illustrate the two corresponding remapped images, respectively; and Fig.8.14(e) demonstrates the thresholded difference, where the overlapping between the two viewing areas indicate the location of the objects; see the red line shown in Fig.8.14(e). The ideal detection and method above would become quite difficult in the real cases, due to texture, irregular shape, and non-homogeneous brightness of real obstacles. To solve the problems, a three step process was adopted to decrease the influence of noise in [48], which was shown to be effective for vehicle detection.

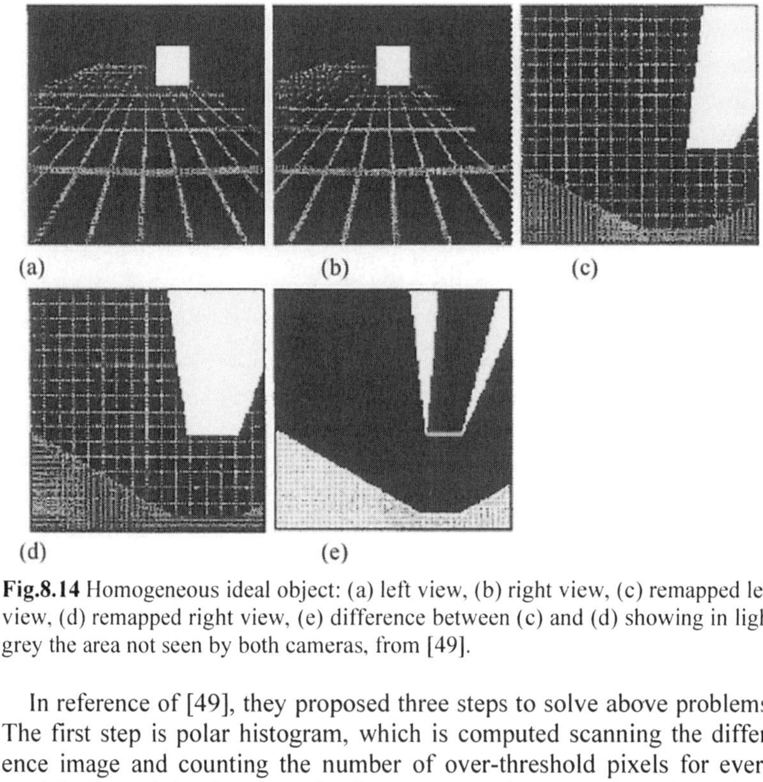

Fig.8.14 Homogeneous ideal object: (a) left view, (b) right view, (c) remapped left view, (d) remapped right view, (e) difference between (c) and (d) showing in light grey the area not seen by both cameras, from [49].

In reference of [49], they proposed three steps to solve above problems. The first step is polar histogram, which is computed scanning the difference image and counting the number of over-threshold pixels for every straight line originating from the focus F , see Fig.8.15. The values of the polar histogram are then normalized using the polar histogram obtained scanning an image where all pixels are set based on reference image. Furthermore, a low-pass filter is applied to decrease the influence of noise.

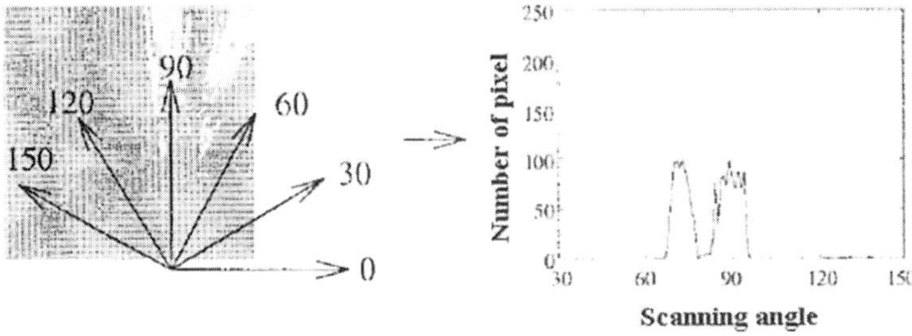

Fig.8.15 Polar histogram, from [49].

The focus of the polar histogram is placed in the middle of $C_{xy}^{(L)}C_{xy}^{(R)}$ in this case the polar histogram presents an appreciable peak corresponding to each triangle. Since the presence of an obstacle produces two disjoint triangles (corresponding to its edges) in the difference image, obstacle detection is reduced to the search for pairs of adjacent peaks; the position of a peak, in fact, determines the angle of view under which the obstacle edge is seen.

As proven in [49], different peaks may have different characteristics, such as amplitude, sharpness, or width, depending on the obstacle distance, the angle of view, and the difference of brightness and texture between the background and the obstacle itself.

The second step is peaks joining, in which two or more peaks can be joined according to different criteria, such as similar amplitude, closeness, or sharpness. Analysis of a large number of different situations showed that it is possible to determine a parameter so as to embed all of the above quantities. Applying peaks joining, it can eliminate the disturbance cause by texture variance occurring in one vehicle. Fig.8.16 shows a typical example, where the two vehicles are correctly detected instead of mistaking for four obstacles on the road.

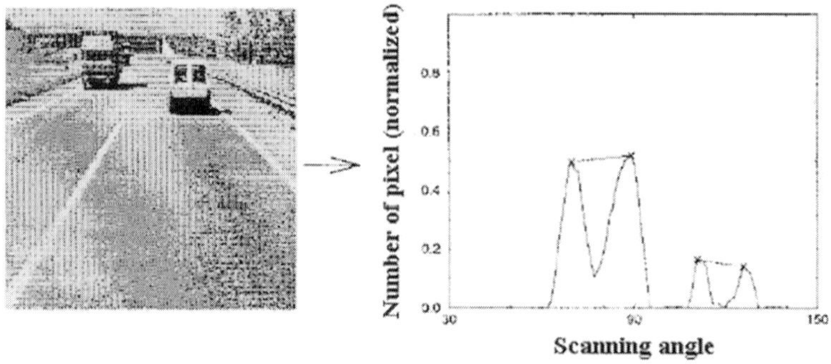

Fig.8.16 Example of peaks join, from [49].

The last proposed step is estimation of obstacle distance, where a radial histogram is computed scanning a specific sector of the difference image for each peak of the polar histogram. The width of the sector is determined as the width of the polar histogram peak in correspondence to the 80% of the peak maximum amplitude as shown in Fig.8.17(a). The number of over-threshold pixels is computed and the result is normalized. The radial histogram is analyzed to detect the corners of triangles, which represents the contact points between obstacles and road plane, thus allowing the determination of the obstacle distance through a simple threshold, see Fig.8.17(b).

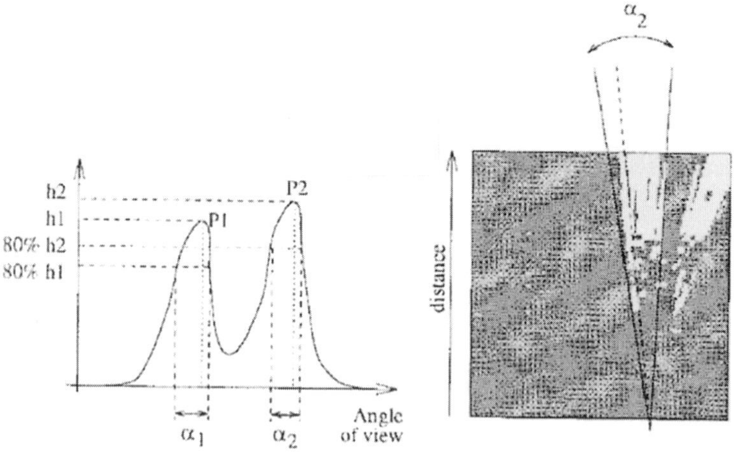

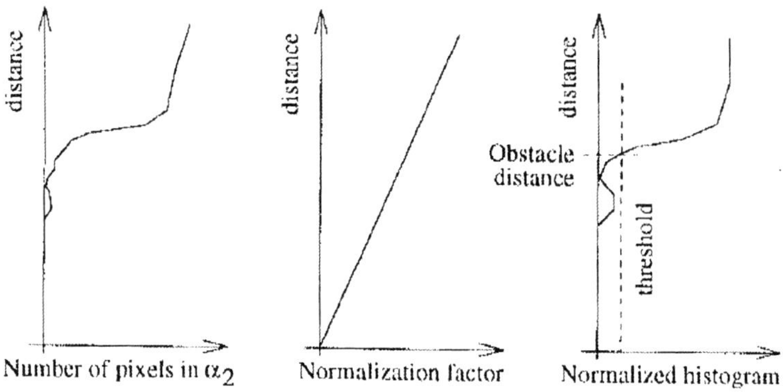

(b)

Fig.8.17 Steps involved in the computation of radial histogram for peak P_2, (a) polar histogram and binary difference image, (b) radial histogram, normalization factor, and normalized radial histogram, from [49].

Motion-based methods are based on the fact that the relative motion between vehicle and background can be detected via the calculation of optical flow, which can be further used to locate the vehicle [117]-[122]. This method is especially useful to detect a vehicle that runs in the opposite direction, since that will yields a diverging flow that can be clearly detected. Any departing or overtaking vehicles will produce a converging flow which is easily to found, too. Vehicle detection using optical flow calculation can be applied to both grey and color images [123], [17]; 2D and 3D scenes [124]. Besides, how to reduce the calculation cost, for example to apply the so called "sparse" optical flow, remained as an interesting topic in this area [15].

2) Hypothesis Verification

In the HV step, the potential locations of vehicles will be further examined, and those hypotheses cannot pass the test will be eliminated. Normally, there are two approaches in this area: explicit rule based methods (template-based) and implicit rule based methods (appearance-based).

The former approaches usually match the potential vehicles with the clearly defined vehicle patterns. For instance, shape and symmetry features are used to identify the vehicle in [125]-[126]. However, explicit rule based methods are not so reliable due to various vehicle size, color and shape. Indeed, vehicle detection depends on mastering the "concepts" of vehicle rather than a specific vehicle pattern set. There is no an easy way to establish an analytical decision rule to separate vehicles from other objects. Recent approaches preferred to learn the generic decision rule for the vehicle class from a set of training examples [127]-[135]. However, as mentioned by [19], in most cases, a large number of features are extracted to compensate for the fact that relevant features are unknown or just a-priori. This naturally leads to notable computation cost on redundant feature information. Recently, how to pick up the predominant features and boost vehicle detection process received more and more discussions [136]-[137]. However, further studies are still expected in the near future.

3) Vehicle Tracking

Tracking algorithms is used to find how an object moves from one frame to the next in the image sequence. General tracking cases usually imply to minimize an error function in terms of the special points existed in two adjacent images. For tracking a particular class of objects such as pedestrians and vehicles, a certain classifier will be employed in advance to distinguish between an object and the background [134]. To track the found vehicles will relief computation burden, since it would be easier find potential vehicles based on their previous location information [138]-[128].

8.3.2 Vehicle Detection Using FMCW Radars

Recently, application of mm-wave radar has received some considerations in vehicle detection area. Most radar equipments are either pulse-modulated radars or continuous wave (CW) radars. Usually, there are two types of CW radar: pure CW radar which relies on the famous Doppler effect to detect moving objects and to measure their speed; and Frequency Modulated CW radar (FMCW) which gives an indication of the distribution range of the target based on the difference in frequency of the reflected wave. The FMCW radar normally use millimeter waves, since a carrier frequency in millimeter wave range allows larger frequency sweeps for better range resolution and limits interface with exiting commercial radio frequent transmitters. Meanwhile, FMCW radar can be robust against interference from other sensors of the same type, because the transmitted signal spreads over a large bandwidth. Since interference can be further reduced by using a coded FM wave form, sometimes the FMCW is called as mm-wave radar. From the investigation of Wenger 1998, most automotive radar projects had chosen FMCW radar as the scanning sensor [151], see Fig.8.18.

The detection theory of FMCW radar can be summarized as follows: radar transmits and receives radar waves, in which the earlier returned signals in power spectra may be use to identify possible obstructions in front of vehicle. If the later signals exceed a certain power threshold selection criteria, they may be verified against a map of special targets. In the meanwhile, the confirmed target observations can be used to determine the vehicle position, simultaneously [143]-[153]. In short, it can be written as:

Year	Company, Institute	Operation frequency [GHz]	Radar type	Output power [mW]	Technical features (No. of beams)
1992	GEC-Plessey	77	FM-CW	10	bistatic lens antenna (3)
	Fujitsu/Fujitsu Ten	60	FM-CW	3	V-shaped WG antenna (1)
	TEMIC&DASA	77	pulsed	<10	MMIC-based, horns and Fresnel lens (3)
	LUCAS Ltd.	77	FM-CW	15	GEC-Plessey sensor
1993	Millitech	77 (60/140)	pulsed or FM-CW	10	folded optics lens antenna, monostatic or bistatic, with MMICs (3/4)
	Technical Univ. Munich/ Germany	61	PN-code modulated, FSK	1.6	digital waveform reconstruction, 5 rect. horns and diel. lens (4/8)
	VORAD Safety Systems	24.725	FM-CW	<1	bias-swept, installed in 1.700 Greyhound buses (1)
	DASA	77	FM-CW	3	low-cost hybrid, monopulse (2)
	Hino	60	FM-CW	60	NRD WG
1994	CelsiusTech	77	FM-CW	10	mechanical scan, Philips sensor
	HIT	77	FM-CW	5	with MMICs (3)
	Philips	77	FM-CW	>10	electronic scanning, slotted WG
	Lucas&Jaguar	77	FM-CW		patch antenna with lens (3)
	Isuzu	60	FM-CW		NRD WG
1995	Toyota&Fujitsu/Fujitsu-Ten	60	FM-CW	3	planar antenna, bistatic
	TRW	94	FM-CW	10	single-chip MMIC transceiver
	TU-Braunschweig/ Germany	77	FM-CW		2 horns, discrete frequencies in 4 modulation segments
	National Academy of Sciences of Ukraine	40	noise radar		2 parabolic antennas
	Nissan	60	pulsed FM-CW	40	NRD WG
	Raytheon	77	FM-CW	2.5	bistatic antenna, with MMICs (4/7)
	Delco	77	FM-CW	10	mechanical/ electrical scan
1996	Furukawa Electric	60	spread-spectrum	0.6	PN coding
	ADC&M/A-Com	77	pulsed	10	lens antenna (3)
	Siemens	77	FM-CW	10	bistatic lens antennas (3)
	Thomson-CSF	77	FM-CW		MMIC-chip set (3)
1997	VORAD Safety Systems	77 (24, 47, 60)	FM-CW, FSK-modulated		fixed beam or monopulse, patch antennas

Fig.8.18 Millimeter-wave automotive radar sensor history from 1992-1997, from [151].

Algorithm 8.VII: vehicle detection using FMCW radar

1) do range-gating filtering for the received FM signal;

2) carry out FFT; save the magnitude and phase of targets, and the gated high resolution range spectrums;

3) detect the vehicle based on the phase and spectrum information.

Fig.8.19 shows the detection process, where only two out of the four receivers are demonstrated. A parallel wave-front reflected from the target is incident at an angle on the receiver array, which also indicates the offset from the major sensor axis.

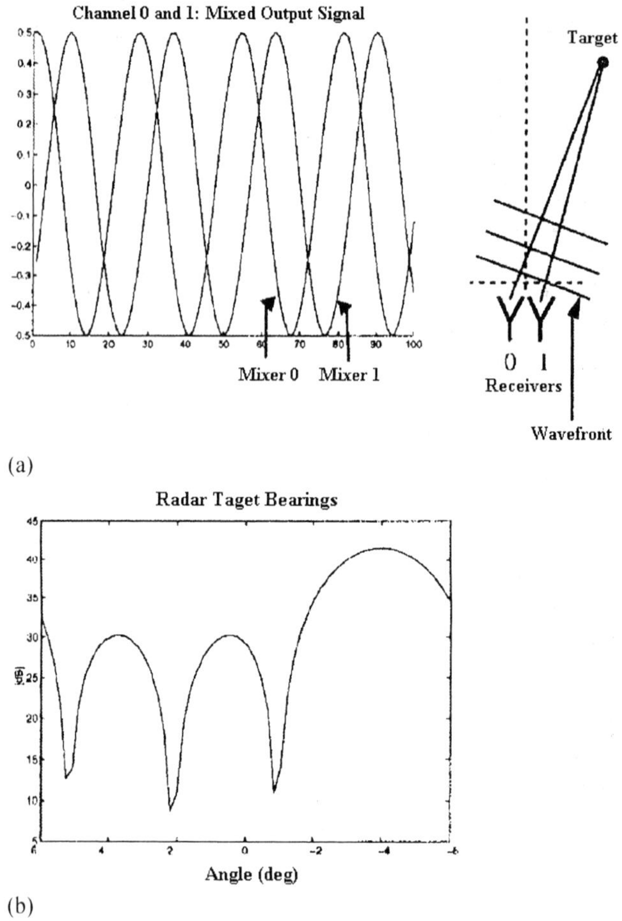

Fig.8.19 Radar simulation of target at 30 m, -4 degree: signal and geometry (a), and bearing in signal phase (b), from [147]. (© [1996] IEEE)

Here, the wave-front arrives slightly earlier in time on receiver 1 than on receiver 0; see Fig.8.20(a). This small difference in range will result in a slightly different frequency output between receivers. This small frequency difference cannot be resolved by the fast Fourier transformation (FFT) in range, but it shows up as a phase difference in the time signal. The first

FFT along time dimension of the receiver output signal now gives target location range and relative phase information. The second FFT along space dimension performs a cross correlation that is a measure of the phase (time) lag between the signals in each receiver channel. The maximum peak in the correlation indicates the phase (time) lag, and thus the bearing of the incoming signal, see Fig.8.20(b).

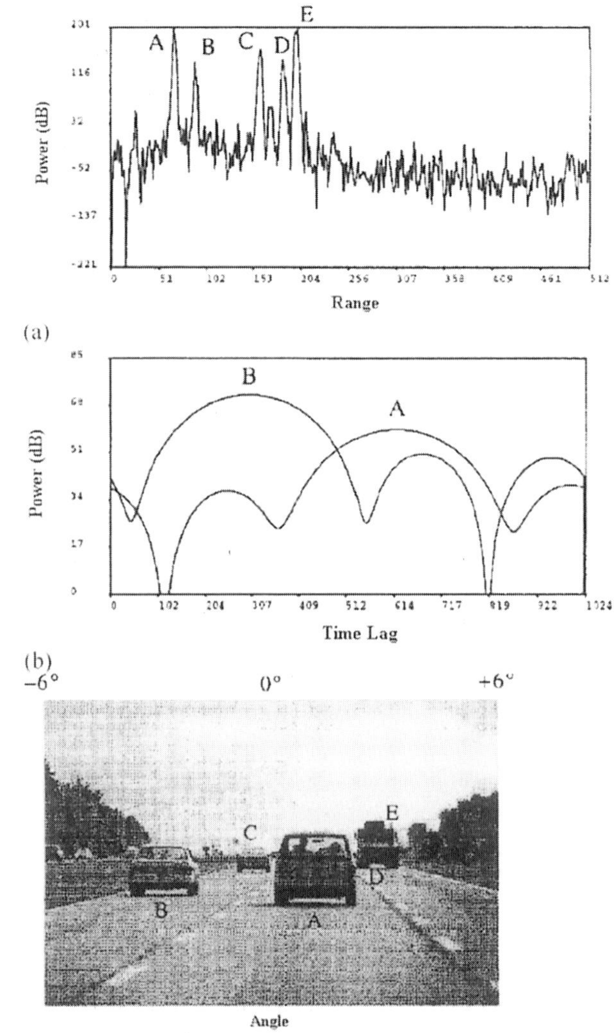

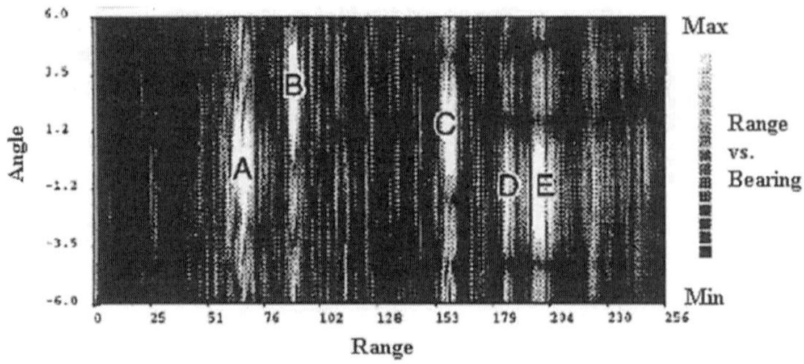

(d)

Fig.8.20 Plot of data: range information (a), bearing of two closest targets (b), traffic scene (c), and intensity map (d); from [147]. (© [1996] IEEE)

Fig.8.20 shows the reconstruction and detection results of range and bearing and the corresponding real scene captured on Highway, where all five vehicles are correctly identified.

The most important performance index is the radar's discern ability. A qualified vehicle-radar should detect the obstacles before it moves into the safety distance.

The target resolution property, which is defined as the ability to distinguish between two adjacent targets here, depends on the number of available sample points and bandwidth. Normally, the range resolution of an FMCW radar could be roughly determined as

$$\delta R = \frac{c}{2B} = \frac{c}{4\Delta f} \tag{8.18}$$

where P_2 c represents the velocity of the light and equals to $3 \times 10^8 m/s^2$ and $2\Delta f$ denotes the FM sweep bandwidth.

Based on (8.18), a resolution of approximately 0.65 meters in range can be achieved due to the FM sweep bandwidth $2\Delta f$ of 240 MHz. If the FM sweep bandwidth increases to 300Mhz, the range resolution could reach 0.5 meters.

In bearing, one resolution cell is $12'' / 4 = 3''$, since there are four receivers in the spatial dimension, see Fig.8.21. It should be noted that two target cannot be resolved anymore and are merged into a single target lobe, if they are at the same range with their bearing less than $3''$ apart.

The accuracy of the resolution also depends on the signal-to-noise ratio (S/N) of the system and the applied algorithms. Through peak approximation in the Fourier Spectrum, it can obtain a relative accuracy of approximately 0.11 meters in range. In bearing, it can obtain a relative accuracy of approximate $0.07''$.

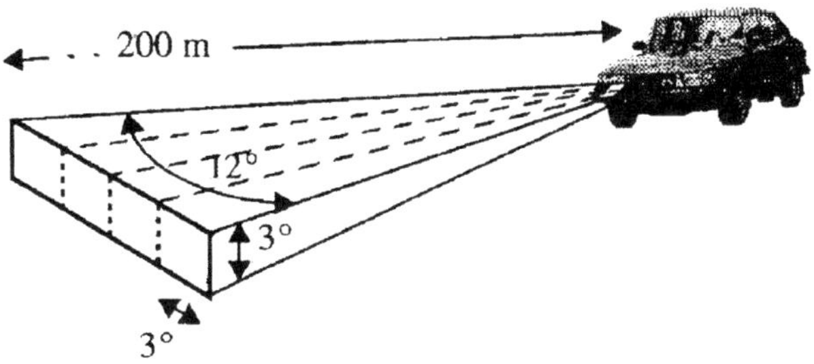

Fig.8.21 Sensor geometry for an mm-wave radar, from [147]. (© [1996] IEEE)

Combining both range and bearing information, the maximum detection distance can be determined with the modulation frequency f_m is given. Suppose a bipolar triangle wave is used, the FM sweep bandwidth $2\Delta f$ is 300Mhz, the maximum center frequency is $f_{c\max} = 500$ kHz, and the modulation frequency $f_m = 625$ Hz, then the maximum detection distance is

$$R = \frac{cf_{c\max}}{8f_m\Delta f} = 200 \tag{8.19}$$

To check whether the above design is feasible, a simplified estimation for the safety detection distance is made as following. In order to stop for a stationary object at high way with an initial speed of 100 km/h (about 65 mph), a detection distance of 55 meters is the minimum span, if it is assumed the acceleration rate $a = -0.7m/s^2$. While in autonomous mode, the vehicle needs longer detection distance, since it needs more time to react for the objects moving towards it. Therefore, neglecting the reaction time, the maximum range of the sensor for detective vehicle should normally be designed to be over 200 meters, which is satisfied by the above radar design.

By experiment, Langer etc. verified that lorries can be detected up to 230 meters using their radar (the nominal maximum range of this radar), cars up to 180 meters, people and animals up to a range of approximately 60 meters, when it runs at 100 km/h (about 65 mph).

The radar cross-section of a person ($0.2m^2 - 1m^2$) is generally an order of magnitude lower than that of a vehicle ($10m^2$ –).This proposed radar is feasible, although the detection range of a small object is correspondingly smaller than a big object. However, this problem could be neglected if humans are assumed only to be present in a road environment where vehicles also move considerably slower, i.e. 50 km/h (32mph). Because the corresponding safe detection distance at low speed would be much shorter than that at high speed.

Another important parameter of the radar is the size of its field of view. The unknown distribution of the obstacles requires wide vertical field of view, while too broad vertical field of view may generate useless environment information and introduce unnecessary difficulty into the obstacle identification process.

As a good attempt, the geometry of the radar sensor proposed by Langer etc. is shown in Fig.8.21. It has a vertical field of view (VFoV) of $3''$, which provides a good compromise between good obstacle coverage in the vertical direction and avoidance of false measurements due to reflections returning from ground, road signs or other structures located overhead. At longer ranges the ground (road) will reflect specularly. The horizontal field of view (HFoV) is $12''$. Since it has four receivers, the HFoV can be further divided into four angular resolution cells. Assuming an average highway lane width of 4 meters in the United States, the sensor covers one lane at a range of 19 meters and three lanes at a range of 57 meters. At a range of 95 meters the sensor covers an area of 20 m by 5 m. Given the geometry requirements of the vehicle-radar, the parameters of radar circuit, including wave forms, modulation frequency, and transmitter power, can be determined.

Practically, the output of mixers on the radar sensor receiver channels is a mixture of different frequencies. Missing or obscured beacons, clutter and false detections are common. Thus, FMCW radar is expected to be further improved by either choosing several different time-frequency transformation algorithms other than simple FFT, or using certain calibration algorithms. More receivers are helpful for improving the resolution performance, too.

Millimeter wave radar satisfies many criteria that are required for an outdoor autonomous vehicle navigation sensor. Its shorter wavelength, compared to the microwave units, provides a high resolution measurement. Moreover, it gives improved performance in inclement weather compared to optical sensors and laser sensors. The system is believed to be able to locate target reflectors at longer distances than other sensors with a relatively high degree of accuracy, if modern digital signal processing (DSP) techniques is used.

As pointed out in [151], mm-wave radar is now continuously upgraded in transceiver technology, antenna development and higher resolution sensors technology. However, mm-wave radar imagery also has some shortcomings such as the higher costs, complexity of the algorithms in the face of backscatter from off-road structures, and considerable power requirements. All these problems should be appropriately solved before it becomes practically used.

8.3.3 Vehicle Detection Using LiDAR or Laser Sensors

Comparing to CMOS/CCD camera approaches, one LiDAR only can provide 3D road information, which contains both object direction and distance. Usually, LiDAR based vehicle detection focuses on how to determine and cluster the potential vehicles' surfaces/edges. There are numerous algorithms that address segmentation of 2D/3D LiDAR data, [154]-[155]. In short, the detection process can be described as:

Algorithm 8.VIII: vehicle detection using LiDAR

1) carry out edge detection using LiDAR data;
2) locate the regions of interests (ROI), i.e. using region growing methods;
3) identify of the objects by bounding object clustering;
4) distinguish the object type using the shape features, see Fig.8.22.

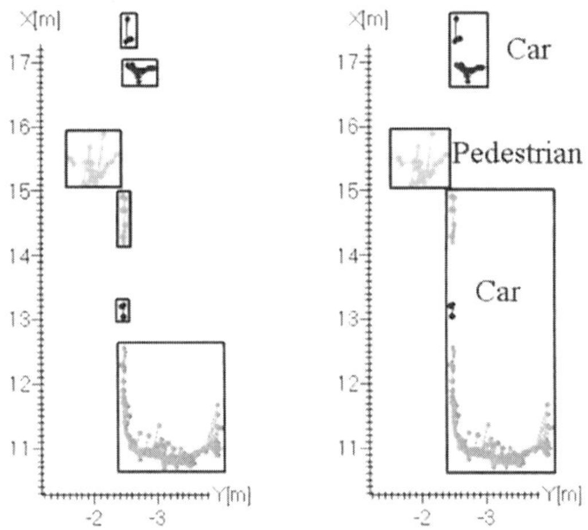

Fig.8.22 Clustering results using the laser scanner, from [159]. (© [2004] IEEE)

One benefit of LiDAR based approach is that two vehicles can be discriminated through their different distances to the camera, although they may have similar angles in the image. However, this leads to more complex clustering calculation, since the dimension of environment data increases [156]-[161]. Fig.8.22 demonstrates a comparison of two segmentation parameters given in [159]. Fig.8.22(a) shows the result using a small segmentation threshold, but to a wrong separation of both objects, while Fig.8.22(b) shows the correct assignment.

8.4 Advances in Vision Based Pedestrian Detection

Pedestrians are vulnerable road users. In European, there are more than 150,000 pedestrians are injured and more than 6000 are killed yearly. In the United States, 12 percent of casualties in traffic accidents are pedestrians. In China, 99,217 persons were killed in traffic accidents in 2004, and one third of them were pedestrians. To conquer this problem, vision-based pedestrian detection techniques have emerged as a hot research topic in the field of driving safety during the last two decades.

Gavrila proposed a classification method in [162]-[163] for human movement detection based on two criteria: the dimensionality of the tracking space (2-D vs. 3-D), and the type of models (e.g., volumetric, statistical, stick figure-based). While this process is useful in smart surveillance, virtual reality, motion analysis, and other applications, it is too general and difficult to apply to pedestrian detection from a moving vehicle. Lombardi [164] presents a new classification. Firstly, many different methods are divided into three groups by whether they include two steps in the process or not: two-step approaches, one-step approaches and those so called "blind" approach. Secondly, various methods are classified in the detection and recognition step respectively. In the detection step, optical flow, background subtraction, range thresholding, and pedestrian shape-based methods are frequently used. In the recognition step, movement analysis and shape analysis attract great interests. However, this classification does not consider the tracking step that is used in many pedestrian detection ways. In the following paragraphs, vision-based pedestrian detection processes are divided into three consecutive steps: pedestrian localization, pedestrian recognition, and pedestrian tracking.

8.4.1 Pedestrian Detection Using CCD/CMOS Cameras

1) Pedestrian Localization

The aim of the detection step is to extract interesting regions where pedestrians may exist, and in these regions, pedestrian recognition and tracking are carried out. Similar to vehicle detection cases, knowledge-based methods, stereo vision based methods and motion-based methods are three major approaches in this field.

Many methods make use of human shape information. These approaches normally do not require temporal information and can overcome the problems caused by changing background. One the other side, because pedestrians have various shapes, the kind of method often results in complex computations [165]-[168]. For example, Curio et. al [167] proposed a method of searching the active regions by local image entropy firstly. They then make shape matching. In their approach, the shape of a pedestrian's legs is represented and detected by a $\wedge$ shape. Besides, they used inverse perspective mapping to detect obstacle in the short distance. After these pre-processing, they developed a temporal dynamic activation (DAF) which directs further processing; see Fig.8.23.

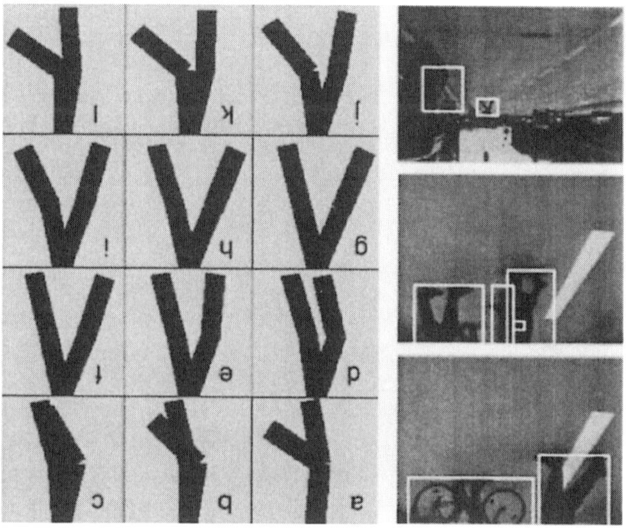

Fig.8.23 Left images show the examples of initial pedestrian detection with the temporal dynamic field where regions extracted using entropy, model contour and stereo cues. Right images show the typical gaits of a human walking model, from left to right and from top to bottom, are shown 12 phases (a)-(l), from [167]. (© [2000] IEEE)

Adopting stereo images is another way to detect the potential regions that may contain pedestrians [169]-[175]. For instance, in [173], the features expressing the shape of a detected 3-D object was calculated and analyzed using the discriminant analysis method of multivariate analysis. The authors found three effective variables in distinguishing pedestrians from other objects. In [122], how to recover the distance between the host vehicle and a character point in the 3-D space is discussed using the positions difference between the coupled images from two cameras mounted on the vehicle. At the same time, they detected the motion of objects using gradient-oriented optical flow method. In order to use the advantages of both stereo vision and motion, the authors proposed a flow/depth constraint approach. This integral method can achieve robust and fast detection of pedestrians.

Motion of pedestrians is a very important cue for extracting regions of interests. One merit of motion based pedestrian localization methods is to detect those persons who are partially occluding other persons [176]-[177]. However, this method needs to analyze a few image frames, which required notable time and simultaneously increases computation cost. Moreover, it cannot detect static pedestrians well. Thus, it is often applied to pedestrian recognition step.

2) Pedestrian Recognition

After the regions where pedestrians possibly exist have been determined, it is necessary to recognize a person among these candidates and eliminate false areas. Recent research shows two main trends: motion-based approaches and shape-based approaches. The former approaches take temporal information into account and try to detect the periodic features of human motion; while the latter ones rely on shape analysis instead of characteristics retrieved from image sequences.

Motion pattern, especially the periodicity of the gait unique to human being is an important cue to recognize the walking pedestrians from other moving objects. For instance, in [178] the Maximum Entropy algorithm was applied to observe the change of image intensity caused by walking pedestrian. Based on the temporal-frequency and spatial-frequency decomposed from the walking rhythm, the authors performed model matching. In [171], Time-Frequency analysis and Short-Time Fourier Transform was used with a Harming window to capture a change in the periodicity. In [179], an adaptable time delay neural network algorithm was employed to extract local spatial- temporal features by processing the input image sequence. Compared with the global methods, this local one was claimed to get better recognition performance with low computational complexity and memory demand.

Shape-based recognition approaches can recognize not only moving pedestrians, but stationary pedestrians. The primary difficulty of this method is how to deal with various appearances of pedestrians caused by lighting, clothing, pose, and occlusion et. al In [161], Gavrila matched the contour of pedestrians with templates and regularizes those matched regions. The regions then are classified by Radial Basis Functions. In [177], the contour matching method is applied, too. Firstly, the binary edge image is obtained by using the edge detection based on Laplacian filtering. Secondly, the edge image is transformed into the DT image by distance transform. Finally, the DT image is compared with the templates and the matching is successful if the difference between them is less than a precisely set threshold. In order to reduce computation cost, active search method [180] was applied for sequent image difference search computation. In [181], a new algorithm was used to detect pedestrians on static images using because the motion information that was mainly resulted from camera movement instead of pedestrians movement. The algorithm mainly considers the features of human legs. AdaBoost training algorithm was employed in [182] to transform a set of "weak" classifiers with success of a little more than 50% to a "strong" classifier, which can detect pedestrians with the success rate close to 100%. In [166], a single frame classification algorithm is proposed. The obtained image is divided into many sub-regions. Running a classifier separately on each sub-region, the local discriminant results are obtained. These results are then integrated by running a classifier on the feature vector comprising of them. Bertozzi et. al [115] searched the image contour in the interesting regions likely to contain pedestrians by means of a shape detection technique on the application of ant colony optimization method [184].

Many one-step methods also used the shape information of human being. In [168], [185]-[186], Papageorgiou et. al defined the shape of objects by wavelet template. The wavelet template consists of a set of regular regions of different scales which correspond to the support of a subset of wavelet functions. The relationships between different regions are expressed as constrains on the values of the wavelet coefficients. Although the intensity of each region changes greatly, the relationship between them remain stabilized. Elzein et. al [175] applied a wavelet-based feature extraction method and a template- matching to detect pedestrians. They used the Haar wavelet transform for each training image and segment the image into some regions, firstly. And secondly, they applied the same approach to calculate the wavelet transform for these regions and obtain feature vector. Finally, the vector is compared with the feature vector of template. If the difference between them is less than an appropriate threshold, a pedestrian is then detected.

3) Pedestrian Tracking

In order to efficiently judge the movement of pedestrians and the corresponding collision prediction, pedestrian tracking becomes another hot topic recently. Normally, the tracking step utilizes the relationship correspond to position, velocity, shape, and other features that can be retrieved from the image sequence. Kalman filtering, Condensation algorithm, Dynamic Bayesian Network and etc. are the most common mathematical algorithms applied for such tasks [187]-[189].

Based on a linear shape model of pedestrians, Philomin et. al [187] employed an efficient variant of the Condensation tracker to track these objectives, which was finally turned to be a high- dimensional problems by using quasi-Monte Carlo methods. In [188], Gavrila et. al used a special alpha-beta ($\alpha-\beta$) tracker to estimate the object state parameters. The tracker is a simplified Kalman filter with a constant velocity model and steady-state gains. Since the shape-variation is handled successfully in the detection step, the tracker only needs to deal with position and bounding box extent. Bertozzi et. al [141] estimated the position of pedestrians in the image, where Kalman filter was applied to compute the process of pedestrians' movement and trajectory.

8.4.2 Pedestrian Detection Using Infrared Cameras

Although numerous methods had been introduced in this field, vision-based pedestrian detection remains as an open challenge because of many difficulties. The outdoor situation is complex and changeable. Pedestrian shapes are countless and variable in dresses, poses, and motions. Nonetheless, sudden changes of background scene are unavoidable, when camera is mounted on a moving vehicle.

To conquer such problems, the thermopile/infrared sensors have been applied to intelligent vehicle systems recently [190]-[197]. In [194], a thermopile sensor was applied to measure the objects exited within the respective field-of-view, because pedestrians almost always have higher temperatures than the backgrounds. Based on the same detection theory, infrared camera is also employed for pedestrian detection now [195]-[197]. It can obtain 2D thermal images, from which far more morphological and thermal characteristics can be retrieved and analyzed.

The most important merit of thermopile/infrared sensor is its capability to detect pedestrian easily. The detection rate is much higher than that of conventional vision-based methods. Fig.8.24 given in [195] vividly shows how pedestrians can be distinguished from environment in the infrared image. Besides, infrared cameras based method seems to be the only reliable solution for nighttime pedestrian detection [196]-[197].

Fig.8.24 Preprocessing phase for a stereo infrared camera system: (a) original input image, (b) focus of attention, from [195]. (© [2005] IEEE)

Another nice advantage of the thermopile/infrared sensor is its ability to detect pedestrians passively without illuminating the environment. Thus, it will not electronically pollute the surroundings and is an environment-friendly sensor. The only shortcoming of infrared cameras that prevent their wide applications is the price. How to realize reliable on-vehicle infrared cameras with low cost should be an interesting topic in the next ten years.

8.5 Advances in Vision Based Traffic Sign Detection

Great efforts had been devoted to the traffic lights and traffic signs recognition problem in the last decade. It is a relatively easy problem since traffic lights and signs have distinct colors and strict shapes than vehicles and pedestrians. Therefore, most approaches in this field are knowledge-based [198]-[199].

Color features are the most important characteristics for traffic lights and road signs [200]-[203]. One most important problem lies in that the color space is very sensitive to lighting changes. To solve it, a modified approach is suggested in [200], in which the color ratio between the intensity of the specified color and the sum of intensity of RGB is used instead. Some special color filters are designed in [201]-[202]. Besides, HSV color space based approaches are also reported in some literals including [203]-[205].

Shape features are frequently used for road signs detection, too [206]-[207]. Besides, symmetry of the road signs is another important feature for detection [208]-[209]. The most difficult problem for those approaches is image distortions due to noise and irregular positions of the road signs. To apply robust filters that are against distortion was widely used in the last decade [210]-[212]. The integrated detection methods which consider the invariant shape features within image sequences are also discussed in [213]-[214].

The road sign detection difficulties lies in several aspects:

(1) the illumination conditions may not be so good;
(2) the position and the orientation of signs are not known;
(3) other objects can block the vision of the road signs;
(4) there exist sign placement and installation problems such as wrong color, shape, or size; improper placement; lack of warning signage; and obsolete signage, etc.

To recognize the contents of road signs, several different methods had been reported, which can be roughly divided into two types as what mentioned for vehicle detection above: explicit rule based methods (template-based symbolic matching) and implicit rule based methods (appearance-based symbolic recognition) [213]-[221]. Gavrila et. al tested some standard pattern classification methods including polynomial classifiers (PCs), multilayer perceptrons (MLPs), radial basis functions (RBFs), and support vector machines (SVMs), when the relevant pixel intensities had been aggregated in a feature vector. It was proven that most methods can yield promising results if the parameters are correctly set.

In order to improve the identification performance, Miura et. al proposed the so called active vision system to obtain zoomed up image for traffic signs identifications [222]. The proposed intelligent vehicle system contains two types of cameras. The first is a wide-camera which has low resolution but wide range. The second is a tele-camera which has high resolution but shallow range. The first camera cannot move, while the second can be moved towards desired direction. Suppose a candidate traffic sign has been detected by the wide-camera, then the intelligent vehicle will guide the tele-camera to direct to it for capturing a zoomed-up image. Since it takes some time to switch the camera and direct the tele-camera, the position in the image of the candidate need to be predicted at the time of the image input by the tele-camera. Assuming that the vehicle moves straight at a constant speed from the time of candidate detection to the end of identification, prediction is performed as what is shown in Fig.8.25.

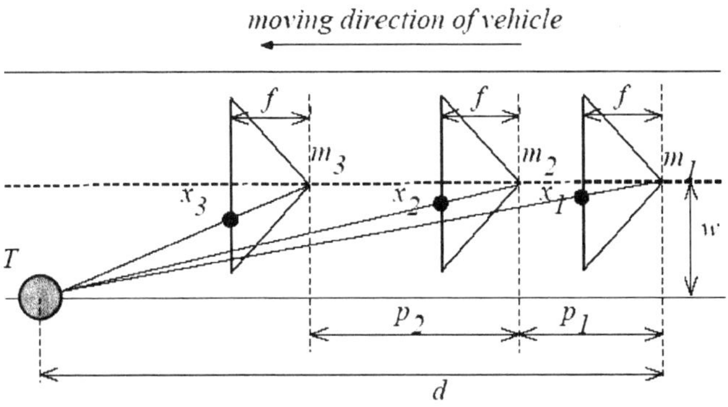

Fig.8.25 Predicting candidate traffic sign's motion, from [222]. (© [2000] IEEE)

Since a sign moves on a line passing the focus of expansion (F.O.E.), it only needs to predict the horizontal position of the sign on the line, see Fig.8.25. In the figure, f is the focal length, m_i is the camera position at the time of image input, x_i is the horizontal position of the target at that time, p_i is the distance between camera positions. d and w are the distances to the target from m_1 in the longitudinal and the lateral direction, respectively. Thus, the position x_3 in the third image can be predicated based on two previous observations x_1 and x_2.

Let t_i be the time to move by distance p_i. t_1 is calculated by measuring the time of input of the first and the second images. t_2 depends on the time needed for camera control, and can be estimated in advance. Let v be the vehicle speed, p_i should be

$$p_i = v \cdot t_i \tag{8.20}$$

From the similarity of triangles in the figure, it has

$$\frac{f}{x_1} = \frac{d}{w}, \quad \frac{f}{x_2} = \frac{d - p_1}{w}, \quad \frac{f}{x_3} = \frac{d - (p_1 + p_2)}{w} \tag{8.21}$$

Solving (8.20)-(8.21) leads to

$$x_3 = \frac{p_1 x_1 x_2}{(p_1 + p_2)x_2 - p_2 x_1} = \frac{x_1 x_2}{(1+\alpha)x_2 - \alpha x_1} \tag{8.22}$$

where $\alpha = p_1 / p_2 \equiv \text{constant}$, since v is assumed to be constant here.

Eq.(8.22) shows that the target position can be predicted without knowing vehicle speed v and focal length f . Thus, by using (8.22) and the record of past navigation position, the direction of the camera (pan and tilt angles) can be calculated.

Fig.8.26 Tracking a complex road guidance sign by the tele-camera (left-top to right-bottom), from [222]. (© [2000] IEEE)

Therefore, during driving process, if the tele-camera does not capture the target in a sufficiently large size in the image, the camera continuously tracks the target to obtain a larger target image using a simple tracking strategy. Using this method, the obtained image quality is significantly boosted, which yields far better performance especially for those complex road signs as shown in Fig.8.26.

8.6 Advances in Vision Based Driver Monitors

Driver/Passenger oriented research gains increasing interests in the last ten years. Conventional research focuses on how to make ride more comfortable. For example, designing advanced suspension and chair systems to avoid injury and implementing smart controller to adjust inside vehicle temperature.

In recent research, it becomes widely believed that advanced driver assistance systems should have at least the following two functions to further ensure driver/passenger's safety.

(1) monitor driver and passengers' positions/gestures so as to employ "smart" air bag;

(2) monitor drivers' behaviors so as to analyze driver's state, study, evaluate and even mimic his/her driving habits, etc.

8.6.1 Driver/Passenger Position/Gesture Detection

No one can deny that airbags have significantly improved the occupant safety in vehicle crashes. However, inappropriate airbag deployment can also cause unnecessary injury or money waste on airbag replacement. NHTSA had pointed out that although airbags had saved over 6,000 lives by the end of 2000, they had also killed over 200 occupants through inappropriate deployment. In response, NHTSA issued a set of regulations mandating smart airbags that can adapt intelligently to the occupant. The study on smart airbag systems focuses on how to intelligently turn on/off or adjust airbags deployment strength according to occupant characteristics. This naturally leads to research on driver/passengers gesture detection, especially for head position monitoring [223]-[225].

In [226]-[227], monocular vision systems were used for driver head detection. More approaches applied stereo vision systems in this area [228]-[234]. In [235], a 3D optical time-of- flight (TOF) sensor, together with active modulated Infrared illumination was employed for driver detection. In their tests, each individual pixel measures not only an intensity grey value but also the distance to a corresponding point in the scene under investigation. In [236], infrared camera was used to help eliminate the disturbances caused by poor lighting conditions. Algorithms can help reject occlusion and the presence of other competing head-like objects in the scene.

Most driver head detection methods are knowledge-based, which normally applied appearance-based algorithms. The following cues are frequently used in this area:

(1) temperatures: human often have higher temperatures than the in-vehicle environments [235]-[236];

(2) face features: usually, it is assumed that human faces may contains more edge information than other items/background in the vehicle. Since the cameras are normally installed near driver/passengers, the potential areas of human faces should have high density of edges [231];

(3) human head and body shape: when all the Points of Interests (POI) are segmented by using their illumination degree, textures and distances to the camera, human shape features can be used to differ driver/passenger's head and background. For instance, [233] described an approach to first find and track the head of the driver and next locate his/her torso in the trinocular stereo vision system.

It is natural that the head position algorithm must be robust to harsh lighting conditions, partial occlusions, and uncontrolled driver postures. Several different methods such as Supporting Vector Machine (SVM) and Principal Component Analysis (PCA) were used to "learn" to check the existence of driver/ passengers from sample images [237]-[240].

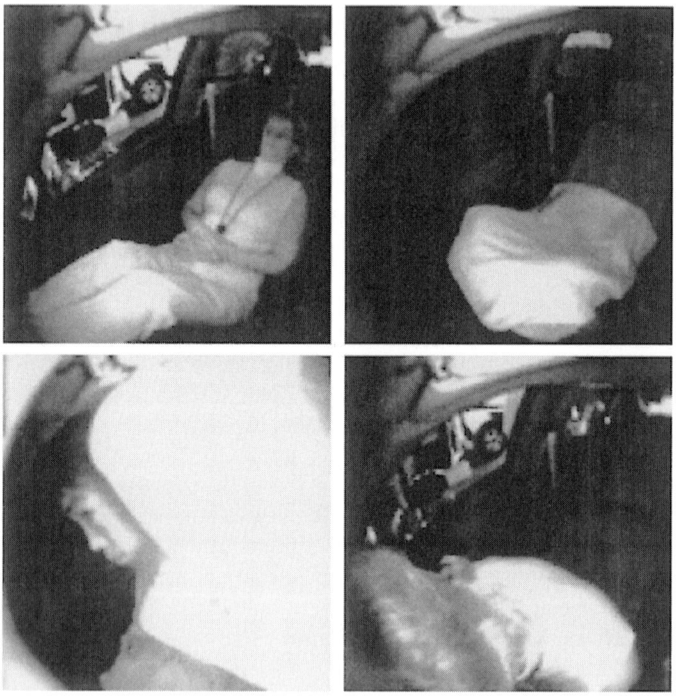

Fig.8.27 Failed detection for some difficult examples: (upper) two similar images, but there is a passenger in (left-top) which nobody in (right-top); (lower) depict two extreme postures of passengers, from [227]. (© [2005] IEEE)

In some recent reports, the competing objects such as hands can be correctly discarded [234]. However, most approaches cannot have 100% detection rate especially for some extreme cases and still needs further studies; see Fig.8.27.

8.6.2 Driver Fatigue Analysis

The diminished driver's vigilance level has become a serious traffic safety problem. The National Highway Traffic Safety Administration (NHTSA) estimates that, in the US, drowsy drivers cause 100,000 accidents each year, resulting in more than 1,500 fatalities and 71,000 injuries. Among different approaches in this field, the driver-eye gaze tracking system has received considerable interests [241]-[245].

Recently, the research interest in this field has shifted to driver fatigue detection using both information of road scene monotony and driver vigilance. Their idea comes from a widely accepted psychology perspective that defines monotony as an exogenous contributing factor of fatigue [246]-[250]. For instance, Fletcher et. al designed an integrated fatigue detection system that uses driver-head-pose and eye-gaze tracking as well as road monotony analysis; Fig.8.28. It was claimed to have good performance, since it combines two different approaches in one system.

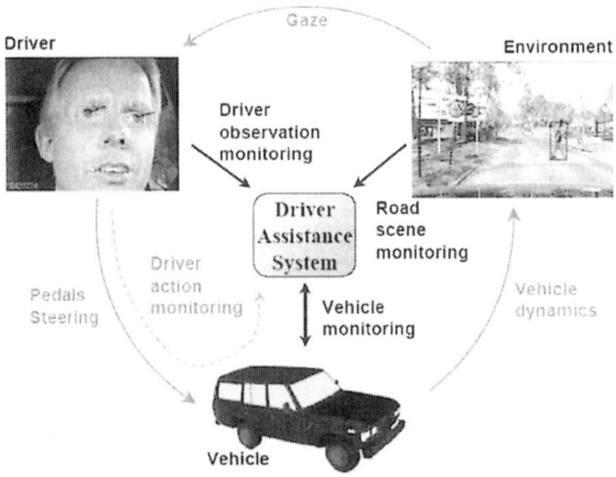

Fig.8.28 Introducing driver observation to supplement driver action monitoring, from [245]. (© [2005] IEEE)

To automatically measure the monotony in a road sequence, the information content of the road image sequence over time is estimated. Practically in [251], the Kolmogorov-Chaitin complexity was chosen to measure the amount of this information source. In real calculations, MPEG encoding was employed, because it exploits the property that in moving pictures only small regions of the image actually change substantially between frames. A simple linear model is built to link a human judged monotony scale with the obtained MPEG file size, Fig.8.29. They claim that this system has better performance than those that focus on driver-head-pose and eye-gaze tracking only.

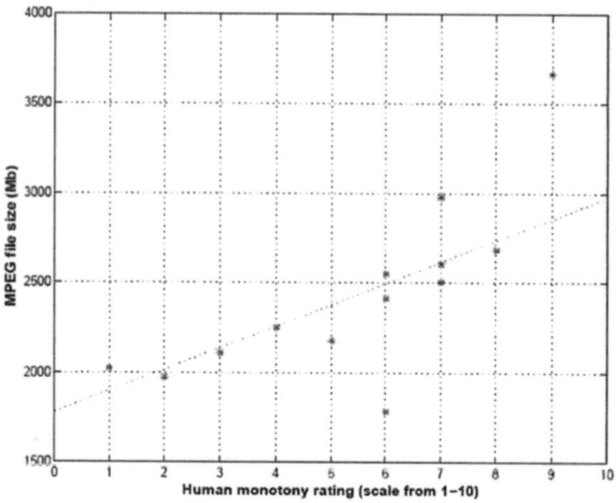

Fig.8.29 Graph of various video file sizes versus a human evaluated monotony scale. 1 = very monotonous, 10 = very stimulating, from [245]. (© [2005] IEEE)

8.6.3 Driving Actions Analysis

In conventional research, only the accelerator and brake pedal pressure signals are recorded and analyzed to recognize the driver driving mode. In some new literals [252]-[255], the drivers' postures/gestures are also tracked and studied. In [253], a hierarchy of driver activity mode is constructed to represent a driver's activity, which is recognized at multiple levels: individual body-part postures/gestures at the low level, single body part action at the middle level, and the driver interaction with the vehicle at the high level. Driving is represented in terms of the interactions among driver, vehicle, and surround, and driver activity is recognized by a rule-based decision tree.

It is clear that this kind of models can also be employed in driver training systems to correct new driver's inappropriate habits. But the attempt in [253] is just the first step to the next generation driver assistance systems. The solid achievements for improving driving safety are still on the path [256]-[257].

8.7 Further Discussions on Intelligent Vision Systems

8.7.1 Multiple Vision Sensor Fusion

Sensor fusions gained more and more efforts during the last decade [258]-[270]. As widely proven over the past years, to appropriately use more than one kind of vision sensors may greatly improve perception performance. In general, multiple vision sensor fusion focuses on the following problems:

(1) scanning range and accuracy of discrimination: because different sensors may measure different areas of environment, how to determine the common space of measurements should be solved first;

(2) data synchronization: basically, sensors are asynchronous, which means they may have different sample rate. Therefore, a synchronization step is necessary before fusion process;

(3) fusion strategy: how to choose the fusion algorithms and strategy is the most important problem. First, the fusion process depends on the representations of measurements and their accuracy. Second, since the accuracies of different sensors depend on the varying environment, the fusion algorithms for the available data should be flexible;

(4) fusion speed: the need for high frame rates in automotive application motivates the investigation into computationally simple methods of combining visible and infrared images. For instance, wavelet transformation is now widely viewed as a good tool for image mapping and merging task.

Here, one interesting emphasis is given to visible and infrared images fusion. An attempt reported in [265] studied how to use wavelet-based image fusion techniques. As shown in Fig.8.30, the fusion result is much better; since color visible image lost most road side environment information; and infrared images missed the color hue of vehicle rear lights and road lights.

Fig.8.30 Applying a computationally simple image fusion technique based on the Discrete Haar Wavelet Transform to combine three images from cameras operating in different wavelength bands: (a) the color visible image, (b) the long wave infrared image (7-14μm), (c) the monochrome visible and near infrared (up to 1100nm) image, and (d) the two level wavelet fusion result, from [265]. (© [2005] IEEE)

8.7.2 Vision Sharing

Although every researcher wants to make an intelligent driving system collect as much environment information as possible, he/she may sometimes fail to do so, since not all kinds of sensors can be installed onto a vehicle because of cost and implementation consideration.

Along with the developments of wireless communication technologies, the concept of multiple vehicles cooperative driving and roadside aid vehicle driving emerged as promising potentials to alleviate traffic congestions recently. The main idea of these newly proposed methods is to share/exchange driving information between vehicles and roadside so as to improve driving safety and efficiency. Therefore, inter-vehicle/vehicle-roadside communication technologies achieved great concerns and efforts in the last two decades.

For example, many divers find it difficult to drive backward, since he/she hardly knows what happens behind their vehicle [271]. For example, in parking garage, the driver sometimes fail to determine whether he/she parks right in the parking bay and still has an enough gap before bumping the curb. To solve this problem, an interesting multi-camera system was built in a parking garage in [272], where vehicle's position information can be easily measured and send to driver in real time. The upper large figure shows the view of the camera, and the lower small figure shows the driver's view, in which the view from the camera is reflected in real-time on the small screen installed before the steering wheel. The driver can easily handle the vehicle to an appropriate position, see Fig.8.31.

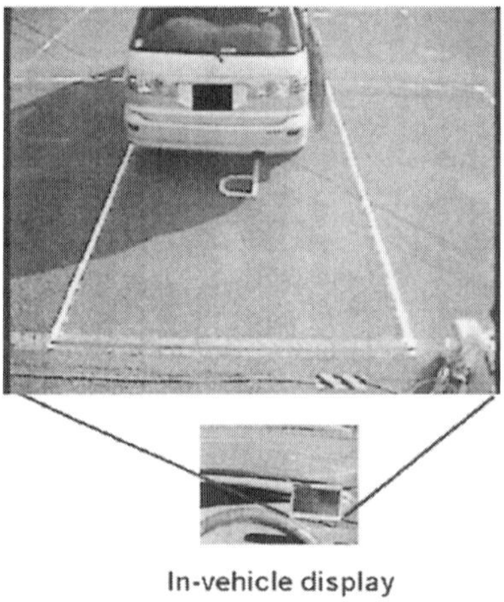

Fig.8.31 Image of parking assistance system using single camera-infrastructure, from [272]. (© [2005] IEEE)

A hot topic that has gained fast increasing attentions recently is driving assistance around intersections with vehicle-road communications. Intersection collision contributes a significant portion of highway accidents. A potential solution is suggested to supply timely alerts to drivers of imminent collisions. In [273], as illustrated in Fig.8.32, an experimental radar is set up and configured to observe the left turn motions of subject vehicles with a trajectory depicted by the yellow turning curve. The subject vehicle in the figure is initially traveling north then turning toward west. As shown in the photograph, there is a left turn pocket for the subject vehicles. The triangle in the figure represents the coverage area of a radar device used for monitoring the movements of other vehicles in the opposite direction. If the system finds it is dangerous to make left turn based on the radar data (a vehicle is coming to the intersection from north with high speed for example), it will notify the driver to stop left turn.

Inter-vehicle communication and its applications undergo significant developments in the last two decades, too. The most important merit of inter-vehicle communication is it does not require infrastructure intelligence [274]-[275]. In some recent reports, i.e. [276]-[279], cooperative vision sensory via inter- vehicle communication were addressed. It is believed that more efforts would be put into in this new field quickly.

Fig.8.32 Diagram of field observation site for the radar at intersection, from [273]. (© [2005] IEEE)

8.7.3 Traffic Infrastructures and Vehicle Vision Systems

Changes of traffic infrastructures affect intelligent vehicle vision systems silently. Variations of lane markings, road lights and traffic signs will naturally change the vision algorithms.

For instance, inspired by a 2001 survey of transportation agencies in the United States, Canada and internationally, [280]-[281] studied whether wider-than-standard longitudinal pavement markings. Subjective opinions that wider lines are "better" may be due in greater part to increased peripheral visibility and consequently, decreased driver workload. Thus, investigations into line width, as well as line brightness, could be improved by the use of new measures of effectiveness related to changes in peripheral vision and driver comfort or workload. Obviously, how the length/width of lane marking affects the vision based lane detection results and what is the optimal length/width is an interesting topic and needs to be analyzed.

A great development in infrastructures achieved recently is to use new Clearview typeface for guide signs [282]-[285]. The most important benefit of Clearview is it eliminates nighttime overglow or haloing. As a result, all the road sign detection and recognition methods should update their sample image library and retest their algorithms. Because Clearview changes letter proportions and the spacing between letters which makes the typeface more readable and words more recognizable, better detection and recognition results are expected.

Besides, many drivers complained that sometimes there are tens of road signs at intersections. It is even impossible for a human driver to read them all. It seems that they have been placed there mainly to comply with legislation, with no thought for the human. Thus, how to properly arrange the placement of road signs so that let human driver or intelligent driving systems easily process the information they're supposed to convey emerges as an interesting problem for the coming decade.

Furthermore, all the collected human driving records and related cognitive research achievements in this field also give some insights of human driver decision process, which can help to improve the performance of vehicle vision system.

8.7.4 Vision Sensor Design, Calibration and Fault Detection

To further improve the performance of intelligent vehicle vision systems, vision sensor design and implementation also attracts considerable attentions during the last ten yeas [286]-[289]. Generally, one important research direction is to increase the detection range and accuracy [289]-[291]. And another direction is to reduce the manufactory cost and make the vision sensor more reliable [11]. It is clear that the price/performance ratio determines when the intelligent vision systems can be applied practically to vehicle industry. From an optimistic viewpoint, a mature intelligent vision system might become possible over the next two decades or three.

Choice of the magnification, resolution, and size of image also needs to be carefully considered according to the smallest feature that needs to be detected, the largest feature that will be encountered and the representativity of the image with respect to the overall sample. Besides, variation of light intensity needs to be checked in the field of view.

For instance, in [201]-[202], De la Escalera et. al analyzed several color filters for RGB color system. The first mentioned color threshold has the following expression:

$$g(x,y)=\begin{cases} k_1 \begin{cases} R_{\min} \le f_{red}(x,y) \le R_{\max} \\ G_{\min} \le f_{green}(x,y) \le G_{\max} \\ B_{\min} \le f_{blue}(x,y) \le B_{\max} \end{cases} \\ k_2, \quad \text{otherwise} \end{cases} \tag{8.23}$$

where $f_{red}(x,y)$, $f_{green}(x,y)$, and $f_{blue}(x,y)$ are the functions that give the red, green and blue levels of each point of the image, respectively.

One of the greatest inconveniences of it lies in that the color space is very sensitive to lighting changes. To solve this problem, a modified approach is suggested by Kamada and Yoshida [200], in which the color ratio between the intensity of the specified color and the sum of intensity of RGB is used instead. Therefore, considering the relation between the components, assuming that the red component is chosen as a reference, the new color filter can be written as

$$g(x,y)=\begin{cases} k_1 \begin{cases} R_{\min} \le f_{red}(x,y) \le R_{\max} \\ TG_{\min} \le \dfrac{f_{green}(x,y)}{f_{red}(x,y)} \le TG_{\max} \\ TB_{\min} \le \dfrac{f_{blue}(x,y)}{f_{red}(x,y)} \le TB_{\max} \end{cases} \\ k_2, \quad \text{otherwise} \end{cases} \tag{8.24}$$

where T is an appropriate weight coefficient.

In [292], two different exposure conditions were used for one single image to achieve better illumination calibration performance. One is to capture images using plural imaging devices with different exposures at a time. The other is to capture images using a single imaging device with a different exposure so as to realize the imaging of moving objects.

Vision sensor calibration is an important task for computer vision, robotics, and vision based manufacturing systems. Over the past few decades, considerable effort has been made on development of effective and accurate procedures and algorithms to identify the internal camera geometric and optical characteristics (intrinsic parameters) and the three-dimensional position and orientation of the camera frame relative to a certain world coordinate system (extrinsic parameters) for various applications [293]-[296]. The so called video stabilization techniques was employed in [297]-[298] to calibrate camera position angle error caused by installment fault and/or vibration.

Besides, fault-tolerant vision systems and vision sensor fault detection gained increasing interests recently, i.e. [299]-[302]. Since detailed discussion on this topic required an individual book chapter, no detailed discussions are made here.

8.7.5 Vision Based Environment Detection and Adoption

One remained important problem for vehicle vision system is how to autonomously applied different algorithms under varied situations. Real-world traffic applications must account for several aspects that accept diverse interpretation, such as the weather, light, static or moving objects on the road, noise, manmade interventions, etc. It is difficult for a vision algorithm to account for all kinds of variations. Moreover, under diverse conditions, different algorithms may yield quite different results. Thus, two major requirements from advanced systems are expected to emerge namely (i) adaptation to environmental changes and (ii) ability of combining (fusing) information from different sources. Apparently, this requires appropriate detection of vehicle driving environments to secure transport facilities safe from accidents and to keep the performance smooth.

For example, an interesting fog detection method is proposed in [303]. Another weather recognition method is proposed in [304], which uses a subspace method to judge rainy weather by detecting raindrops on the windshield. First, the concept of "Eigendrops" is defined to represent the principal components extracted from raindrop images in the learning stage. Then the method detects raindrops by template matching; see Fig.8.33. In addition to determination of rainy or fair weather, location of these rainy drops could also be helpful for rainy brush controls and pedestrian detection.

In [305], the road condition was monitored since it is one of the most important factors toward detection of vehicle driving environments. In their approach, features related to water were extracted by the ratio of horizontal polarization image intensity to vertical polarization image intensity for each pixel. Features related to snow were extracted by texture analysis using the co-occurrence matrix. A multivariate analysis was employed to discriminate five kinds of the road conditions: "Dry," "Wet," "Slushy," "Icy," and "Snowy," on the basis of these features extracted from the road images as well as temperature. It was claimed that this algorithm can obtain favorable discrimination accuracy rate of 92.3% on the average.

In the Ground Challenge competition organized by DARPA [306], the involved vehicles attempted to run across a 250-mile (400 km) desert within a fixed time period. The encountered driving environment varies from ordinary highway to dusty construction road, which naturally requires robust vision sensors and flexible algorithms. It is really promising that 5 vehicles passed the desert within the time limit. And optimistically predicting, smarter vehicles will appear in the next twenty years.

Fig.8.33 Raindrops detection and change of number of detection by wiping, (left) before wiping, (right) during wiping; from [304]. (© [2005] IEEE)

8.8 Reference

1. C. Thorpe, M. Hebett, and T. Kanade, et. al, "Vision and navigation for the Camegie-Mellon Navlab," IEEE Transactions on Pattern Analysis and Machine Intelligence, vol. 10, no. 3, pp. 401-412, 1988.
2. K. Chen and B. A. Galler, "An overview of intelligent vehicle-highway systems (IVHS) activities in North America," Proceedings of the 5th Jerusalem Conference on Information Technology, 'Next Decade in Information Technology', pp. 694-701, 1990.
3. S. Tsugawa, "Vision-based vehicles in Japan: machine vision systems and driving control systems," IEEE Transactions on Industrial Electronics, vol. 41, no. 4, pp. 398-405, 1994.
4. R. Larsen, "AVCS: an overview of current applications and technology," Proceedings of Intelligent Vehicles Symposium, pp. 152-157, 1995.
5. S. Hahn, "Automation of driving functions - future development, benefits and pitfalls," Proceedings of IEEE Intelligent Vehicles Symposium, pp. 309-312, 1996.
6. I. Masaki, "Machine-vision systems for intelligent transportation systems," IEEE Intelligent Systems, vol. 13, no. 6, pp. 24–31, 1998.

7. E. D. Dickmanns, "Vehicles capable of dynamic vision: a new breed of technical beings?" Artificial Intelligence, vol. 103, no. 1-2, pp. 49-76, 1998.
8. R. Bishop, "Intelligent vehicle applications worldwide," IEEE Intelligent Systems and Their Applications, vol. 15, no. 1, pp.78-81, 2000.
9. D. M. Gavrila, U. Franke, and C. Wohler, et. al., "Real time vision for intelligent vehicles," IEEE Instrumentation and Measurement Magazine, vol. 4, no. 2, pp. 22-27, 2001.
10. E. D. Dickmanns, "Vision for ground vehicles: history and prospects," International Journal of Vehicle Autonomous Systems, vol. 1, no. 1, pp. 1-44, 2002.
11. E. D. Dickmanns, "The development of machine vision for road vehicles in the last decade," Proceedings of IEEE Intelligent Vehicle Symposium, vol. 1, pp. 268-281, 2002.
12. W. Jones, "Building safer cars," IEEE Spectrum, vol. 39, no. 1, pp. 82-85, 2002.
13. J. Miura, M. Itoh, and Y. Shirai, "Toward vision-based intelligent navigator: its concept and prototype," IEEE Transactions on Intelligent Transportation Systems, vol.3, no. 2, pp. 136- 146, 2002.
14. M. Bertozzi, A. Broggi, and M. Cellario, et. al, "Artificial vision in road vehicles," Proceedings of the IEEE, vol. 90, no. 7, pp. 1258-1271, 2002.
15. F.-Y. Wang, P. B. Mirchandani, Z. Wang, "The VISTA project and its applications," IEEE Intelligent Systems, vol. 17, no. 6, pp. 72-75, 2002.
16. T. Bucher, C. Curio, and J. Edelbrunner, et. al, "Image processing and behavior planning for intelligent vehicles," IEEE Transactions on Industrial Electronics, vol. 50, no. 1, pp. 62-75, 2003.
17. L. Fletcher, N. Apostoloff, and L. Petersson, et. al, "Vision in and out of vehicles," IEEE Intelligent Systems. vol. 18, no. 3, pp. 12-17, 2003.
18. F.-Y. Wang, X. Wang, and L. Li, et. al, "Creating a digital-vehicle proving ground," IEEE Intelligent Systems, vol. 18, no. 2, pp. 12-15, 2003.
19. Z. Sun, G. Bebis, and R. Miller, "On-road vehicle detection using optical sensors: a review," Proceedings of International IEEE Conference on Intelligent Transportation Systems, pp. 585-590, 2004.
20. W. Hu, T. Tan, and L. Wang, et. al, "A survey on visual surveillance of object motion and behaviors," IEEE Transactions on Systems, Man and Cybernetics, Part C, vol. 34, no. 3, pp. 334-352, 2004.
21. N.-N. Zheng, S. Tang, and H. Cheng, et. al, "Toward intelligent driver-assistance and safety warning system," IEEE Intelligent Systems, vol. 19, no. 2, pp. 8-11, 2004.
22. L. Li, J. Song, and F.-Y. Wang, et. al, "New developments and research trends for intelligent vehicles," IEEE Intelligent Systems, vol. 20, no. 4, pp. 10-14, 2005.
23. R. M. Inigo, "Traffic monitoring and control using machine vision: a survey," IEEE Transactions on Industrial Electronics, vol. 32, no. 3, pp. 177-185, 1985.

24. V. Kastrinaki, M. Zervakis, and K. Kalaitzakis, "A survey of video processing techniques for traffic applications," Image and Vision Computing, vol. 21, no. 4, pp. 359-381, 2003.
25. O. Sidla, L. Paletta, and Y. Lypetskyy, et. al, "Vehicle recognition for highway lane survey," Proceedings of IEEE International Conference on Intelligent Transportation Systems, pp. 531-536, 2004.
26. P. Hough, "Method and means for recognizing complex patterns," in US Patent, 1962.
27. J. Canny, "A computational approach to edge detection," IEEE Transactions on Pattern Analysis and Machine Intelligence, vol. 8, no. 6, pp. 679-698, 1986.
28. J. Illingworth and J. Kittler, "The adaptive Hough transform," IEEE Transactions on Pattern Analysis and Machine Intelligence, vol. 9, no. 5, pp. 690-698, 1987.
29. V. Leavers, "Which Hough transform?" CVGIP: Image Understanding, vol. 58, pp. 250-264, 1993.
30. J. Matas, C. Galambos, and J. Kittler, "Robust detection of lines using the progressive probabilistic Hough transform," Computer Vision and Image Understanding, vol. 78, pp. 119-137, 2000.
31. C. R. Jung and C. R. Kelber, "A robust linear-parabolic model for lane following," Proceedings of 17th Brazilian Symposium on Computer Graphics and Image Processing, pp. 72-79, 2004.
32. S. Lakshmanan and D. Grimmer, "A deformable template approach to detecting straight edges in radar images," IEEE Transactions on Pattern Analysis and Machine Intelligence, vol. 18, no. 4, pp. 438-443, 1996.
33. B. Ma, S. Lakshmanan, and A. O. Hero, "Simultaneous detection of lane and pavement boundaries using model-based multi-sensor fusion," IEEE Transactions on Intelligent Transportation Systems, vol. 1, no. 3, pp. 135-147, 2000.
34. C. Kreucher and S. Lakshmanan, "LANA: a lane extraction algorithm that uses frequency domain features," IEEE Transactions on Robotics and Automation, vol. 15, no. 2, pp. 343-350, 1999.
35. K. Kaliyaperumal, S. Lakshmanan, and K. Kluge, "An algorithm for detecting roads and obstacles in radar images," IEEE Transactions on Vehicular Technology, vol. 50, no. 1, pp. 170-182, 2001.
36. U. Y. Young and S.-Y. Oh, "Three-feature based automatic lane detection algorithm (TFALDA) for autonomous driving," IEEE Transactions on Intelligent Transportation Systems, vol. 4, no. 4, pp. 219-225, 2003.
37. Y. Wang, E. Teoh, and D. Shen, "Lane detection and tracking using B-snake," Image and Vision Computing, vol. 22, pp. 269–280, 2004.
38. M. S. von Trzebiatowski, A. Gern, and U. Franke, et. al, "Detecting reflection posts - lane recognition on country roads," Proceedings of IEEE Intelligent Vehicles Symposium, pp. 304-309, 2004.
39. Y. Wang, D. G. Shen, E. K. Teoh, "Lane detection using spline model," Pattern Recognition Letters, vol. 21, no. 8, pp. 677-689, 2000.

40. J. Park, J. Lee, and K. Jhang, "A lane-curve detection based on an lcf," Pattern Recognition Letters, vol. 24, pp. 2301-2313, 2003.
41. Q. Li, N. Zheng, and H. Cheng, "Springrobot: a prototype autonomous vehicle and its algorithms for lane detection," IEEE Transactions on Intelligent Transportation Systems, vol. 5, no. 4, pp. 300-308, 2004.
42. S. T. Park, S. Y. Yang, and J. H. Jung, "Real-time lane recognition by simulated annealing algorithm," Proceedings of The 4th Korea-Russia International Symposium on Science and Technology, vol. 3, pp. 95-98, 2000.
43. T. Liu, N. Zheng, and H. Cheng, et. al, "A novel approach of road recognition based on deformable template and genetic algorithm," Proceedings of IEEE Intelligent Transportation Systems Conference, vol. 2, pp. 1251-1256, 2003.
44. D. J. Kang and M. H. Jung, "Road lane segmentation using dynamic programming for active safety vehicles," Pattern Recognition Letters, vol. 24, pp. 3177–3185, 2003.
45. H. Kim, S. Hong, and H. Son, et. al, "High speed road boundary detection on the images for autonomous vehicle with the multi-layer CNN," Proceedings of International Symposium on Circuits and Systems, vol. 5, pp. 769-772, 2003.
46. A. Takahashi, Y. Ninomiya, and M. Ohta, et. al, "Rear view lane detection by wide angle camera," Proceedings of IEEE Intelligent Vehicle Symposium, vol. 1, pp. 148-153, 2002.
47. K. Ishikawa, K. Odagiri, and K. Kobayashi, et. al, "Lane detection by using omni-directional camera for outdoor terrains," Proceedings of the 41st SICE Annual Conference, vol. 2, pp. 1334-1335, 2002.
48. M. Bertozzi, A. Broggi, and A. Fascioli, "Stereo inverse perspective mapping: theory and applications," Image and Vision Computing Journal, vol. 8, no. 16, pp. 585-590, 1998.
49. M. Bertozzi and A. Broggi, "GOLD: a parallel real-time stereo vision system for generic obstacle and lane detection," IEEE Transactions on Image Processing, vol. 7, no. 1, pp. 62-81, 1998.
50. R. Labayrade, D. Aubert, and J.-P. Tarel, "Real time obstacle detection in stereovision on non flat road geometry through 'v-disparity' representation," Proceedings of IEEE Intelligent Vehicle Symposium, vol. 2, pp. 646-651, 2002.
51. D. Charnley and R. Blissett, "Surface reconstruction from outdoor image sequences," Image and Vision Computing, vol. 7, no. 1, pp. 10-16, 1989.
52. R. Chapuis, R. Aufrere, and F. Chausse, "Accurate road following and reconstruction by computer vision," IEEE Transactions on Intelligent Transportation Systems, vol. 3, no. 4, pp. 261-270, 2002.
53. S. Nedevschi, R. Schmidt, and T. Graf, et. al, "3D lane detection system based on stereovision," Proceedings of International IEEE Conference on Intelligent Transportation Systems, pp. 161-166, 2004.

54. J. C. McCall and M. M. Trivedi, "An integrated, robust approach to lane marking detection and lane tracking," Proceedings of IEEE Intelligent Vehicles Symposium, pp. 533–537, 2004.
55. A. L. Yuille and J. M. Coughlan, "Fundamental limits of Bayesian inference: order parameters and phase transitions for road tracking," IEEE Transactions on Pattern Analysis and Machine Intelligence, vol. 22, no. 2, pp. 160-173, 2000.
56. J. P. Gonzalez and U. Ozguner, "Lane detection using histogram-based segmentation and decision trees," Proceedings of IEEE Intelligent Transportation Systems Conference, pp. 346-351, 2000.
57. M. Hu, W. Yang; and M. Ren, et. al, "A vision based road detection algorithm," IEEE Conference on Robotics, Automation and Mechatronics, vol. 2, pp. 846-850, 2004.
58. J. D. Crisman and C. E. Thorpe, "SCARF: a color vision system that tracks roads and intersections," IEEE Transactions on Robotics and Automation, vol. 9, no. 1, pp. 49-58, 1993.
59. Y. He, H. Wang, and B. Zhang, "Color-based road detection in urban traffic scenes," IEEE Transactions on Intelligent Transportation Systems, vol. 5, no. 4, pp. 309-318, 2004.
60. A. Broggi and A. Fascioli, "Artificial vision in extreme environments for snowcat tracks detection," IEEE Transactions on Intelligent Transportation Systems, vol. 3, no. 3, pp. 162-172, 2002.
61. M. A. Sotelo, F. J. Rodriguez, and L. Magdalena, "VIRTUOUS: vision-based road transportation for unmanned operation on urban-like scenarios," IEEE Transactions on Intelligent Transportation Systems, vol. 5, no. 2, pp. 69-83, 2004.
62. F. Meyer and S. Beucher, "Morphological segmentation," Journal of Visual Communication and Image Representations, vol. 1, no. 1, pp. 121-146, 1990.
63. S. Beucher and M. Bilodeau, "Road segmentation and obstacle detection by a fast watershed transformation," Proceedings of the Intelligent Vehicles Symposium, pp. 296-301, 1994.
64. J. Sparbert, K. Dietmayer, and D. Streller, "Lane detection and street type classification using laser range images," Proceedings of IEEE International Conference on Intelligent Transportation Systems, pp. 454-459, 2001.
65. H. Cramer and G. Wanielik, "Road border detection and tracking in non cooperative areas with a laser radar system," Proceedings of 11th International Symposium of Advanced Technologies for Adas Systems, 2002.
66. C. Rasmussen, "Combining laser range, color, and texture cues for autonomous road following," Proceedings of IEEE International Conference on Robotics and Automation, pp. 4320-4325, 2002.
67. B. Fardi, U. Scheunert, and H. Cramer, et. al, "Multi-modal detection and parameter-based tracking of road borders with a laser scanner," Proceedings of IEEE Intelligent Vehicles Symposium, pp. 95-99, 2003.

68. W. S. Wijesoma, K. R. S. Kodagoda, and A. P. Balasuriya, et. al, "Laser and camera for road edge and mid-line detection," Proceedings of the Second International Workshop on Robot Motion and Control, pp. 269-274, 2001.
69. W. S. Wijesoma, K. R. S. Kodagoda, and A. P. Balasuriya, "Road-boundary detection and tracking using ladar sensing," IEEE Transactions on Robotics and Automation, vol. 20, no. 3, pp. 456-464, 2004.
70. A. von Reyher, A. Joos, and H. Winner, "A lidar-based approach for near range lane detection," Proceedings of IEEE Intelligent Vehicles Symposium, pp. 147-152, 2005.
71. C. Mertz, J. Kozar, and J.R. Miller, "Eye-Safe laser line striper for outside use," Proceedings of IEEE Intelligent Vehicle Symposium, pp. 507-512, 2002.
72. R. Aufrere, C. Mertz, and C. Thorpe, "Multiple sensor fusion for detecting location of curbs, walls, and barriers," Proceedings of the IEEE Intelligent Vehicles Symposium, pp. 126-131, 2003.
73. C. Mertz, D. Duggins, and J. Gowdy, "Collision warning and sensor data processing in urban areas," Proceedings of the 5th international conference on ITS telecommunications, 2005, pp. 73-78.
74. H. Zhao, R. Shibasaki, "Reconstruction of Textured Urban 3D Model by Ground-Based Laser Range and CCD Images," IEICE Transactions on Information and Systems, vol.E83-D, No.7, pp.1429-1440, 2000.
75. H. Zhao and R. Shibasaki, "Reconstructing textured CAD model of urban environment using vehicle-borne laser range scanners and line cameras," Machine Vision and Applications, Special issue on computer vision system, vol. 14, no. 1, pp. 35-41, 2003.
76. H. Zhao and R. Shibasaki, "A vehicle-borne urban 3-D acquisition system using single-row laser range scanners," IEEE Transactions on Systems, Man and Cybernetics, Part B, vol. 33, no. 4, pp. 658 - 666, 2003.
77. E. D. Dickmanns and B. D. Mysliwetz, "Recursive 3-D road and relative ego-state recognition," IEEE Transactions on Pattern Analysis and Machine Intelligence, vol. 14, no. 2, pp. 199-213, 1992.
78. D. J. LeBlanc, R. D. Ervin, and G. E. Johnson, et. al, "CAPC: an implementation of a road-departure warning system," IEEE Control Systems Magazine, vol. 16, no. 6, pp. 61-71, 1996.
79. D. J. LeBlanc, R. D. Ervin, and G. E. Johnson, et. al, "Warning and intervention system to prevent road-departure accidents," Vehicle System Dynamics, vol. 25. Suppl., pp. 383-396, 1996.
80. J. Lee, "A machine vision system for lane-departure detection," Computer Vision and Image Understanding, vol. 86, pp. 52–78, 2002.
81. J. W. Lee, C.-D. Kee, and U. K. Yi, "A new approach for lane departure identification," Proceedings of IEEE Intelligent Vehicles Symposium, pp. 100–105, 2003.
82. J. W. Lee and U. K. Yi, "A lane-departure identification based on LBPE, Hough transform, and linear regression," Computer Vision and Image Understanding, vol. 99, no. 3, pp. 359-383, 2005.

83. W. Travis, A. T. Simmons, and D. M. Bevly, "Corridor Navigation with a LiDAR/INS Kalman Filter Solution," Proceedings of IEEE Intelligent Vehicles Symposium, pp. 343–347, 2005.
84. R. Muradore, M. Assom, and C. Spagnol, et. al, "Model based GPS/INS integration for high accuracy land vehicle applications," Proceedings of IEEE Intelligent Vehicles Symposium, pp. 400-405, 2005.
85. J. C. McCall and M. M. Trivedi, "Performance evaluation of a vision based lane tracker designed for driver assistance systems," Proceedings of IEEE Intelligent Vehicles Symposium, 2005.
86. F. Paetzold and U. Franke, "Road recognition in urban environment," Image and Vision Computing, vol. 18, no. 5, pp. 377-387, 2000.
87. C.-Y. Chan, "Magnetic sensing as a position reference system for ground vehicle control," IEEE Transactions on Instrumentation and Measurement, vol. 51, no. 1, pp. 43-52, 2002.
88. J. I. Hernandez and C.-Y. Kuo, "Steering control of automated vehicles using absolute positioning GPS and magnetic markers," IEEE Transactions on Vehicular Technology, vol. 52, no. 1, pp. 150-161, 2003.
89. N. Mudaliar, D. LeBlanc, and H. Peng, "Linear estimator for road departure warning systems," Proceedings of American Control Conference, vol. 3, pp. 2104-2109, 2004.
90. F. Chausse, J. Laneurit, and R. Chapuis, "Vehicle localization on a digital map using particles filtering," Proceedings of IEEE Intelligent Vehicles Symposium, pp. 243–248, 2005.
91. R. Labayrade and D. Aubert, "A single framework for vehicle roll, pitch, yaw estimation and obstacles detection by stereovision," Proceedings of IEEE Intelligent Vehicles Symposium, pp. 31-36, 2003.
92. P. Coulombeau and C. Laurgeau, "Vehicle yaw, pitch, roll and 3D lane shape recovery by vision," IEEE Intelligent Vehicle Symposium, vol. 2, pp. 619-625, 2002.
93. J. Manigel and W. Leonhard, "Vehicle control by computer vision," IEEE Transactions on Industrial Electronics, vol. 39, no. 3, pp. 181-188, 1992.
94. L. Li and F.-Y. Wang, "Research advances in vehicle lateral motion monitoring and control," International Journal of Intelligent Control and Systems, vol. 9, no. 3, 2004.
95. N. D. Matthews, P. E. An, and C. J. Harris, "Vehicle detection and recognition in grey-scale imagery," The Second International Workshop on Intelligent Autonomous Vehicles, pp. 1-6, 1995.
96. Z. Sun, R. Miller, and G. Bebis, "A real-time precrash vehicle detection system," Proceedings of the Sixth IEEE Workshop on Applications of Computer Vision, pp. 171-176, 2002.
97. U. Handmann, T. Kalinke, and C. Tzomakas, et. al, "Computer vision for driver assistance systems," Proceedings of SPIE, vol. 3364, 1998.
98. C. Hoffman, T. Dang, and C. Stiller, "Vehicle detection fusing 2D visual features," IEEE Intelligent Vehicles Symposium, pp. 280-285, 2004.

99. M. B. van Leeuwen and F. C. A. Groen, "Vehicle detection with a mobile camera: spotting midrange, distant, and passing cars," IEEE Robotics and Automation Magazine, vol. 12, no. 1, pp. 37-43, 2005.
100. T. Kalinke, C. Tzomakas and W. V. Seelen, "A texture-based object detection and an adaptive model-based classification," IEEE International Conference on Intelligent Vehicles, pp. 143–148, 1998.
101. T. K. ten Kate, M. B. van Leewen, and S. E. Moro-Ellenberger, et. al, "Mid-range and distant vehicle detection with a mobile camera," Proceedings of IEEE Intelligent Vehicles Symposium, pp. 72-77, 2004.
102. T. Zielke, M. Brauckmann, and W. von Seelen, "Intensity and edge-based symmetry detection with an application to car-following," Computer Vision, Graphics, and Image Processing: Image Understanding, vol. 58, no. 2, pp. 177-190, 1993.
103. A. Broggi, P. Cerri, and P. C. Antonello, "Multi-resolution vehicle detection using artificial vision," Proceedings of IEEE Intelligent Vehicles Symposium, pp. 310-314, 2004.
104. J. M. Collado, C. Hilario, and A. de la Escalera, "Model based vehicle detection for intelligent vehicles," Proceedings of IEEE Intelligent Vehicles Symposium, pp. 572-577, 2004.
105. D. Guo, T. Fraichard, M. Xie and C. Laugier, "Color modeling by spherical influence field in sensing driving environment," Proceedings of IEEE Intelligent Vehicle Symposium, pp. 249–254, 2000.
106. T. Xiong and C. Debrunner, "Stochastic car tracking with line- and color-based features," IEEE Transactions on Intelligent Transportation Systems, vol. 5, no. 4, pp. 324-328, pp. 324-328, 2004.
107. P. Meer, D. Mintz, and A. Rosenfeld, et. al, "Robust regression methods for computer vision: a review," International Journal of Computer Vision, vol. 6, no. 1, pp. 59–70, 1991.
108. J. Weber, D. Koller, and Q.-T. Luong, et. al, "New results in stereo-based automatic vehicle guidance," Proceedings of IEEE Intelligent Vehicles Symposium, pp. 530–535, 1995.
109. U. Franke and I. Kutzbach, "Fast stereo based object detection for stop & go traffic," Proceedings of Intelligent Vehicles Symposium, pp. 339-344, 1996.
110. M. Bertozzi and A. Broggi, "Vision-based vehicle guidance," Computer Vision, vol. 30, no. 7, 1997.
111. C. Sun, "Fast stereo matching using rectangular subregioning and 3D maximum-surface techniques," International Journal of Computer Vision, vol.47, No.1/2/3, pp. 99-117, 2002
112. R. Labayrade, D. Aubert, and J.-P. Tarel, "Real time obstacle detection in stereovision on non flat road geometry through 'V-disparity' representation," Proceedings of IEEE Intelligent Vehicle Symposium, 2002.
113. T. Bucher, C. Curio, and J. Edelbrunner, et. al, "Image processing and behavior planning for intelligent vehicles," IEEE Transactions on Industrial Electronics, vol. 50, no. 1, pp. 62-75, 2003.

114. A. Broggi, M. Bertozzi, and A. Fascioli, et. al, "Visual perception of obstacles and vehicles for platooning," IEEE Transactions on Intelligent Transportation Systems, vol. 1, no. 3, pp. 164-176, 2000.
115. M. Bertozzi, A. Broggi, and A. Fascioli, et. al, "Vision-based pedestrian detection: will ants help?" Proceedings of IEEE Intelligent Vehicle Symposium, vol. 1, pp. 1-7, 2002.
116. R. Labayrade, D. Aubert, and J.-P. Tarel, "Real time obstacle detection in stereovision on non flat road geometry through 'v-disparity' representation," Proceedings of IEEE Intelligent Vehicle Symposium, vol. 2, pp. 646-651, 2002.
117. R. Nelson and J. Aloimonos, "Using flow field divergence for obstacle avoidance: towards qualitative vision," Proceedings Of International Conference on Compute Vision, pp.188-196, 1988.
118. W. Enkelmann, "Obstacle detection by evaluation of optical flow fields form image sequence," Image and Vision Computing, pp. 160-168, 1991.
119. M. H. Young, T. H. Hong, and A. Yang, "Obstacle detection for a vehicle using optical flow," SAE Intelligent Vehicle, 1992.
120. W. Kruger, W. Enkelmann and S. Rossle, "Real-time estimation and tracking of optical flow vectors for obstacle detection," Proceedings of IEEE Intelligent Vehicle Symposium, pp. 304–309, 1995.
121. A. Giachetti, M. Campani, and V. Torre, "The use of optical flow for road navigation," IEEE Transactions on Robotics and Automation, vol. 14, no.1, pp. 34-48, 1998.
122. U. Franke and S. Heinrich, "Fast obstacle detection for urban traffic situations," IEEE Transactions on Intelligent Transportation Systems, vol. 3, no. 3, pp. 173-181, 2002.
123. B. Heisele and W. Ritter, "Obstacle detection based on color blob flow," Proceedings of Intelligent Vehicles Symposium, pp. 282-286, 1995.
124. M. Irani and P. Anandan, "A unified approach to moving object detection in 2D and 3D scenes," IEEE Transactions on Pattern Analysis and Machine Intelligence, vol. 20, no. 6, pp. 577-589, 1998.
125. J. Wu, R. Rink, T. Caelli, and V. Gourishankar, "Recovery of the 3-d location and motion of a rigid object through camera image (an extended kalman filter approach)," International Journal of Computer Vision, vol. 2, pp. 373-394, 1989.
126. U. Regensburger and V. Graefe, "Visual recognition of obstacles on roads," Intelligent Robots and Systems, pp. 73-86, Academic Press, Inc., 1995.
127. M. Betke, E. Haritaoglu, and L.S. Davis, "Real-time multiple vehicle detection and tracking from a moving vehicle," Machine Vision and Applications, vol.12, no. 2, pp. 69–83, 2000.
128. C. Goerick, D. Noll and M. Werner, "Artificial neural networks in real time car detection and tracking applications," Pattern Recognition Letters, vol. 17, pp. 335–343, 1996.

129. M. Weber, M. Welling and P. Perona, "Unsupervised learning of models for recognition," Proceedings of European Conference on Computer Vision, vol. 1, pp. 18–32, 2000.
130. U. Handmann, T. Kalinke, and C. Tzomakas, et. al, "An image processing system for driver assistance," Image and Vision Computing, vol. 18, pp. 367-376, 2000.
131. C. Papageorgiou and T. Poggio, "A trainable system for object detection," International Journal of Computer Vision, vol. 38, no. 1, pp. 15-33, 2000.
132. T. Kato, Y. Ninomiya, and I. Masaki, "Preceding vehicle recognition based on learning from sample images," IEEE Transactions on Intelligent Transportation Systems, vol. 3, no. 4, pp. 252-260, 2002.
133. N. Srinivasa, "Vision-based vehicle detection and tracking method for forward collision warning," Proceedings of IEEE Intelligent vehicle symposium, pp. 626-631, 2002.
134. S. Avidan, "Support vector tracking," IEEE Transactions on Pattern Analysis and Machine Intelligence, vol. 26, no. 8, pp. 1064-1072, 2004.
135. Z. Sun, G. Bebis, and R. Miller, "On-road vehicle detection using evolutionary Gabor filter optimization," IEEE Transactions on Intelligent Transportation Systems, vol. 6, no. 2, pp. 125-137, 2005.
136. Y. Zhu, D. Comaniciu, and M. Pellkofer, "An integrated framework of vision-based vehicle detection with knowledge fusion," Proceedings of IEEE Intelligent Vehicle Symposium, pp.199-204, 2005.
137. P. Viola and M. Jones, "Rapid object detection using a boosted cascade of simple features," Proceedings of IEEE Conference on Computer Vision and Pattern Recognition, pp. 511-518, 2001.
138. Z. Sun, G. Bebis, and R. Miller, "Boosting object detection using feature selection," Proceedings of IEEE International Conference on Advanced Video and Signal Based Surveillance, pp. 290-296, 2003.
139. G. L. Foresti, V. Murino, and C. Regazzoni, "Vehicle recognition and tracking from road image sequences," IEEE Transactions on Vehicular Technology, vol. 48, no. 1, pp. 301-318, 1999.
140. Z. Kim and J. Malik, "Fast vehicle detection with probabilistic feature grouping and its application to vehicle tracking," Proceedings of the Ninth IEEE International Conference on Computer Vision, vol. 1, pp. 524-531, 2003.
141. A. Broggi, M. Bertozzi, and R. Chapuis, et. al, "Pedestrian localization and tracking system with Kalman filtering," Proceedings of IEEE Intelligent Vehicles Symposium, 2004.
142. C. Demonceaux, A. Potelle, and D. Kachi-Akkouche, "Obstacle detection in a road scene based on motion analysis," IEEE Transactions on Vehicular Technology, vol. 53, no. 6, pp. 1649-1656, 2004.
143. R. Tribe, "Automotive applications of microwave radar," IEE Colloquium on Consumer Applications of Radar and Sonar, pp. 1/1-1/5, 1993.
144. L. H. Eriksson, "A radar sensor for automatic AICC," Proceedings of IEEE Vehicular Technology Conference, vol. 1, pp. 434-437, 1994.

145. P. Ganci, S. Potts, and F. Okurowski, "A forward looking automotive radar sensor," Proceedings of IEEE Intelligent Vehicles Symposium, pp. 321-325, 1995.
146. H. H. Meinel, "Commercial applications of millimeterwaves: history, present status, and future trends," IEEE Transactions on Microwave Theory and Techniques, vol. 43, no. 7, pp. 1639-1653, 1995.
147. D. Langer and T. Jochem, "Fusing radar and vision for detecting, classifying and avoiding roadway obstacles," Proceedings of IEEE Intelligent Vehicles Symposium, pp. 333-338, 1996.
148. W. Nagy and J. Wilhelm, "System and parametric tradeoffs of forward looking automotive radar systems," Proceedings of the 1996 IEEE National Radar Conference, pp. 19-26, 1996.
149. H. H. Meinel, "Automotive radar and related traffic applications of millimeterwaves," 1997 Topical Symposium on Millimeter Waves, pp. 151-154, 1997.
150. D. Langer and B. Kumar, "Integrating radar and carrier phase GPS for classifying roadway obstacles," Proceedings of IEEE Conference on Intelligent Transportation System, pp. 835-840, 1997.
151. J. Wenger, "Automotive mm-wave radar: status and trends in system design and technology," IEE Colloquium on Automotive Radar and Navigation Techniques, ref. no. 1998/230, pp. 1-7, 1998.
152. G. M. Rebeiz, G.-L. Tan, and J. S. Hayden, "RF MEMS phase shifters: design and applications," IEEE Microwave Magazine, vol. 3, no. 2, pp. 72-81, 2002.
153. M. Ulm, J. Schobel, and M. Reimann, et. al "Millimeter-wave microelectromechanical (MEMS) switches for automotive surround sensing systems," 2003 Topical Meeting on Silicon Monolithic Integrated Circuits in RF Systems, Digest of Papers, pp. 142-149, 2003.
154. A. Hoover, G. Jean-Baptiste, and X. Jiang, et. al, "An experimental comparison of range image segmentation algorithms," IEEE Transactions of Pattern Analysis and Machine Intelligence, vol. 18, no. 7, pp. 673–689, 1996.
155. O. R. P. Bellon and L. Silva, "New improvements to range image segmentation by edge detection," IEEE Signal Processing Letters, vol. 9, no. 2, pp. 43–45, 2002.
156. A. Kirchner and C. Ameling, "Integrated obstacle and road tracking using a laser scanner," Proceedings of IEEE Intelligent Vehicles Symposium, pp. 675-681, 2000.
157. N. Shimomura, K. Fujimoto, and T. Oki, et. al, "An algorithm for distinguishing the types of objects on the road using laser radar and vision," IEEE Transactions on Intelligent Transportation Systems, vol. 3, no. 3, pp. 189-195, 2002.
158. K. C. Fuerstenberg and K. Dietmayer, "Object tracking and classification for multiple active safety and comfort applications using a multilayer laser scanner," Proceedings of IEEE Intelligent Vehicles Symposium, pp. 802-807, 2004.

159. D. Streller and K. Dietmayer, "Multiple hypothesis classification with laser range finders," Proceedings of International IEEE Conference on Intelligent Transportation Systems, pp. 195-200, 2004.
160. N. Kaempchen, M. Buehler, and K. Dietmayer, "Feature-Level fusion for free-form object tracking using laserscanner and video," Proceedings of IEEE Intelligent Vehicle Symposium, pp. 453-458, 2005.
161. A. Kapp, "Robust object segmentation and parametrization of 3D Lidar data," Proceedings of IEEE Intelligent Vehicle Symposium, pp. 694-699, 2005.
162. D. M. Gavrila, "The visual analysis of human movement: a survey," Computer Vision and Image Understanding, vol. 73, no. 1, pp. 82-98, 1999.
163. D. M. Gavrila, "Sensor-based pedestrian protection," IEEE Intelligent Systems, vol. 16, no. 6, pp. 77-81, 2001.
164. P. Lombardi, "A survey on pedestrian detection for autonomous driving systems," Technical Report, 2001.
165. A. Broggi, M. Bertozzi, and A. Fascioli, et. al, "Shape-based pedestrian detection," Proceedings of IEEE Intelligent Vehicle Symposium, pp. 215-220, 2000.
166. D. M. Gavrila, "Pedestrian detection from a moving vehicle," Proceedings of the European Conference on Computer Vision, pp. 37-49, 2000.
167. C. Curio, J. Edelbrunner, T. Kalinke, and C. Tzomakas, et. al, "Walking pedestrian recognition," IEEE Transactions on Intelligent Transportation Systems, vol. 1, no. 3, pp. 155-163, 2000.
168. A. Mohan, C. Papageorgiou, and T. Poggio, "Example-based object detection in images by components," IEEE Transactions on Pattern Analysis and Machine Intelligence, vol. 23, no. 4, pp. 349-361, 2001.
169. H. Mori, N. M. Charkari, and T. Matsushita, "On-line vehicle and pedestrian detections based on sign pattern," IEEE Transactions on Industrial Electronics, vol. 41, no. 4, pp. 384-391, 1994.
170. C. Wohler and J. K. Anlauf, "An adaptable time-delay neural-network algorithm for image sequence analysis," IEEE Transactions on Neural Networks, vol. 10, no. 6, pp. 1531-1536, 1999.
171. R. Cutler and L. Davis, "Robust real-time periodic motion detection, analysis and applications," IEEE Transactions on Pattern Analysis and Machine Intelligence, vol. 22, no. 8, pp. 781-796, 2000.
172. L. Zhao and C. E. Thorpe, "Stereo- and neural network-based pedestrian detection," IEEE Transactions on Intelligent Transportation Systems, vol.1, no. 3, pp. 148-154, 2000.
173. K. Fujimoto, H. Muro, and N. Shimomura, et. al. "A study on pedestrian detection technology using stereo images," JSAE Review, vol. 23, pp. 383-385, 2002.
174. C. Bregler, "Learning and recognizing human dynamics in video sequences," Proceedings of IEEE Computer Society Conference on Computer Vision and Pattern Recognition, pp. 568-574, 1997.

175. H. Elzein, S. Lakshmanan, and P. Watta, "A motion and shape-based pedestrian detection algorithm," Proceedings of the IEEE Intelligent Vehicle Symposium, pp. 500–504, 2003.
176. S. A. Niyogi, E. H. Adelson, "Analysing and recognizing walking figures in XYT," Proceedings of IEEE Conference on Computer Vision and Pattern Recognition, pp. 469-474, 1994.
177. T. Hashiyama, D. Mochizuki, and Y. Yano, et. al, "Active frame subtraction for pedestrian detection from images of moving camera," Proceedings of IEEE International Conference on Systems, Man and Cybernetics, pp. 480-485, 2003.
178. S. Yasutomi and H. Mori, "A method for discriminating of pedestrian based on rhythm," Proceedings of IEEE International Conference on Intelligent Robots and Systems, pp. 988-995, 1994.
179. C. Wohler, U. Kressler, J. K. Anlauf, "Pedestrian recognition by classification of image sequences global approaches vs local spatio- temporal processing," Proceedings of IEEE International Conference on Pattern Recognition, pp. 540-544, 2000.
180. H. Murase and V. V. Vinod, "Fast visual search using color matching-active search," The Transactions of the Institute of Electronics, Information and Communication Engineers, vol. J81-D-II, no. 9, pp. 2035-2042, 1998.
181. Y. Abramson and B. Steux, "Hardware-friendly pedestrian detection and impact prediction," Proceedings of Intelligent Vehicles Symposium, 2004.
182. Y. Freund and R. E. Schapire. "A decision-theoretic generalization of on-line learning and an application to boosting," Computational Learning Theory: Eurocolt '95, pp. 23–37, 1995.
183. A. Shashua, Y. Gdalyahu and G. Hayun, "Pedestrian detection for driving assistance systems: single-frame classification and system level performance," Proceedings of IEEE Intelligent Vehicle Symposium, 2004.
184. M. Dorigo and G. Di Caro, "The ant colony optimization meta-heuristic," New Ideas in Optimization, eds., pp. 11–32, London, UK: McGraw-Hill, 1999.
185. P. Sinha, E. Osuna, and M. Oren, et. al, "Pedestrian detection using wavelet templates," Proceedings of IEEE Conference on Computer Vision and Pattern Recognition, pp. 193–199, 1997.
186. P. Papageorgiou and T. Poggio, "Trainable pedestrian detection," Proceedings of International Conference on Image Processing, vol. 4, pp. 35-39, 1999.
187. V. Philomin, R. Duraiswami and L. Davis, "Pedestrian tracking from a moving vehicle," Proceedings of IEEE Intelligent Vehicles Symposium, pp. 350–355, 2000.
188. D. Gavrila and J. Giebel, "Shape-based pedestrian detection and tracking," Proceedings of IEEE Intelligent Vehicle Symposium, pp. 8-14, 2002.

189. B. Leibe, E. Seemann and B. Schiele, "Pedestrian detection in crowded scenes," Proceedings of International Conference on Computer Vision and Pattern Recognition, 2005.
190. Y. L. Guilloux and J. Lonnoy, "PAROTO Project: the benefit of infrared imagery for obstacle avoidance," Proceedings of IEEE Intelligent Vehicles Symposium, 2002.
191. Y. Fang, K. Yamada, and Y. Ninomiya, et. al, "Comparison between infrared-image-based and visible-image-based approaches for pedestrian detection," Proceedings of IEEE Intelligent Vehicles Symposium, 2003.
192. Y. Fang, K. Yamada, and Y. Ninomiya, et. al, "A shape-independent method for pedestrian detection with far-infrared images," IEEE Transactions on Vehicular Technology, vol. 53, no. 5, pp. 1679-1697, 2004.
193. M. Bertozzi, A. Broggi, and A. Fascioli, "Pedestrian detection for driver assistance using multiresolution infrared vision," IEEE Transactions on Vehicular Technology, vol. 53, no. 6, pp. 1666-1678, 2004.
194. D. T. Linzmeier, D. Vogt, and R. Prasanna, et. al, "Probabilistic signal interpretation methods for a thermopile pedestrian detection system," Proceedings of IEEE Intelligent Vehicles Symposium, pp. 12–17, 2005.
195. M. Bertozzi, A. Broggi, and A. Lasagni, "Infrared stereo vision-based pedestrian detection," Proceedings of IEEE Intelligent Vehicles Symposium, pp. 24–29, 2005.
196. X. Liu and K. Fujimura, "Pedestrian detection using stereo night vision," IEEE Transactions on Vehicular Technology, vol. 53, no. 6, pp. 1657-1665, 2004.
197. F. Xu, X. Liu, and K. Fujimura, "Pedestrian detection and tracking with night vision," IEEE Transactions on Intelligent Transportation Systems, vol. 6, no. 1, pp. 63-71, 2005.
198. L. Priese, J. Klieber, and R. Lakmann, et. al, "New results on traffic sign recognition," Proceedings of the Intelligent Vehicles '94 Symposium, pp. 249-254, 1994.
199. S. Estable, J. Schick, and F. Stein, "A real-time traffic sign recognition system," Proceedings of IEEE Intelligent Vehicles Symposium, pp. 213-218, 1994.
200. H. Kamada and M. Yoshida, "A visual control system using image processing and fuzzy theory," Vision Based Vehicle Guidance, I. Masaki ed. Berlin, Germany, Springer-Verlag, pp: 111-128, 1992.
201. A. de la Escalera, L. E. Moreno, and M. A. Salichs, et. al "Road traffic sign detection and classification," IEEE Transactions on Industrial Electronics, vol. 44, no. 6, pp. 848-859, 1997.
202. A. de la Escalera, J. M. Armingol, and J. M. Pastor, et. al, "Visual sign information extraction and identification by deformable models for intelligent vehicles," IEEE Transactions on Intelligent Transportation Systems, vol. 5, no. 2, pp. 57-68, 2004.
203. G. Piccioli, E. De Micheli, and P. Parodi, et. al, "Robust method for road sign detection and recognition," Image and Vision Computing, vol. 14, no. 3, pp. 209-223, 1996.

204. S. Vitabile, G. Pollaccia, and G. Pilato, et. al, "Road signs recognition using a dynamic pixel aggregation technique in the HSV color space," Proceedings of International Conference on Image Analysis and Processing, pp. 572-577, 2001.
205. A. de la Escalera, J. M Armingol and M. Mata, "Traffic sign recognition and analysis for intelligent vehicles," Image and Vision Computing, vol. 21, no. 3, pp. 247-258, 2003.
206. C.-Y. Fang, S.-W. Chen, and C.-S. Fuh, "Road-sign detection and tracking," IEEE Transactions on Vehicular Technology, vol. 52, no. 5, pp. 1329-1341, 2003.
207. C. Y. Fang, C. S. Fuh, and P. S. Yen, et. al. "An automatic road sign recognition system based on a computational model of human recognition processing," Computer Vision and Image Understanding," vol. 96, no. 2, pp. 237–268, 2004.
208. G. Loy and A. Zelinsky, "Fast radial symmetry for detecting points of interest," IEEE Transactions on Pattern Analysis and Machine Intelligence, vol. 25, no. 8, pp. 959-973, 2003.
209. N. Barnes and A. Zelinsky, "Real-time radial symmetry for speed sign detection," Proceedings of IEEE Intelligent Vehicles Symposium, pp. 566-571, 2004.
210. Y. Aoyagi and T. Asakura, "A study on traffic sign recognition in scene image using genetic algorithms and neural networks," Proceedings of IEEE International Conference on Industrial Electronics, Control, and Instrumentation, vol. 3, pp. 1838-1843, 1996.
211. T. Asakura, Y. Aoyagi, and O. K. Hirose, "Real-time recognition of road traffic sign in moving scene image using new image filter," Proceedings of the 39th SICE Annual Conference on International Session Papers, pp. 13-18, 2000.
212. E. Perez and B. Javidi, "Nonlinear distortion-tolerant filters for detection of road signs in background noise," IEEE Transactions on Vehicular Technology, vol. 51, no. 3, pp. 567-576, 2002.
213. G. Piccioli, E. De Micheli, and P. Parodi, "Robust road sign detection and recognition from image sequences," Proceedings of IEEE Intelligent Vehicles Symposium, pp. 278-283, 1994.
214. G. Piccioli, E. D. Micheli, and P. Parodi, et. al, "A robust method for road sign detection and recognition," Image Vision Computing, vol. 14, pp. 208–223, 1996.
215. P. Paclik, J. Novovicova, and P. Somol, et. al, "Road sign classification using Laplace kernel classifier," Pattern Recognition Letter, vol. 21, no. 13–14, pp. 1165–1173, 2000.
216. S. H. Hsu and C. L. Huang, "Road sign detection and recognition using matching pursuit method," Image and Vision Computing, vol. 19, no. 3, pp. 119–129, 2001.
217. S. Vitabile, A. Gentile, and F. Sorbello, "A neural network based automatic road signs recognizer," Proceedings of International Joint Conference on Neural Networks, vol. 3, pp. 2315-2320, 2002.

218. A. Farag and A. E. Abdel-Hakim, "Detection, categorization and recognition of road signs for autonomous navigation," Proceedings of Advanced Concepts for Intelligent Vision Systems, pp. 125-130, 2004.
219. C. Bahlmann, Y. Zhu, and V. Ramesh, "A system for traffic sign detection, tracking, and recognition using color, shape, and motion information," Proceedings of IEEE Intelligent Vehicles Symposium, pp. 255-260, 2005.
220. S. Lafuente-Arroyo, P. Gil-Jimenez, and R. Maldonado-Bascon, et. al, "Traffic sign shape classification evaluation I: SVM using Distance to Borders," Proceedings of IEEE Intelligent Vehicles Symposium, pp. 557-562, 2005.
221. P. Gil-Jimenez, S. Lafuente-Arroyo, and H. Gomez-Moreno, et. al, "Traffic sign shape classification evaluation II: FFT applied to the signature of blobs," Proceedings of IEEE Intelligent Vehicles Symposium, pp. 607-612, 2005.
222. J. Miura, T. Kanda, and T. Shirai, "An active vision system for real-time traffic sign recognition," Proceedings of IEEE Intelligent Transportation Systems, pp. 52-57, 2000.
223. P. Faber, "Seat occupation detection inside vehicles," Proceedings of the 4th IEEE Southwest Symposium on Image Analysis and Interpretation, pp. 187-191, 2000.
224. S. Boverie, M. Devy, and G. Kirk, "3D perception for new airbag generations," IFACWorld Congress on Automatic Control, pp. 21-26, 2002.
225. M. E. Farmer and A. K. Jain, “"ccupant classification system for automotive airbag suppression," Proceedings of IEEE International Conference on Computer Vision and Pattern Recognition, vol. 1, pp. 756-761, 2003.
226. J. Krumm and G. Kirk, "Video occupant detection for airbag deployment," Proceedings of the 4th IEEE Workshop on Applications of Computer Vision, pp. 30-35, 1998.
227. Y. Zhang, S. Kiselewich, and W. A. Bauson, "A monocular vision-based occupant classification approach for smart airbag deployment," Proceedings of IEEE Intelligent Vehicle Symposium, pp. 632-637, 2005.
228. M. Devy, A. Giralt, and A. Marin-Hernandez, "Detection and classification of passenger seat occupancy using stereovision," Proceedings of IEEE Intelligent Vehicles Symposium, pp. 714-719, 2000.
229. W. Huang and Y. Guo, "Real-time stereo tracking of multiple moving heads," Proceedings of IEEE ICCV Workshop on Recognition, Analysis, and Tracking of Faces and Gestures in Real-Time Systems, pp. 55-60, 2001.
230. Y. Owechko, N. Srinivasa, and S. Medasani, et. al., "Vision-based fusion system for smart airbag applications," Proceedings of IEEE Intelligent Vehicle Symposium, vol. 1, pp. 245–250, 2002.
231. Y. Owechko, N. Srinivasa, and S. Medasani, "High performance sensor fusion architecture for vision-based occupant detection," Proceedings of IEEE Intelligent Transportation Systems, vol. 2, pp. 1128-1133, 2003.

232. B. Alefs, M. Clabian, and H. Bischof, et. al, "Robust occupancy detection from stereo images," Proceedings of IEEE Intelligent Transportation Systems Conference, pp. 1-6, 2004.
233. T. D. Schoenmackers and M. M. Trivedi, "Real-time stereobased vehicle occupant posture determination for intelligent airbag deployment," Proceedings of IEEE Intelligent Vehicles Symposium, pp. 570-574, 2003.
234. S. J. Krotosky, S. Y. Cheng, and M. M Trivedi, "Real-time stereo-based head detection using size, shape and disparity constraints," Proceedings of IEEE Intelligent Vehicles Symposium, pp. 550–556, 2005.
235. M. Fritzsche, C. Prestele, and G. Becker, et al., "Vehicle occupancy monitoring with optical range-sensors," Proceedings of IEEE Intelligent Vehicles Symposium, pp. 90-94, 2004.
236. M. Trivedi, S. Y. Cheng, and E. Childers, et. al, "Occupant posture analysis with stereo and thermal infrared video: Algorithms and experimental evaluation," IEEE Transactions on Vehicle Technology, vol. 53, no. 6, pp. 1968–1712, 2004.
237. R. Reyna, A. Giralt, and D. Esteve, "Head detection inside vehicles with a modified SVM for safer airbags," Proceedings of IEEE Intelligent Transportation Systems Conference, pp. 268-272, 2001.
238. A. Santana Diaz, B. Jammes, and D. Esteve, et. al, "Driver hypovigilance diagnosis using wavelets and statistical analysis," Proceedings of IEEE International Conference on Intelligent Transportation Systems, pp. 162-167, 2002.
239. Y. Luo, Y. L. Murphy, and F. Khairallah, "Human head detection using multi-modal object features," Proceedings of International Joint Conference on Neural Networks, pp. 2134-2139, 2003.
240. Y. Zhu and K. Fujimura, "Head pose estimation for driver monitoring," Proceedings of IEEE Intelligent Vehicles Symposium, pp. 501-506, 2004.
241. T. Pilutti and A. G. Ulsoy, "Identification of driver state for lane-keeping tasks," IEEE Transactions on Systems, Man and Cybernetics, Part A, vol. 29, no. 5, pp. 486-502, 1999.
242. X. Liu, F. Xu, and K. Fujimura, "Real-time eye detection and tracking for driver observation under various light conditions," IEEE Intelligent Vehicle Symposium, vol. 2, pp. 344-351, 2002.
243. R.-B. Wang, K.-Y. Guo, and S.-M. Shi, et. al, "A monitoring method of driver fatigue behavior based on machine vision," Proceedings of IEEE Intelligent Vehicles Symposium, pp. 110-113, 2003.
244. Q. Ji, Z. Zhu and P. Lan, "Real-time nonintrusive monitoring and prediction of driver fatigue," IEEE Transactions on Vehicular Technology, vol. 53, no. 4, pp. 1052-1068, 2004.
245. L. Fletcher, L. Petersson, and A. Zelinsky, "Road scene monotony detection in a fatigue management driver assistance systeme," Proceedings of Intelligent Vehicles Symposium, pp. 484-489, 2005.

246. P. Thiffault and J. Bergeron, "Monotony of road environment and driver fatigue: a simulator study," Accident Analysis and Prevention, vol. 35, pp. 381–391, 2003.
247. N. Apostoloff and A. Zelinsky, "Vision in and out of vehicles: integrated driver and road scene monitoring," International Journal of Robotics Research, pp. 5–28, 2003.
248. J. Roge, T. Pebayle, and E. Lambilliotte, et. al, "Influence of age, speed and duration of monotonous driving task in traffic on the driver's useful visual field," Vision Research, vol. 44, no. 23, pp. 2737-2744, 2004.
249. T. Brandt, R. Stemmer, and A. Rakotonirainy, "Affordable visual driver monitoring system for fatigue and monotony," Proceedings of IEEE International Conference on Systems, Man and Cybernetics, vol. 7, pp. 6451-6456, 2004.
250. L. Fletcher, L. Petersson, A. Zelinsky, "Road scene monotony detection in a fatigue management driver assistance system," Proceedings of IEEE Intelligent Vehicles Symposium, pp. 484–489, 2005.
251. G. J. Chaitin, "Information-theoretic limitations of formal systems," Journal of the ACM, vol. 21, no. 3, pp. 403-424, 1974.
252. J. McCall, O. Achler, and M. M. Trivedi, et. al, "A collaborative approach for human-centered driver assistance systems," Proceedings of IEEE Conference on Intelligent Transportation Systems, 2004.
253. S. Park and M. Trivedi, "Driver activity analysis for intelligent vehicles: issues and development framework," Proceedings of IEEE Intelligent Vehicles Symposium, 2005.
254. G. McAllister, S. J. McKenna and I. W. Ricketts, "Hand tracking for behaviour understanding," Image and Vision Computing, vol. 20, no. 12, pp. 827-840, 2002.
255. M. Donath, "Designing human centered systems for crash avoidance," Keynote Speech in the IEEE Intelligent Vehicles Symposium, 2005.
256. D. Pomerleau and T. Jochem, "Rapidly adapting machine vision for automated vehicle steering," IEEE Expert, vol. 11, no. 2, pp. 19-27, 1996.
257. R. D. Ellis, T. J. Meitzler, and G. Witus, et. al, "Computational modeling of age-differences in a visually demanding driving task: vehicle detection," IEEE Transactions on Systems, Man and Cybernetics, Part A, vol. 30, no. 3, pp. 336-346, 2000.
258. C. Stiller, J. Hipp, and C. Rssig, et. al, "Multisensor obstacle detection and tracking," Image and Vision Computing, vol. 18, pp. 389–396, 2000.
259. D. M. Gavrila, M. Kunert, and U. Lages, "A multi-sensor approach for the protection of vulnerable traffic participants the PROTECTOR project," Proceedings of IEEE Instrumentation and Measurement Technology Conference, vo. 3, pp. 2044-2048, 2001.
260. T. Kato, Y. Ninomiya, and I. Masaki, "An obstacle detection method by fusion of radar and motion stereo," IEEE Transactions on Intelligent Transportation Systems, vol. 3, no. 3, pp. 182-188, 2002.

261. B. Steux, C. Laurgeau, and L. Salesse, "Fade: a vehicle detection and tracking system featuring monocular color vision and radar data fusion," Proceedings of IEEE Intelligent Vehicle Symposium, vol. 2, pp. 632-639, 2002.
262. J. Laneurit, C. Blanc, and R. Chapuis, et. al, "Multisensorial data fusion for global vehicle and obstacles absolute positioning," Proceedings of IEEE Intelligent Vehicles Symposium, pp. 138-143, 2003.
263. C. Blanc, L. Trassoudaine, and Y. Le Guilloux, et. al, "Track to track fusion method applied to road obstacle detection, Proceedings of The 7th International Conference on Information Fusion, pp. 775-782, 2004.
264. B. Fardi, U. Schuenert, G. Wanielik, "Shape and motion-based pedestrian detection in infrared images: a multi sensor approach, Proceedings of IEEE Intelligent Vehicles Symposium, 2005.
265. W. F. Herrington, B. K. P. Horn, and I. Masaki, "Application of the discrete Haar wavelet transform to image fusion for nighttime driving," Proceedings of IEEE Intelligent Vehicles Symposium, pp. 273–277, 2005.
266. R. Labayrade, C. Royere, and D. Aubert, "A collision mitigation system using laser scanner and stereovision fusion and its assessment," Proceedings of IEEE Intelligent Vehicles Symposium, pp. 441-446, 2005.
267. R. R. Murphy, "Sensor and information fusion improved vision-based vehicle guidance," IEEE Intelligent Systems and Their Applications, vol. 13, no. 6, pp. 49-56, 1998.
268. Y. Fang, I. Masaki, and B. Horn, "Depth-based target segmentation for intelligent vehicles: fusion of radar and binocular stereo," IEEE Transactions on Intelligent Transportation Systems, vol. 3, no. 3, pp. 196-202, 2002.
269. E. D. Dickmanns, "Expectation-based, multi-focal, saccadic (EMS) vision for dynamic scene understanding," Proceedings of Norwegian Signal Processing Symposium, 2002.
270. E. D. Dickmanns, "Three-Stage visual perception for vertebrate-type dynamic machine vision," Engineering of Intelligent Systems, 2004.
271. M. Wada, K. S. Yoon, and H. Hashimoto, "Development of advanced parking assistance system," IEEE Transactions on Industrial Electronics, vol. 50, no. 1, pp. 4-17, 2003.
272. Y. Suzuki, T. Fujii, and M. Tanimoto, "Parking assistance using multi-camera infrastructure," Proceedings of IEEE Intelligent Vehicles Symposium, pp. 106–100, 2005.
273. C. Chan and B. Bougler, "Evaluation of cooperative roadside and vehicle-based data collection for assessing intersection conflicts," Proceedings of IEEE Intelligent Vehicles Symposium, pp. 165–170, 2005.
274. Y. Liu, U. Ozguner, and E. Ekici, "Performance evaluation of intersection warning system using a vehicle traffic and wireless simulator," Proceedings of IEEE Intelligent Vehicles Symposium, pp. 171–176, 2005.

275. S. Tsugawa, "Inter-vehicle communications and their applications to intelligent vehicles: an overview," Proceedings of IEEE Intelligent Vehicle Symposium, vol. 2, pp: 564-569, 2002.
276. S. Tsugawa and S. Kato, "Evaluation of incident information transmission on highways over inter-vehicle communications," Proceedings of IEEE Intelligent Vehicle Symposium, pp. 12-16, 2003.
277. N. Navet, Y. Song, and F. Simonot-Lion, "Trends in automotive communication systems," Proceedings of the IEEE, vol. 93, no. 6, pp. 1204-1223, 2005.
278. L. Li and F.-Y. Wang, "Cooperative driving and lane changing at blind Crossings," Proceedings of IEEE Intelligent Vehicles Symposium, pp. 435–439, 2005.
279. K. Tischler, and B. Hummel, "Enhanced environmental perception by inter-vehicle data exchange," Proceedings of IEEE Intelligent Vehicles Symposium, pp. 313-318, 2005.
280. T. J. Gates and H. G. Hawkins, "The use of wider longitudinal pavement markings," Texas Transportation Institute Report, College Station, TX, 2002.
281. T. J. Gates, S. T. Chrysler, and H. G. Hwakins, "Innovative visibility-based measures of effectiveness for wider longitudinal pavement markings," Texas Transportation Institute Report, College Station, TX, 2002.
282. R. K. Lens and R. Y. Tsai, "Techniques for calibration of the scale factor and image center for high accuracy 3-D machine vision metrology," IEEE Transactions on Pattern Analysis and Machine Intelligence, vol. 10, no. 5, pp. 713-720, 1987.
283. Manual on Uniform Traffic Control Devices, Federal Highway Administration, U.S. Department of Transportation, 2003.
284. P. Baines and C. Dixon, Signs: Lettering in the Environment, Laurence King Publishing, 2003.
285. ClearviewHwy, http://clearviewhwy.com/
286. C. L. Wan, K. W. Dickinson and T. D. Binnie, "A cost-effective image sensor system for transport applications utilising a miniature CMOS single chip camera," Proceedings of IFAC Transportation Systems, 1994.
287. R. Gregor, M. Lutzeler, and M. Pellkofer, et. al, "EMS-Vision: a perceptual system for autonomous vehicles," IEEE Transactions on Intelligent Transportation Systems, vol. 3, no. 1, pp. 48-59, 2002.
288. K. Yamada and M. Soga, "A compact integrated visual motion sensor for ITS applications," IEEE Transactions on Robotics and Automation, vol. 4, no. 1, pp. 35-42, 2003.
289. K. Beier and H. Gemperlein, "Simulation of infrared detection range at fog conditions for Enhanced Vision Systems in civil aviation," Aerospace Science and Technology, vol. 8, no. 1, pp. 63-71, 2004.
290. C. Mertz, S. McNeil, and C. Thorpe, "Side collision warning systems for transit buses," Proceedings of IEEE Intelligent Vehicle Symposium, pp. 344-349, 2000.

291. M. Sergi and M. Donath, "Bus rapid transit technologies: a virtual mirror for eliminating vehicle blind zones," Final Research Reports of University of Minnesota, Center for Transportation Studies, vol. 2, report no. CTS 04-12, 2005.
292. K. Yamada, T. Nakano, and S. Yamamoto, "A vision sensor having an expanded dynamic range for autonomous vehicles," IEEE Transactions on Vehicular Technology, vol. 47, no. 1, pp. 332-341, 1998.
293. S. De Ma, "A self-calibration technique for active vision systems," IEEE Transactions on Robotics and Automation, vol. 12, pp. 114–120, 1996.
294. T. A. Clarke and J. G. Fryer, "The development of camera calibration methods and models," Photogrammetric Record, vol. 16, no. 91, pp. 51–66, 1998.
295. A. Broggi, M. Bertozzi, and A. Fascioli, "Self-calibration of a stereo vision system for automotive applications," Proceedings of 2001 IEEE International Conference on Robotics and Automation, vol. 4, pp. 3698–3703, 2001.
296. F.-Y. Wang, "A simple and analytical procedure for calibrating extrinsic camera parameters," IEEE Transactions on Robotics and Automation, vol. 20, no. 1, pp. 121-124, 2004.
297. J. S. Jin, Z. Zhu and G. Xu, "A stable vision system for moving vehicles," IEEE Transactions on Intelligent Transportation Systems, vol. 1, no. 1, pp. 32-39, 2000.
298. Y.-M. Liang, H.-R. Tyan, and S.-L. Chang, et. al, "Video stabilization for a camcorder mounted on a moving vehicle," IEEE Transactions on Vehicular Technology, vol. 53, no. 6, pp. 1636-1648, 2004.
299. M. R. Kabuka, S. Harjadi, and A. Younis, "A fault-tolerant architecture for an automatic vision-guided vehicle," IEEE Transactions on Systems, Man and Cybernetics, vol. 20, no. 2, pp. 380-394, 1990.
300. W. Sohn and N.D. Kehtarnavaz, "Analysis of camera movement errors In vision-based vehicle tracking," IEEE Transactions on Pattern Analysis and Machine Intelligence, vol. 17, no. 1, pp. 57-61, 1995.
301. N. D. Kehtarnavaz and W. Sohn, "Error analysis of camera movements in stereo vehicle tracking systems," Computer Vision and Image Understanding, vol. 62, no. 3, pp. 347-359, 1995.
302. J. A. Fayman, E. Rivlin, and D. Moss, "Real-time active vision with fault tolerance," Proceedings of International Conference on Pattern Recognition, vol. 3, pp. 7276-7276, 2005.
303. N. Hautire and D. Aubert, "Driving assistance: automatic fog detection and measure of the visibility distance," Proceedings of Intelligent Transport Systems World Congress, 2003.
304. H. Kurihata, T. Takahashi, I. Ide, "Rainy weather recognition from in-vehicle camera images for driver assistance," Proceedings of IEEE Intelligent Vehicles Symposium, pp. 205-210, 2005.
305. M. Yamada, K. Ueda, and I. Horiba, et. al, "Discrimination of the road condition toward understanding of vehicle driving environments," IEEE

Transactions on Intelligent Transportation Systems, vol. 2, no. 1, pp. 26-31, 2001.

306. Grand Challenge Home, http://www.darpa.mil/grandchallenge/

Intelligent Vehicle Tire Inspection and Monitoring

9.1 Introduction

The comfort and safety of driving tightly depends on the good working states of the vehicle tires. However, it is not easy for people to estimate tire pressure, temperature and tire/road friction force etc. correctly, especially when driving on road. As a result, numerous novel tire sensors and inspection devices were developed in the last two decades to make driving more safe and comfortable. In addition, these approaches relates to resolving scrap tire pollution and highway noise reduction. This Chapter presents a brief survey of the recent advances in this field with a special focus on online tire sensors.

According to National Highway Traffic Safety Administration's investigation data, United States Department of Transportation (NHTSA), tire failures raise great safety concerns in the last decade, because a significant percentage of tire failures were caused by manufacturing defects or improper maintenance. Among them, under-inflated tires are known as one prime reason for roads accidents now, extremely for some really bad ones. Moreover, under-inflated tires can also cause: quicker wear at tire surface, bad car handling, and lower fuel mileage. However, a great number of people failed to realize they had under-inflated tires. Based on recent survey data, it is primarily because that many drivers check their vehicles' tire pressure infrequently. Other contributing factors include the difficulty of visually tire-pressure estimation when a tire is notably under-inflated or deflated due to natural leakage or seasonal climatic changes [1]-[3].

Besides tire pressure, the working condition of a tire is affected by numerous other factors including tire surface, tire belt and bearing, tire temperature, etc. For instance, tire bearing damage causes increased fuel costs, abnormal wheel vibration and noise, which will lead to increased stopping distance or decreased braking power. Thus, in recent approaches, different tire sensors are developed to take an integrated inspecting/monitoring of all these factors to make driving more safe and comfortable [4]-[7].

Fig.9.1 shows structure of a general radial tire of automobiles [8]-[9]. If the wheel bearing part is also considered, offline tire inspection can be categorized from the outside to the inside of a tire as follows:

(1) Tire tread condition analysis, which aims to detect tire surface wear condition;

(2) Tire ply/belt inspection, which examine tire ply/belt structure;

(3) Tire bearing examination, which check out the possible damage of tire bearing.

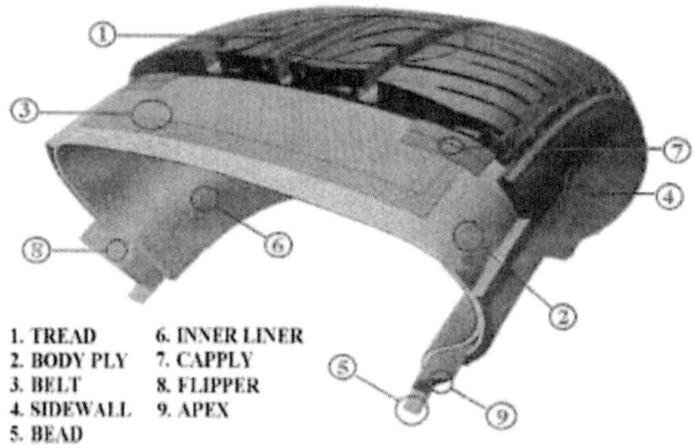

Fig.9.1 Diagram of automobile radial tire structure, from [9].

And online tire monitoring mainly includes:

(1) Tire pressure monitoring, since an under-inflated tire may cause great accident especially when steering or breaking;

(2) Tire temperature monitoring, since too high temperature may lead to tire explosion and severer tire wear;

(3) Tire beard force and deformation monitoring, which is important to tire dynamics modeling, tire manufactory [10] and riding safety examination [7];

(4) Tire/road friction monitoring, which is directly related to vehicle motion control.

The new terms "Intelligent Tire" and "Smart Tire", which mean online tire monitoring, are now achieving increasingly popularity among automotive and automotive component manufacturers. Now, from the new viewpoint, "tire" is regarded as an integral element of an intelligent vehicle: as link between road and vehicle body. It not only supports the vehicle's weight, but supplies automotive control systems with valuable real-time information. In other words, tire is taken as a virtual sensor of the whole vehicle.

More specifically, different approaches of "Intelligent Tire" can be divided into two kinds: direct and indirect tire monitor from their different techniques. In direct approaches, electronic sensors are employed to measure the desired tire characteristics, such as classic deformation-voltage sensors and magnetic field sensors. However, in this Chapter, prominence is given to a newly developing tire sensor using SAW technique.

Indirect approaches are based on the effect of tire deflation on the vehicle movement [11]. Notice the abnormal variation of tire/road friction coefficient also indicates potential leakage or severer tire wear, to monitor the friction force characteristics at the tire/road interface has also received a great deal of attention in automotive literatures recently [12]-[14]. Because these approaches were well addressed in *Chapter 2*, they will not be further mentioned here.

In the rest of this Chapter, a rough survey is organized to review the recent developments and trends in this field. The references presented here is surely incomplete, but it tries to give a comprehensive representation of different ways on tire inspection/monitoring.

9.2 Advances in Offline Tire Inspection

9.2.1 Tire Surface Inspection

Tire surface textures and surface temperature inspection are the two frequently mentioned directions in this field. Tire surface textures at least greatly affect two things: tire/road friction coefficients [15] and tire noise [6], [16]-[20]. Since the wear degree of a tire is relatively easy to estimate by human eyes, automatic tire surface texture inspection received little attention in the past thirty years. More efforts are put into tire surface temperature inspection, since too high temperature will damage tire/wheel structure [21]-[25]. Furthermore, the obtained tire surface temperature data can be used for tire road friction modeling [26].

Usually, infrared thermo-meters are used for non-contact tire temperature measurement. This kind of sensor is based on the famous Planck's black body radiation law, which quantitatively ascertains the quantity of electromagnetic radian power of various wavelengths at different temperatures. One advantage of it is of being able to measure surface temperature distribution of the measure object.

A typical tire temperature variation process is shown in Fig.9.2. It is clear that tire temperature rapidly rises at the initiation of 15 minutes. After a certain time of period, i.e. 45 minutes, from the start, the tire temperature will arrives at a steady state. At this moment, heat generation is almost in equilibrium with heat dissipation. Nevertheless, these curves also indicate that temperature rose at every section of the tire should has the similar temperature rise rate.

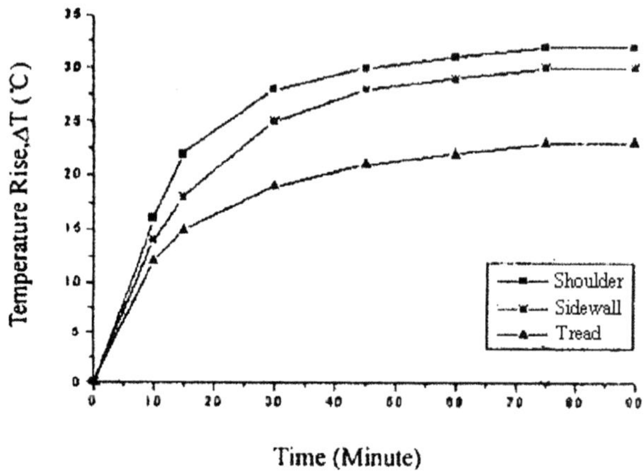

Fig.9.2 Surface temperature rise of a bias truck tire during warm-up at 70km/h, 480kPa, and 18kN load, form [24]. (© [1999] IEEE)

Apparently, these principles can be applied to tire fault detection. Any detected surface temperature that is abnormally higher than equilibrium threshold can be taken as tire failure, and any temperature variation inconsistence between different tire surface sections might indicate a potential failure too. However, this method is sensitivity to environment variations. For example, if the vehicle suddenly passes a wet road section and gets its tires wet, a false tire defect alarm may be triggered due to sudden temperature decrease caused by evaporation.

Furthermore, an infrared image of the tire surface can reveal how the tire is gripping the road surface. Notice that a warmer band around the tire indicates that this section of the tire works harder to the road. Thus, an abnormal hot spot/tread on the infrared tire image can mean an unexpected lubrication, wear or misalignment anomaly exists, when a real infrared image of a rolling tire is comparing to a predicted temperature distribution model, see Fig.9.3. Infrared thermography was first used to measure tire surface temperatures during aircraft taxi tests and laboratory tests [23]. It is also shown to be an effective method for racing cars recently, where the suspension and balancing systems can be correctly adjusted based on infrared image information.

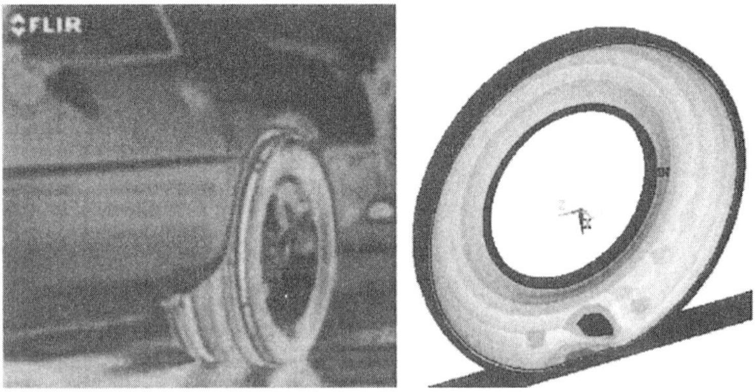

Fig.9.3 (left) from http://www.flirthermography.com/, infrared image of a rolling tire, (right) predicated temperature distribution of a rolling tire, Reprinted from [26] with permission from Elsevier.

9.2.2 Tire Ply/Belt Inspection

Under-inflation will cause high tire deflection and tire heat build-up, which leads to ply separation and may also cause uneven wearing of the tread and rapid wearing of tire shoulder. To detection tire ply/belt failure can reduce accidences.

In another point, to retreat tires based on tire tread inspection attracts great interest, since it is desired for price, energy and environmental perspectives. It is an important problem to solve, because, for example, in 1996, 266 million scrap tires were generated with an approximate weight of 3.3 million tons [5].

A novel approach using ultrasonic inspection of tires had been proposed for tire tread failure detection in [27]. The basic theory of this technique could be briefly described by a typical model given in [28].

As shown in Fig.9.1, a tire could have several different layers, which exist because of variations in rubber compounding and belt configuration. Existence of these layers results in variation in the acoustic impedance of the material between layers. In addition, the tire has significant attenuation due to the viscoelasticity of rubber and the inhomogeneity of structure.

It is assumed that the material properties or the defect properties are known and the received ultrasonic signal is unknown. The kernel part of this method is to first determine and then monitor the uncertainty in the ultrasonic transmission coefficient during the tire inspection process. Any unexpected variation of the estimated ultrasonic transmission coefficient found in the inspection will be considered as a defect.

In [27], the material was initially assumed to be a layered homogeneous elastic material with material properties that vary between the layers because of the density and structure of the belts. And viscoelasticity has been implemented as a complex modulus. The Voight model Downs used for linear viscoelasticity is shown in Fig.9.4. For a plane harmonic incident wave, the complex moduli from the Voigt model have the form

$$\lambda^* = \lambda + \frac{2}{3} i\omega\eta \text{, } \mu^* = \mu - i\omega\eta \tag{9.1}$$

where λ and μ are Lame constants and the real part of the complex modulus is the standard elastic modulus and the imaginary part changes with frequency and the viscosity coefficient.

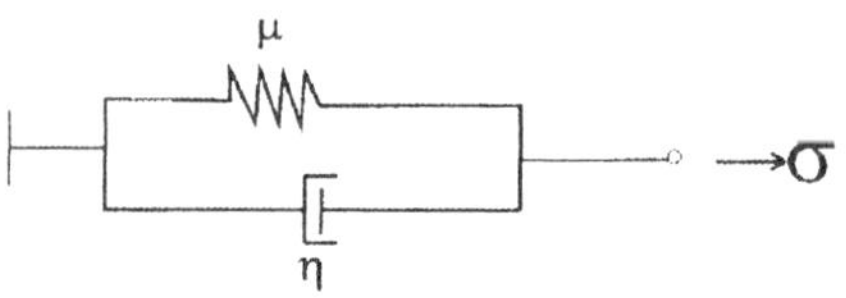

Fig.9.4 Rubber model two-element Voigt model, from [27]. (© [1999] IEEE)

The tire being inspected was modeled as shown in Fig.9.5. The tire casing was represented by five solid layers with fluid loading (the water coupling) on either side of the sample. Layers 2–6 are solid rubber and medium 1 and 7 are fluid water. Angle θ_7 is the incident angle and θ_1 is the transmitted angle. The scalar and vector potentials of waves are

$$\varphi^{(j)} = \varphi_1^{(j)} \exp(ia(z - z_{j-1})) + \varphi_2^{(j)} \exp(-ia(z - z_{j-1})) \tag{9.2}$$

and

$$\phi^{(j)} = \phi_1^{(j)} \exp(ia(z - z_{j-1})) + \phi_2^{(j)} \exp(-ia(z - z_{j-1})) \tag{9.3}$$

where $z_j \le z \le z_{j-1}$ · $\alpha = k_l \cos\theta_l$, $\beta = k_t \cos\theta_t$ and $\xi = k_i \cos\theta_i$. Here, terms φ_1 and φ_2 are the reflection amplitude and the incident amplitude for the scalar potential. ϕ_1 and ϕ_2 are the reflection amplitude and the incident amplitude for the y component of the vector potential. The vector potential can then be written as $\vec{\varphi} = [\varphi_1 \quad \varphi_2 \quad \phi_1 \quad \phi_2]^T$.

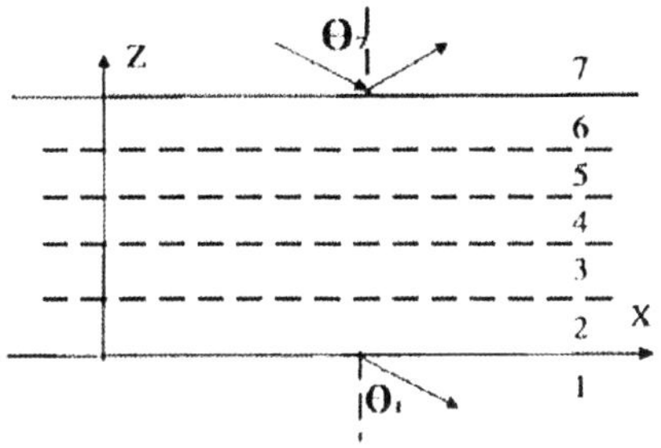

Fig.9.5 The proposed five-layers model of tire material, from [27]. (© [1999] IEEE)

By estimating the vector potential, a comparison of the sensitivity of the continuous boundary conditions, interlayer slip in solids, and reversion layer with angle is shown in Fig.9.6. These results show that, under all angles, attenuation due to the two defects should exist. The attenuation for the slip condition at normal incidence is less than the attenuation that would be expected to be detectable in experiments, which creates some concerns. In addition, a large variation in amplitude between the portions of the tire casing due to abnormal incidence clearly indicates the need for model-based gain control. In this context, it is useful to consider how the measurements are to be made and then to look at experimental results.

9.2.3 Tire Bearing Inspection

Tire/wheel non-uniformities or damage is another problem that should be carefully handled. It can cause lots of troubles including increased fuel costs and abnormal wheel vibration/noise. Risks of higher maintenance costs may increase if no warnings are fired.

In [32], Gustafsson et. al. analyzed this problem and claimed that the offset error of each tooth can be identified individually and compensated with an adaptive algorithm. Fig.9.7(a) illustrates an offset error for a tooth of the tire; where solid line represents the un-ideal toothed wheel and dash line ideal toothed wheel. The peak of normal vehicle vibration is clearly weakened by the peaks introduced by the disturbance from the toothed wheel.

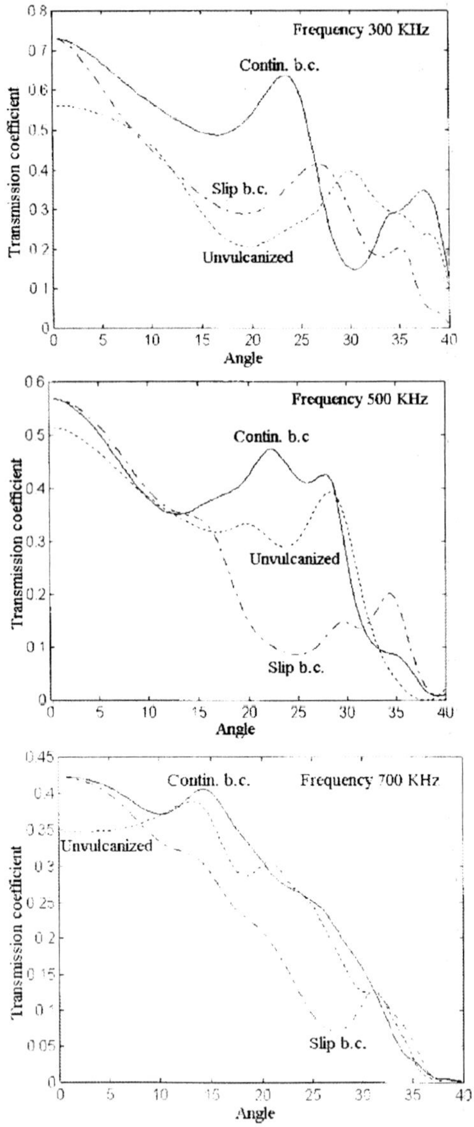

Fig.9.6 Transmission coefficients versus angles at the nominal center frequency of the transducer, curves represent an acceptable in (top) and two expected types of defects in (middle), (bottom). (top) shows the three cases at 500 kHz, (middle) and (bottom) show the same sensitivity to angle at 300 and 700 kHz, respectively, from [27]. (© [1999] IEEE)

With a recursive least squares adaptation algorithm (RLS), Gustafsson et. al. showed that the wheel offset errors can be explicitly estimated online based on the wheel rolling dynamics. Fig. 9.7(b) shows the obtained estimation of offset errors. Fig.9.8 shows that the signal to noise ratio has been much improved. The obtained spectrum clearly shows a vibration mode around 45 Hz.

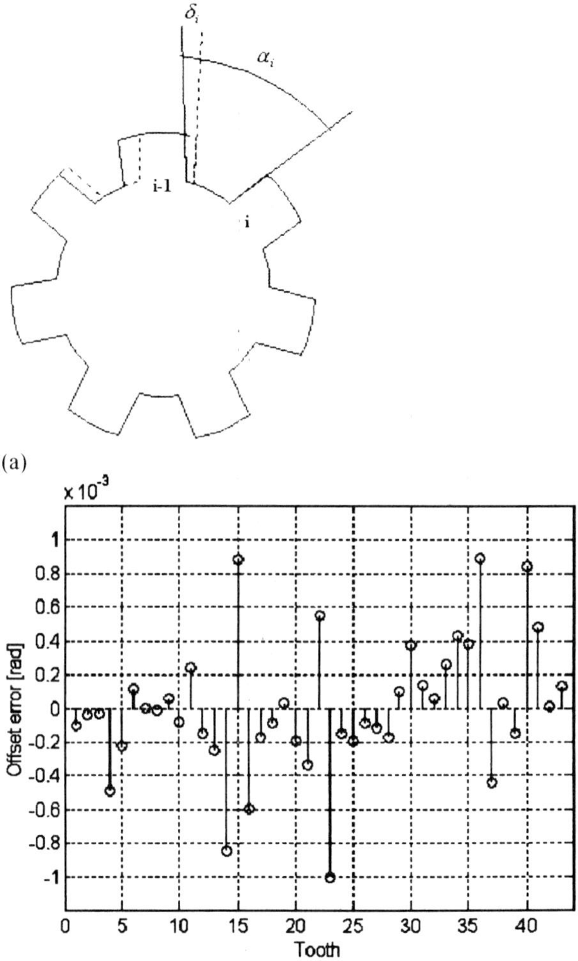

(a)

(b)

Fig.9.7 Offset error of wheel: diagram of wheel offset error (a), and estimation results for offset error of a 43 teeth wheel (b), from [32].

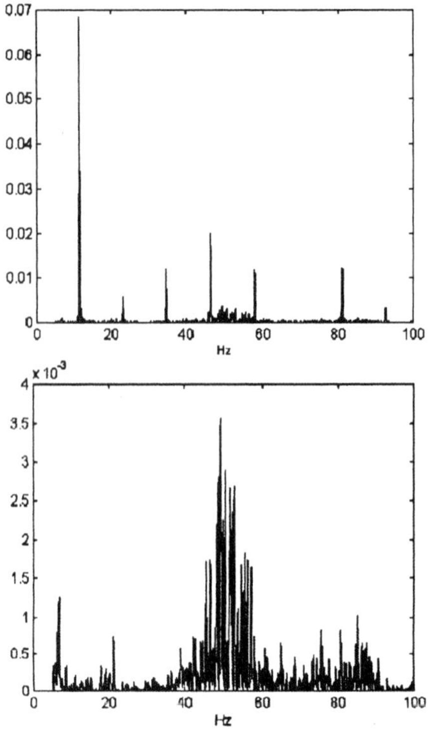

Fig.9.8 Spectrum for uncorrected spectrum for uncorrected signal: the periodic disturbances stemming from the un-ideal toothed wheel distorts the spectrum (upper), the periodic disturbances are eliminated using the intelligent algorithm (lower), from [32].

Besides RLS, Kalman filter and Wavelet transformation methods are also applied to identify wheel offset errors [7]. However, it should be pointed out hat the disturbance is assumed to be narrow-banded and well-behaved in the simulation shown here. In a real driving process, vehicle velocity may change, which make the energy leak out to other frequencies. Thus, no peak can be clearly found as the simulation results shown above.

9.3 Advances in Online Tire Monitoring

During the last twenty years, real-time tire tread force, pressure and temperature monitoring got consistent attentions because of their importance. For instance, DOT estimates that the total quantified safety benefits will be 124 fatalities prevented and 8,722 injuries prevented or reduced in severity each year, if all light vehicles meet the four-tire, 25 percent compliance option. If all light vehicles meet the one-tire, 30 percent compliance option, 79 fatalities will be prevented and 5,176 injuries prevented or reduced in severity each year [2].

9.3.1 SAW Tire Sensors

How to obtain tire pressure and temperature directly through equipped sensors had been studied for about two decades. The most important problem that needs to be solved is how to provide the power to tire sensors. An alternative approach is to transfer power to the wheel from the vehicle supply via some forms of sliding contact. This approach was once seriously discussed, but finally rejected for both cost and reliability consideration. The new technique is to use the so-called radio sensor technology [33]-[34].

Generally, all the radio sensors could be divided into active devices, which are powered by battery; semi-active devices, which are energized by inductive coupling or strong radio frequency signals; and passive transponder devices. In the experiments, people found that classic measurement systems using active or semi active sensors will be quite expensive for tire products. An applicable tire sensor should be low-cost, reliable and durable. In contrast, a sensor system that consists of a battery powered sensor has short lifetime due to the unavoidable battery recharge. It's reported that Ford Motor Company concluded in 1987 that an acceptable maximum cost for a tire pressure monitoring system is around 15 dollars per wheel (at 1987 prices) [33]. This price will immediately rule out almost all the active and semi-active radio sensors that need to recharge battery nearly once per year.

In late 1990s, a novel passive radio sensor, surface acoustic wave (SAW) sensor, emerged and offered exciting perspectives for remote monitoring. This is a purely passive radio sensor that could work in harsh environments. The first industrial application is a road pricing system for the Norwegian highways around Oslo. But it was soon developed and used as successful tire pressure and temperature sensors. Before discussing the tire sensor applications of SAW, it would better to review the basic knowledge of radio sensor technology first.

A radio sensor system normally employs a radio request unit and one or more distant (usually less than 5 meters for tire sensors) sensor units. The radio interrogation consists of the radio request and the sensor response and its evaluation. The radio request signal is transmitted from the transmitter section of the interrogation system to the sensor (often cited as downlink). The sensor responds with a radio signal (often called as uplink) that is received by the receiver section of the interrogation system. Up- and down-links are used to separate the request signal and the sensor's response. A separation in frequency (frequency domain division; FDD) requires a frequency conversion or at least a nonlinear device in the sensor unit. Comparing to FDD, a separation in space is difficult for linear passive or semi-active devices; because a separation between request and response signal in time (time domain division; TDD) is necessary, which usually requires a certain energy storage mechanism.

Specially, SAW devices retransmit an approximately linearly-distorted radio request signal. The distortion is affected by the measurand. The energy of the RF radio signal is stored in the SAW, which yields the required time delay between request signal and sensor response for TDD. The RF operation frequency is limited by the substrate size and the photolithographic process. Usually, SAW devices are manufactured in the frequency range between 30MHz and 3 GHz.

A SAW transponder works in the following way. It picks up an electromagnetic request signal and stores it until all echoes generated by multipath propagation have faded away, see Fig.9.9. Then, a characteristic response signal is beamed back to the interrogator unit. Any physical or chemical influences in the nearby environment will consequently change the propagation properties of SAW and the response pattern of the device [28]-[45].

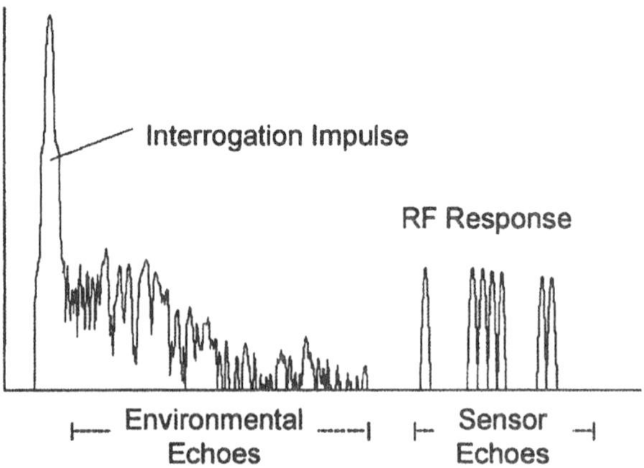

Fig.9.9 Interrogation pulse, environmental echoes, and RF response of a SAW reflective delay line, from [37]. (© [1998] IEEE)

The most important merit of SAW devices is that a low propagation velocity of SAWs allows a long delay time to be realized within a small chip. And the "reading" operation only takes several microseconds, it is possible to execute over than 105 interrogations per second.

Nevertheless, SAW transponders are small, robust, inexpensive, and can withstand extreme conditions (e.g., weather, radiation, temperature, shock, etc.). By using standard fabrication processes and packages, the operating range of a SAW transponder can vary from $-195\,^{o}C$ (liquid nitrogen) up to $200\,^{o}C$. With special crystals, adhesives and electrode materials, the temperature range can be extended to higher temperatures. Based on the combinations of new material, SAW devices have already been shown to function at temperatures as high as $1000\,^{o}C$. Since the design of wireless SAW devices on these new materials poses no problem, sensors working at even higher temperatures are expected to be realized in the near future.

The application spectrum of SAW includes reference devices and sensors. In reference applications, the identification information is either hard coded by special arrangement of reflectors and transducers within the propagation path of the SAW or controlled by connecting different loads to the transducers.

Fig.9.10(upper) shows a passive SAW tag with several transducers, whereas Fig.9.10(lower) shows a passive SAW tag with reflectors. Using this technique, each tire could be labeled uniquely. This is essential for Intelligent Unit to determine which tire is fault.

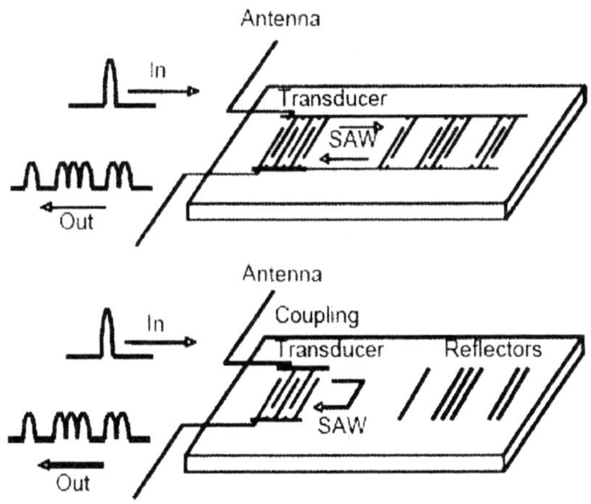

Fig.9.10 Schematic layout of a passive SAW tag consisting of several transducers connected together with a common bus bar (top), and a passive SAW tag consisting of one transducer and several reflectors (bottom), from [58].

Different from SAW reference devices, SAW sensory employ the sensitivity of special crystal cuts of different substrate materials to strain, temperature and mechanical stress, or their stability against such environment effects. The value of the associated physical parameter could be converted and measured by the change of sensor's surface length or surface acoustic wave's velocity, respectively.

One of the most important performance indices of SAW sensors is their temperature variation coefficient (TK). Apart from temperature, the propagation of the SAW depends on the geometry of the substrate and the material parameters subject to environment conditions.

Suppose delay of the responses for reflector i is denoted by τ_i in the following of this section. Thus τ_i should be the ratio of SAW propagation length L_i on the substrate's surface and propagation velocity v_{si}. The effect of the measurand on the sensor will cause a scaling of the sensor's response, which would be detected as individual delay shifts $\Delta\tau_i$ of the response signals s_i from reflector i. Mechanical measurands can be collected by loading the sensor mechanically.

In the delay line case, the interrogator transmits a burst signal, the sensor responds with a chain of burst, one for every reflector arranged at the substrate's surface. The differential delay between two or more response signal is evaluated. If the distance between the interrogation unit and the SAW transponder is unknown or varying, differential arrangements are used to evaluate differences in amplitude, time, frequency, or phase change of the RF signal.

Suppose the delay τ_i of the corresponding delay line i is affected by the measurand y. Generally the sensitivity S_x^y which describes the dependence of y from parameter x can be calculated with

$$S_x^y = \lim_{x \to 0} \frac{1}{x} \cdot \frac{\Delta x}{\Delta y} = \frac{1}{x} \cdot \frac{dx}{dy} \tag{9.4}$$

A linear approximation of the resulting delay yields

$$\tau_{iy} = \tau_{i0}(1 + S_x^y \cdot y) = \tau_{i0}(1 + \xi) \tag{9.5}$$

where ξ is an introduced scale variable.

From (9.5), it is apparent that the sensor effect yields a scaling of the delay time by a factor $1 + \xi$. The variation $\Delta\tau_{iy}$ of a time delay $\Delta\tau_{i0}$ is

$$\Delta\tau_{iy} = \tau_{iy} - \tau_{i0} = \tau_{i0}\xi = \tau_{i0} \cdot S_x^y \cdot y \tag{9.6}$$

Generally, to avoid errors caused by error sensitivity for temperature, etc., differential measurements of delay or center frequency are performed. Some typical linearized coefficients for physical effects to SAW substrate material sensors can be found in [40].

Temperature variation will simultaneously generates crystal deformation and SAW propagation velocity change. Suppose a change $\Delta\vartheta$ in the environmental temperature will result in a variation of the path length Δl and a variation of the SAW velocity Δv. Thus the propagation time $\Delta\tau$ changes according to

$$\frac{\Delta\tau}{\tau}=\left(\frac{1}{l}\frac{\Delta l}{\Delta\vartheta}-\frac{1}{v}\frac{\Delta v}{\Delta\vartheta}\right)\Delta\vartheta=TCD_1\Delta\vartheta \tag{9.7}$$

where TCD_1 represents the first order temperature coefficient of delay.

Normally, on the high coupling YZ-Lithiumniobate substrate, the temperature coefficient is one order smaller compared to the temperature coefficient of velocity. However, on special cuts of Quartz like STX-Quartz the variation in the path length is compensated by the variation of the SAW velocity in the first order. Therefore Quartz substrates are traditionally used for mechanical sensors Seifert and $LiNbO_3$ for temperature sensors.

SAW tire temperature sensors are based on the following facts: surface wave velocities are temperature dependent and are affected by the orientation and type of crystalline material used to fabricate the sensor. Fig.9.11. shows a sample measurement conducted on a transponder. When the phase difference between two pulses stemming from widely separated reflectors (i.e. Num. 1 and 11) is evaluated, a large phase shift per unit temperature is observed. This results in a high accuracy: with a phase resolution of the request unit of $\pm 1''C$, the temperature resolution amounts to $0.02''C$ if the SNR is large enough. Moreover, when the phase shift exceeds $360''C$, the raised ambiguity can be resolved by simultaneously evaluating the phase difference between two pulses stemming from adjacent reflectors (i.e. Num. 1 and 2).

Since SAW sensors are very sensitive to mass loading, they normally are sealed in a hermetic package. Usually, temperature sensors using SAW delay line oscillators have millidegree resolution, good linearity, and low hysteresis [39].

SAW velocities are strongly affected by stresses applied to the piezoelectric substrate on which the wave is propagating. A SAW pressure sensor can be therefore created by monitoring the variations of the scattering parameter S_{11}. In [58], Stelzer et. al. 2001 demonstrated the measured phase response and for the measured magnitude response due to tire pressure variation, see Fig.9.13-15. It is clear that in the most interesting range between 200 and 300 kPa, the phase change is a maximum. This naturally allows high precision in normal operational conditions combined with a great measurement range for the fault case.

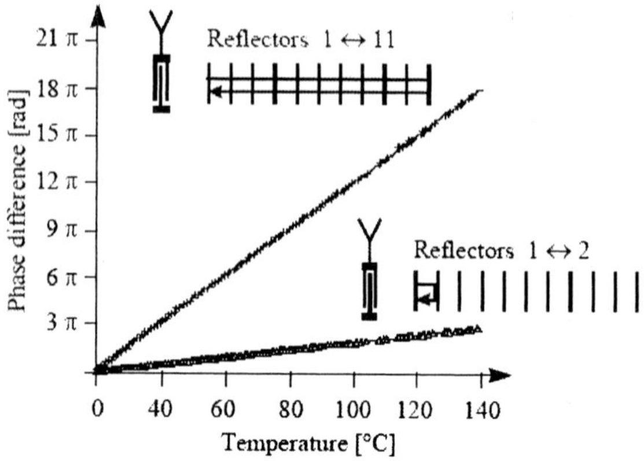

Fig.9.11 Temperature-dependent phase difference between two selected pulses in the time response of a SAW ID tag, the tag works at 434 MHz and was requested from a distance of 3 m, the time distance between the first and the last reflector is 4.55 μs, from [43]. (© [2001] IEEE)

Usually an electrical matching network is used to connect the external sensor to the SAW device. This leads to the following complex scattering parameter S_{11} discussed with the well-known P-matrix formulation (9.8), which is initialized by Reindl and Ruile. By using the matching network, a transformation of the impedance variation of the sensor to either a maximum variation in the amplitude or a maximum variation in the phase of the detectable reflected electrical wave can be performed. It depends on the principle of operation of the interrogation unit if only amplitude or also the phase can be evaluated. This can be viewed as a prerequisite for state of the art wireless sensors.

$$S_{11}(Z_{sensor}) = P_{11} + \frac{2P_{13}^2}{P_{33} + \dfrac{1}{Z_{sensor} + Z_{match}}} \tag{9.8}$$

where the short circuit reflection P_{11} as a function of a complex termination impedance Z_{load} at its electrical port. P_{13} denotes the electro-acoustic transfer coefficient, and P_{33} denotes the input admittance of the transducer.

Thus, it has

$$Z_{load} = Z_{sensor} + Z_{match} \tag{9.9}$$

The externally loaded sensor is frequently used in tire pressure monitoring applications. As shown in Fig.9.12, in a schematic drawing of the tire pressure monitoring system, the interrogation unit consists of an RF front-end, a DSP for the base band evaluation and a CAN bus interface. Regarding the sensor unit, the pressure sensor itself is connected via a matching network to the SAW chip. Usually, the temperature drifts that tend to interfere with SAW pressure sensors can be minimized by placing a reference SAW device close to the measuring SAW on the same substrate and mixing the two signals. More specifically, the hybrid tire sensors are used to isolate and compensate temperature effects from the stresses that the pressure experiences. A SAW pressure sensor weighing <1 g, with a resolution of 0.73 psi, was recently integrated into a car tire with excellent results. Clearly, this technology is particularly interesting for the new run-flat (also called zero pressure or extended mobility) tire market.

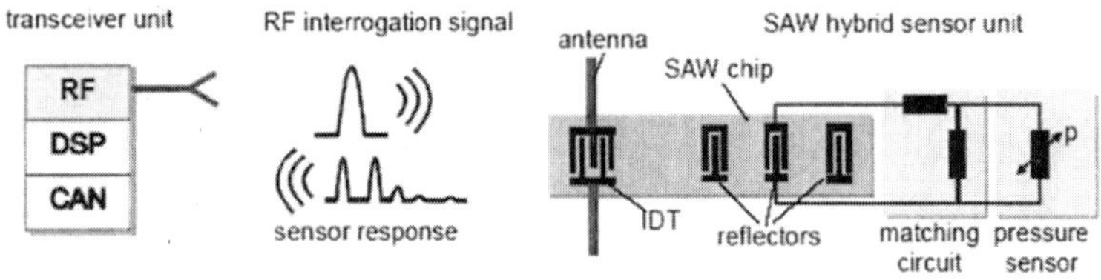

Fig.9.12 Schematic of a SAW sensor system for tire pressure monitoring, from [58].

Apparently Eq.(9.9) indicates that the acoustic reflection coefficient is a complex parameter and the corresponding sensor response is affected in both magnitude and phase. The following two simulation results show the behavior of the whole sensor with respect to a single matching element and the pressure sensor capacity. The pressure sensor itself is modeled by the pressure depending capacity C and an ohmic series resistor. For the matching element, a lossless series inductor with the inductivity L is used. Fig.9.13(a) shows the magnitude of the reflection coefficient S_{11} with lines of constant magnitude. In Fig. 9.13(b), the phase of S_{11} is plotted with lines of constant phase.

Eq.(9.9) also reveals the advantage of this new indirectly affected SAW DL sensor. The separation between time delaying RF transponder and actual sensor make the measurement more accurate. In traditional designs, the amplitude response is evaluated with the drawback of a strong dependency of the achievable amplitude resolution from the actual signal-to-noise ratio which also limits the useable dynamic range. As presented in [54] and [57], this can be overcome by using a phase evaluation instead.

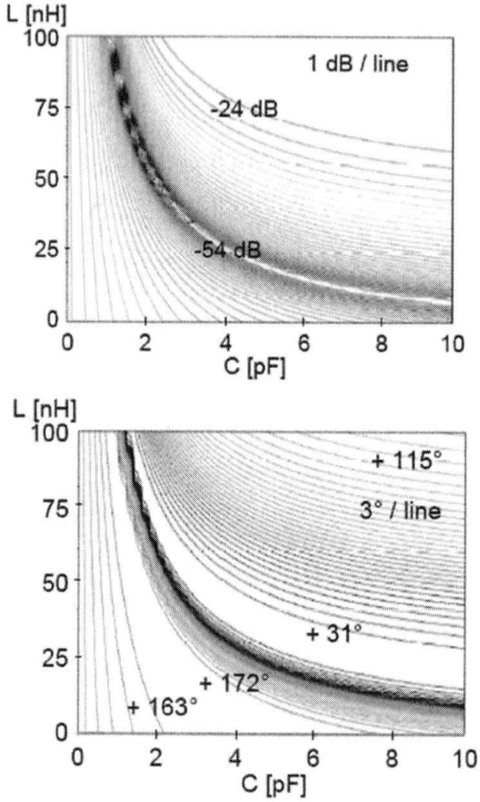

Fig.9.13 Simulated magnitude of the scattering parameter S_{11} (upper) and simulated phase difference $\Delta\varphi_{S_{11}}$ (lower), based on measured SAW device parameters, as a function of the sensor capacitance, from [58].

Stelzer et. al. 2001 showed the finally achieved results of the tire pressure sensor in Fig.9.14 for the measured phase response and in Fig.9.15. For the measured magnitude response, it is obvious that, in the most interesting range between 200 and 300 kPa, the phase change is a maximum as a result of a proper matching network. This allows high precision in normal operational conditions combined with a great measurement range for the fault case.

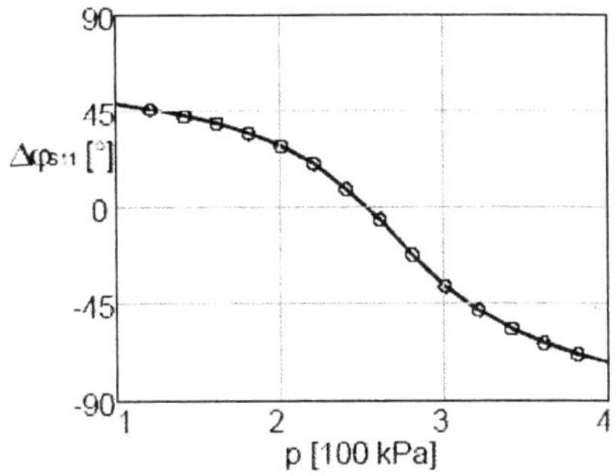

Fig.9.14 Measured phase difference signal $\Delta\varphi_{S_{11}}$ as a function of tire pressure, from [58].

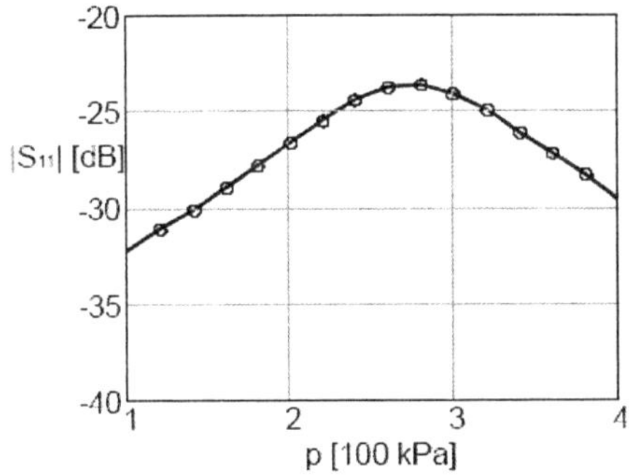

Fig.9.15 Magnitude of the measured scattering parameter S_{11} representing the insertion loss of the sensor unit, from [58].

SAW tire deformation sensor research also received notable considerations recently, because to employ such kind of "real" sensors in online tire/road friction modeling, monitoring and control issues is expected to significantly benefit current research, se reference [53], [15]. Fig.9.16 shows the Bristle model for tire deformation and the resulting tire force. As shown in [53], if a tire is running on a relatively flat surface, the bristle will be bent opposite to the cruising direction by a stress growing with position in contact length, which deforming the bristle pin will obey the resulting tangential stress schematically drawn in Fig.16(b). As a proof, Fig.17 shows the obtained experiment tire force data. Fig.18 further demonstrates the installation method of the SAW tire deformation sensor. It is obvious that this information can be used to test and developed those analytical or practical tire models, i.e. [60]-[62], [15].

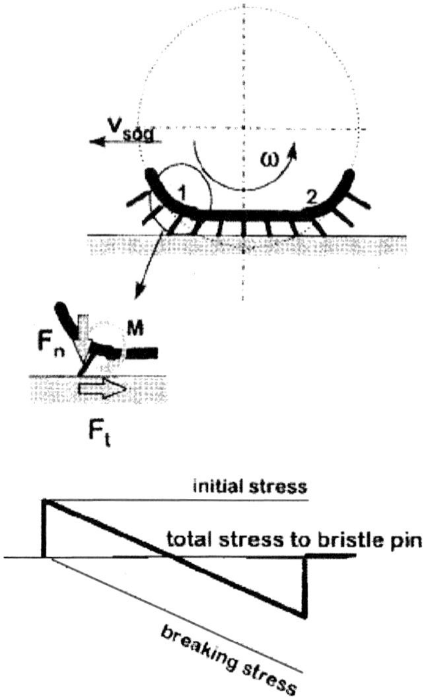

Fig.9.16 Deformation of tread element during road contact. (upper) Bristle model and (lower) resulting stress in a bristle pin, from [53]. (© [1999] IEEE)

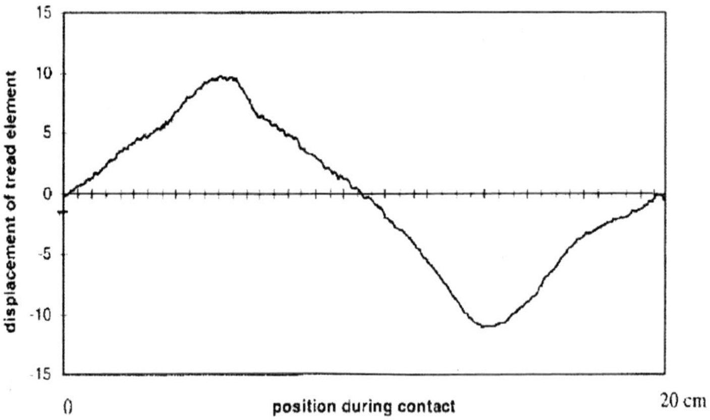

(a)

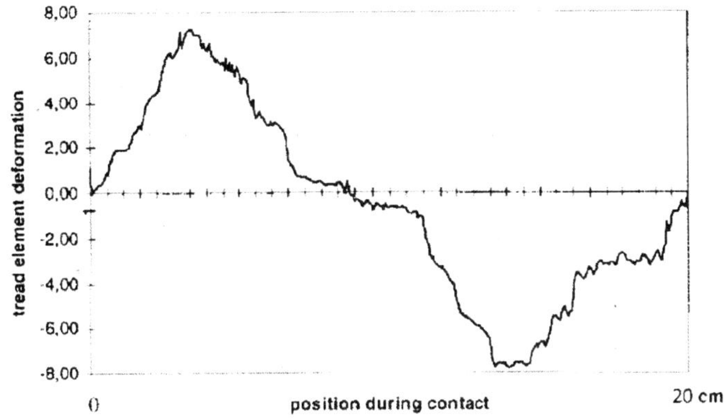

(b)

Fig.9.17 Displacement of tread element versus position during contact (a) and tread element's deformation versus position during contact (b), from [53]. (© [1999] IEEE)

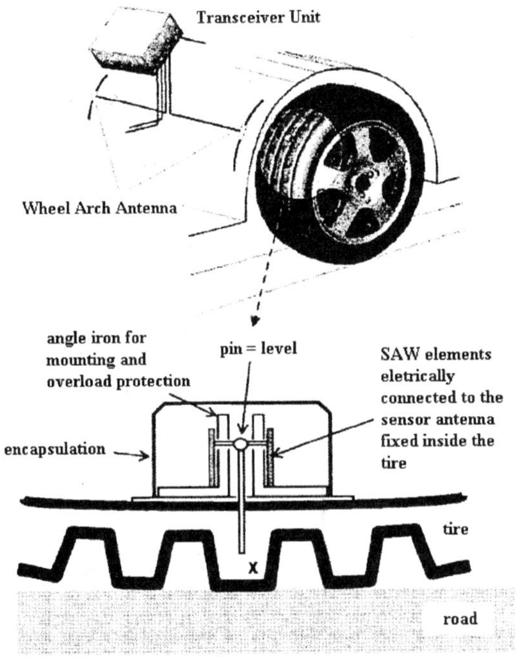

Fig.9.18 Experimental setup for tread element deformation measurement with SAW radio requestable sensors, from [53]. (© [1999] IEEE)

Besides, SAW tire sensor research also emphasizes on varied directions i.e. antenna [63], the relative distances between the two reflector arrays [64], and quartz substrate optimization [65]-[70].

9.3.2 Tire Rolling/Rotation Analysis and Pressure Monitoring

One fact that might be used fro tire pressure detection is that tire pressure will affect the tire rolling/vibration behavior. For instance, Fig.9.19 illustrates the power spectrum density of the rotational speed of a free rolling wheel of a vehicle running at a speed of 60 km/h. The peak around 10 Hz is a reflection of vertical resonance in suspension system. The peaks around 34-40 Hz are the tire resonances which have main components of torsional vibrations. When the tire becomes under-inflated, pressure changes 200 to 120 KPa, and the resonance frequency reduces from 40 to 34 Hz. The resonance characteristics change with the tire pressure.

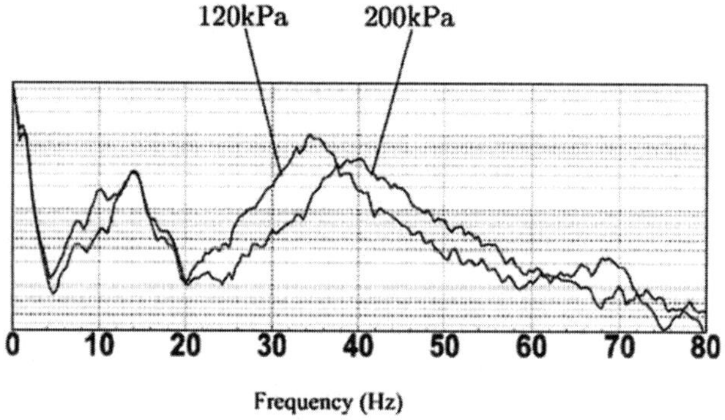

Fig.9.19 Tire rolling power spectrums under different pressures, from [74].

Therefore, several approaches are carried out to monitor tire pressure by detecting tire vibration behavior changes [71]-[78]. For instance, a tire model was presented in [73]-[74] as shown in Fig.9.20. Here, J_1 is a moment of inertia at the rim side, J_2 a moment of inertia at the belt side, K the torsional spring constant of the tire, D a damping of the tire, ω_1 a rotational speed (as can be detected) at the rim side, ω_2 a rotational speed at the belt side, θ_s a torsional angle, and T_d the various disturbances from the road surface assuming the white noise.

Thus, the state equation of the tire can be expressed from the tire model as follows:

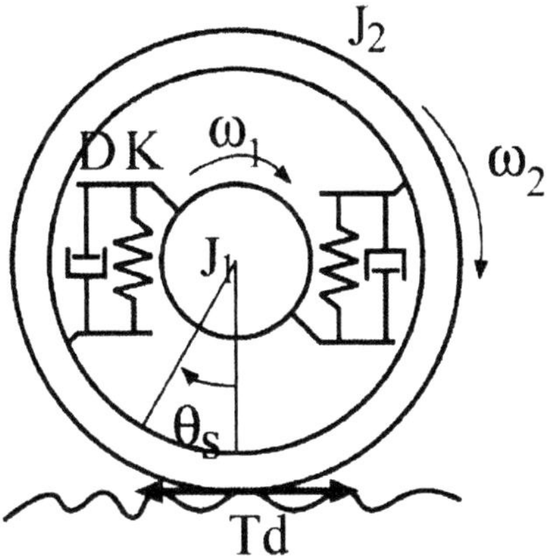

Fig.9.20 A tire model presented in [74].

$$\begin{bmatrix} \dot{\omega}_1 \\ \dot{\omega}_2 \\ \dot{\theta}_s \end{bmatrix} = \begin{bmatrix} -\frac{D}{J_1} & \frac{D}{J_1} & -\frac{K}{J_1} \\ \frac{D}{J_1} & -\frac{D}{J_1} & \frac{K}{J_1} \\ 1 & -1 & 0 \end{bmatrix} \begin{bmatrix} \omega_1 \\ \omega_2 \\ \theta_s \end{bmatrix} + \begin{bmatrix} 0 \\ -\frac{1}{J_2} \\ 0 \end{bmatrix} T_d \tag{9.10}$$

If the air pressure changes so that the spring constant K and the damping constant D change, a disturbance observer to estimate the portion of the equivalent disturbances can be designed as:

$$w = -T_d + \Delta K \theta_s + \Delta D \dot{\theta}_s \tag{9.11}$$

After compensate the wheel signal that is influenced by the vibration of the drive system components, the output of this observer can be used to detect abnormal tire power spectrums under different pressures. Fig.9.21 shows the experimental result under normal driving when the front left and rear right tires have constant pressure, while front rear and rear right tires get leakage. It is clear that the tire fault can be detected.

There were some more complex tire vibration/rolling model that had been established and employed. For instance, without losing too much generality, the tire was modeled as a spring-damper system both in vertical and torsional direction, see Fig.9.22.

The vertical spring-damper in the tire model leads to two conclusions. First, the effective rolling radius is dependent on the inflation pressure. If the pressure is decreased, the radius will decrease as well which causes the tire to rotate faster, too. Otherwise, vice verse.

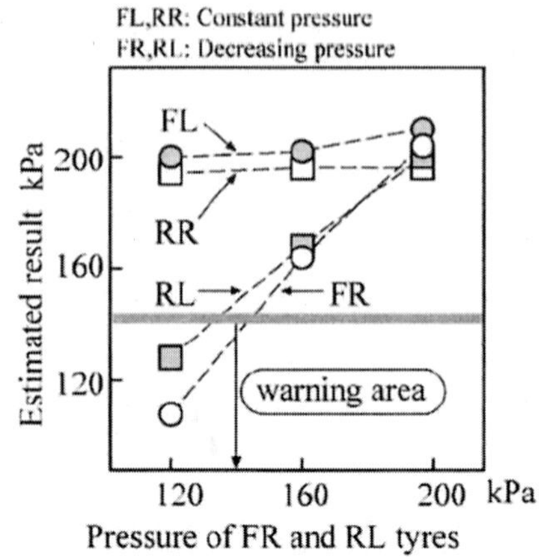

Fig.9.21 Experimental result under normal driving, from [74].

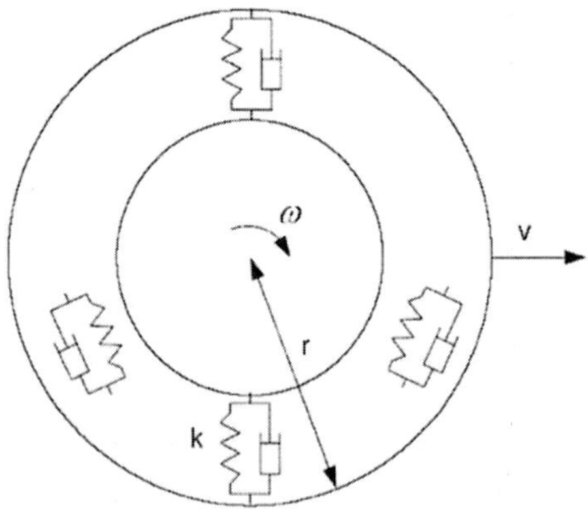

Fig.9.22 Tire modeled as spring-damper system in both vertical and torsional direction, from [77].

Secondly, the tire vibrates in the vertical direction excited by road roughness and causes the effective rolling radius to fluctuate with a specific resonance frequency. Indirectly this results in a fluctuation of the wheel speed at the same resonance frequency. In [77], Persson et al. 2002 showed that the spring-constant will decrease when the tire inflation pressure decreases, which yields a lower resonance frequency. Particularly, the frequency content of the wheel speed signal for three different test runs (with different pressure) is shown in Fig.9.23, where batches of 30 seconds test runs are examined. As is evident from Fig.9.23, the correlation between resonance peak frequency and inflation pressure can be used for tire pressure monitoring.

However, these approaches still need further discussions, since in general cases the effects of ground (especially soil), suspension devices, vehicle load distributions cannot be easily isolated from tire rolling/vibrations analysis [79]-[81].

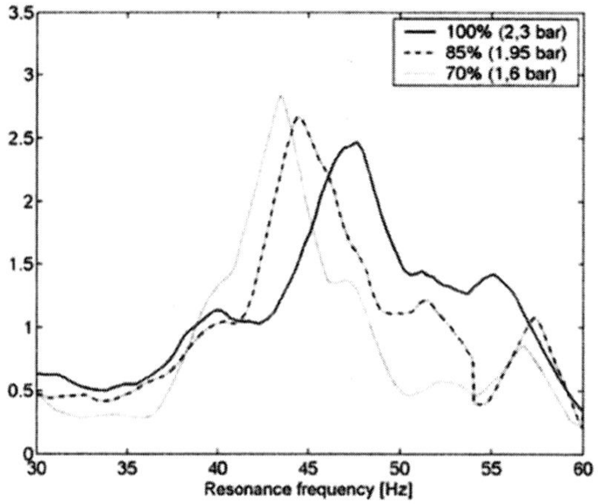

Fig.9.23 Smoothed FFT of wheel speed signal for three different test runs with 100%, 85% and 70% of the nominal inflation pressure, respectively, from [77].

9.3.3 Other Tire Deformation/Pressure Monitoring Sensors

Besides SAW sensor based approaches, there are several other tire deformation monitoring methods. For example, an ultrasonic tire sensor was mounted in [82] on the base of the wheel rim inside the tire. It can measure the distance to the opposite inner wall of the tire and provides the information of actual tire rolling radius, which indirectly measure the tire deformation status, see Fig.9.24.

In [83], a novel approach was proposed to continuously measure the mechanical deformations of a tire by embedding capacitive-resistive sensors on it. In this way, the sensing area is pushed towards the tread interface (the part of the tire in direct contact with asphalt) where the information concerning tire state is actually gotten. Tire deformation causes a change of the spacing between the steel wires inside the tire carcass and this change is translated into an impedance change of that region of the tire. By measuring such an impedance change, it enables to determine the deformation of the tire. And it is not sensitive to tire temperature changes.

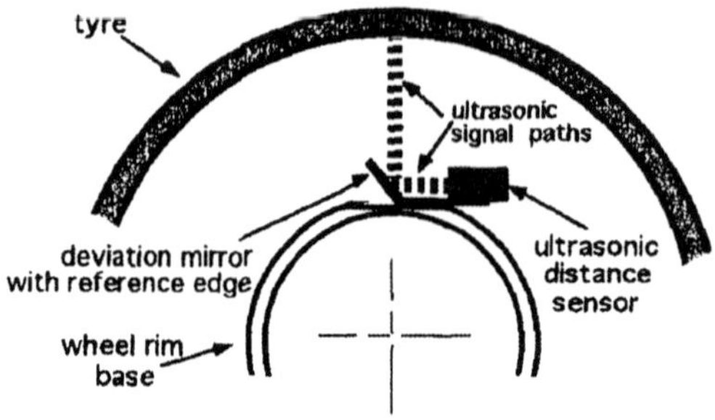

Fig.9.24 Schemas of ultrasonic sensor inside tire, from [82]. (© [1998] IEEE)

In [83]-[84], the wireless tire strain sensors based on electrical capacitance changes with an oscillating circuit were discussed. As shown in Fig.9.25, the steel wire is considered as a straight electrically conductive material in the steel-wire layer, where the rubber is a dielectric material. The steel wires are placed face-to-face, and the dielectric rubber. Thus, any deformation will lead to distance change, which results as capacity changes between two wires. Finally, the capacity variation will be transformed to strain changes, after the disturbance caused by temperature changes is compensated, see fig.9.26.

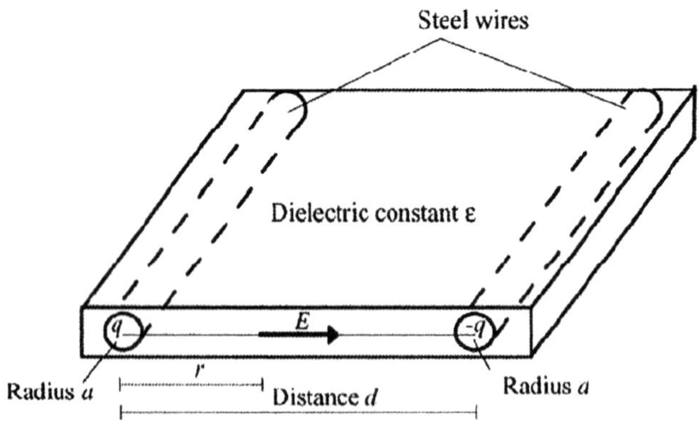

Fig.9.25 The condenser model of a steel wire belt of a tire, Reprinted from [84] with permission from Elsevier.

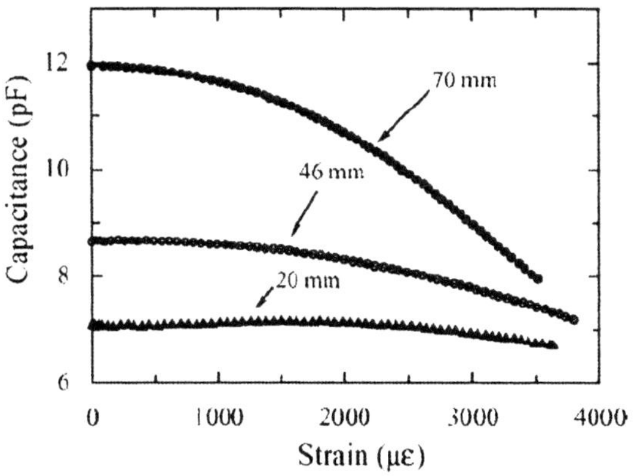

Fig.9.26 Measured capacitance change for various distances: 20, 46 and 70 mm, Reprinted from [84] with permission from Elsevier.

In [85]-[87], different touch mode tire pressure sensors were developed and tested. Varied 3D magnetic field tire tread deformation sensors were presented in [88]-[90]. In [91], tire pressure is indirectly measured by checking tire actual rolling radius. However, since tire pressure can only slightly affect the tire's rolling radius, namely about 1 mm per bar of pressure change, this method seems to be too sensitive to vehicle load variation and road disturbance.

9.4 Further Discussions

Generally, the integrated workflow of "Intelligent Tire" can be briefly depicted as Fig.9.27. Besides, tire inspection devices and monitoring sensors design, tire sensor related intra-vehicle communication devices/protocol design, warning information display method, and tire failure decision should gained notable attentions recently [7], [91]-[96].

For example, every tire sensor should have a build-in ID to distinguish each when communicating with the "intelligent center" of a vehicle. As pointed out in [95], to realize such an ID-tag using SAW is quite straightforward and easy.

To correctly determine tire status is not a simple task even the latest tire sensors are applied, since tire conditions can be affected by several environment factors. For instance, Fig.9.28 shows the different vibration peaks of a front tire working in summer and winter. The "intelligent center" of a vehicle should be smart enough to deal with this point.

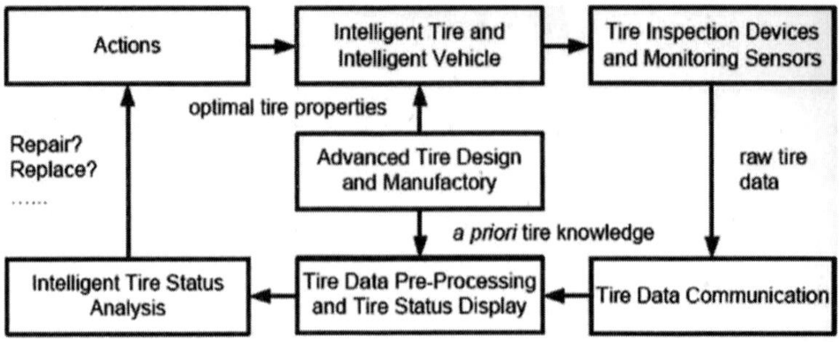

Fig.9.27 The integrated workflow of "Intelligent Tire", partly from [7].

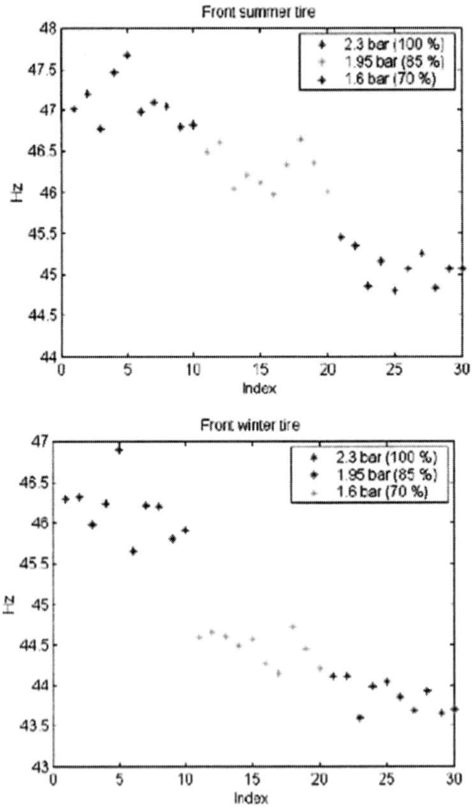

Fig.9.28 Measured tire vibration peak frequencies under different tire pressure and different seasons, from [32].

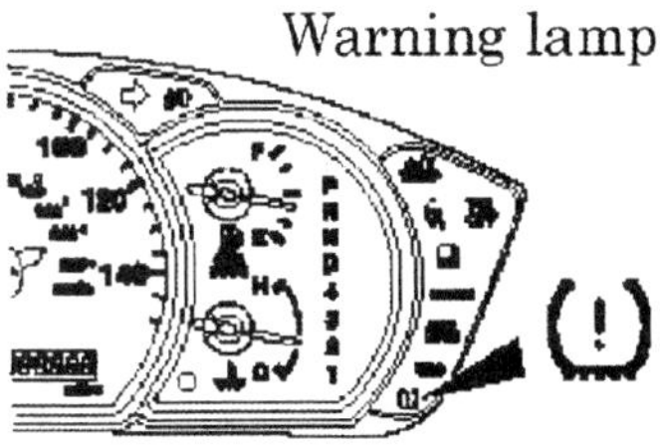

(a)

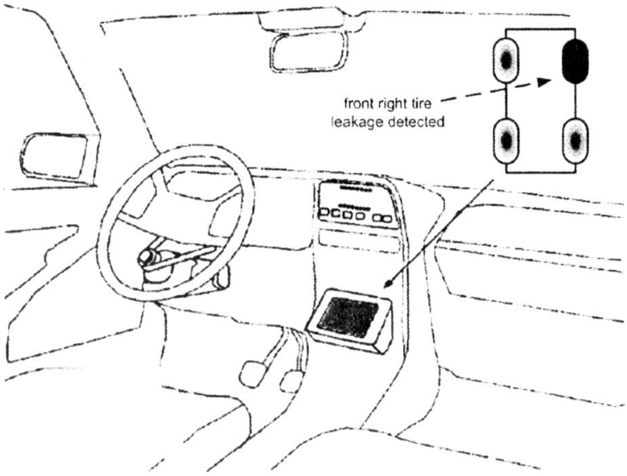

(b)

Fig.9.29 (a) The warning symbol suggested in [94]; (b) advanced display panel for tire sensors indicating that the front right tire is leaking.

How to appropriately display tire information to driver is also an interesting topic. Fig.9.29(a) shows the warning symbol suggested in [10]. However, many recent approaches prefer to design more complex but clear display panel, from which driver can easily find which tire is leaking, see Fig.9.29(b). As pointed out in [96], more efforts in this field can be seen in the new future.

9.5 Reference

1. Tyre Pressure Sensor Markets, III-Vs Review, vol. 16, no. 6, pp. 9, 2003.

2. Federal Motor Vehicle Safety Standards; Tire Pressure Monitoring Systems; Controls and Displays, National Highway Traffic Safety Administration, US Department of Transportation, 2004.
3. M. Kowalewski, "Monitoring and managing tire pressure," IEEE Potentials, vol. 23, no. 3, pp. 8-10, 2004.
4. F. S. Conant, "Tire temperature," Rubber Chemistry and Technology, vol. 44, pp. 398-409, 1971.
5. C. Clark, K. Meardon, and D. Russell, Scrap Tire Technology and Markets, Park Ridge, NJ: Noyes Data, 1993.
6. Highway Traffic Noise in the United States - Problem and Response, Federal Highway Administration, US Department of Transportation, 2000.
7. F.-Y. Wang, G.-S. Shan, and L. Li, et. al, "The research of smart tire and correlative core techniques," Tire Industry Sinica, vol. 22, no. 12, pp. 713-719, 2002.
8. S. K. Clark, Mechanics of pneumatic tires, U.S. Government Printing Office, Washington D. C., 1982.
9. J. R. Cho, H. S. Jeong, W. S. Yoo, "Multi-objective optimization of tire carcass contours using a systematic aspiration-level adjustment procedure," Computational Mechanics, vol. 29, pp. 498–509, 2002.
10. J. R. Cho, S. W. Shin, and W. S. Yoo, "Crown shape optimization for enhancing tire wear performance by ANN," Computers and Structures, vol. 83, no. 12-13, pp. 920-933, 2005.
11. R. J. Grogan, "The effect of tyre deflation on vehicle behaviour," Journal of Forensic Science Society, vol. 12, pp. 285-302, 1972.
12. Li, L., F.-Y. Wang, and G. Shan, et. al, "Design of tire fault observer based on estimation of tire/road friction conditions," Automatica Sinica, vol. 28, no. 5, pp. 689-694, 2003.
13. J. Wang, L. Alexander, R. Rajamani, "Friction estimation on highway vehicles using longitudinal measurements," ASME Journal of Dynamic Systems, Measurement, and Control, vol. 126, no. 2, pp: 265-275, 2004.
14. J. Stephant, A. Charara, and D. Meizel, "Virtual sensor: application to vehicle sideslip angle and transversal forces," IEEE Transactions on Industrial Electronics, vol. 51, no. 2, pp: 278-289, 2004.
15. Li, L., F.-Y. Wang, and Q. Zhou, "Integrated longitudinal and lateral tire/road friction modeling and monitoring for vehicle motion control," IEEE Transactions on Intelligent Transportation Systems, 2005.
16. Surface Texture for Asphalt and Concrete Pavements, U.S. Department of Transportation, Federal Highway Administration, Technical Advisory T5040.36, 2005.
17. O. Bennerhult, "Acoustical and mechanical impedances of road surfaces and the influence on tire noise," Proceedings of International Tire Noise Conference, pp. 185-197, 1979.
18. U. Sandberg and G. Descornet, "Road surface influence on tire/road noise," Internoise, pp. 259-272, 1980.

19. U. Sandberg, "Tyre/road noise—myths and realities," Proceedings of the Internoise, pp. 35-56, 2001.
20. K. Seki, S. Shin, and T. Tabaru, "Analysis of wavelet correlation between tyre sounds and tread patterns," Proceedings of IEEE Networking, Sensing and Control Conference, pp. 241-246, 2005.
21. F. S. Conant, G. L. Hall and J. D. Walter, "Surface temperature of running tires using infrared scanning," SAE# 700475, pp. 1-6, 1970.
22. N. M. Trivisionno, "Thermal analysis of a rolling tire," SAE#700474, pp. 1-5, 1970.
23. R. K. Kiminecz, "Laboratory simulation tests and infrared thermography of A-10 main landing gear tires," Final Report, Jul. 1980 - Dec. 1981.
24. Z. Zhao, Y. Li, and Q. Wang, et. al, "Application of infrared thermometer in testing of temperature field for a rolling tire," Proceedings of the IEEE International Vehicle Electronics Conference, pp. 98-101, 1999.
25. J. Li, Q.-N. Wang, and Z.-L. Zhao, "Steady characteristic of surface temperature of tire of middle duty truck," Journal of Jilin University: Engineering and Technology Edition, vol. 33, no. 2, pp. 1-5. 2003.
26. Y.-J. Lin and S.-J. Hwang, "Temperature prediction of rolling tires by computer simulation," Mathematics and Computers in Simulation, vol. 67, no. 3, pp. 235-249, 2004.
27. J. III Downs, P. Zhang, and M. L. Peterson, "A high-speed high-resolution ultrasonic inspection machine," IEEE/ASME Transactions on Mechatronics, vol. 4, no. 3, pp. 301-311, 1999.
28. N. W. Tschoegl, The Phenomenological Theory of Linear Viscoelastic Behavior, New York: Springer-Verlag, 1989.
29. M. Snow, "Laser triangulation sensors in the tire industry," Sensors Magazine, vol. 19, no. 3, 2002.
30. G. F. Blackwell, "Machine vision in the tire industry," IEEE Conference Record of 1989 Forty-First Annual Conference of Electrical Engineering Problems in the Rubber and Plastics Industries, pp. 67-79, 1989.
31. L. Wang, W. Zhang, and W. Zhu, "Study on profile measurement of extruding tire tread by laser," Proceedings of SPIE, Vol. 2899, p. 22-26, 1996.
32. F. Gustafsson, M. Drevo, and U. Forssell, et. al, "Virtual sensors of tire pressure and road friction," SAE# 2001-01-0796, 2001.
33. M. Hill and J. D. Turner, "Automotive tyre pressure sensing," Proceedings of IEE Colloquium on Automotive Sensors, vol. 5, pp. 1-6, 1992.
34. C. Kolle, W. Scherr, and D. Hammerschmidt, et. al, "Ultra low-power monolithically integrated, capacitive pressure sensor for tire pressure monitoring," Proceedings of IEEE Sensors, vol.1, pp. 244-247, 2004.
35. R. Grossman, J. Michel, and T. Sachs, et. al, "Measurement of mechanical, quantities using quartz sensors," Proceedings of IEE European Frequency and Time Forum Conference, pp. 376-381, 1996.
36. G. Scholl, F. Schmidt, and T. Ostertag, et. al, "Wireless passive SAW sensor systems for industrial and domestic applications," Proceedings of

the IEEE International Frequency Control Symposium, pp. 595-601, 1998.
37. L. Reindl, G. Scholl, and T. Ostertag, et. al, "Theory and application of passive SAW radio transponders as sensors," IEEE Transactions on Ultrasonics, Ferroelectrics and Frequency Control, vol. 45, no. 5, pp. 1281-1292, 1998.
38. A. Pohl and F. Seifert, "New applications of wirelessly interrogable passive SAW sensors," IEEE Transactions on Microwave Theory and Techniques, vol. 46, no. 12, pp: 2208-2212, 1998.
39. B. Drafts, "Acoustic Wave Technology Sensors," Sensors Magazine, vol. 17, no. 10, 2000.
40. A. Pohl, "A review of wireless SAW sensors," IEEE Transactions on Ultrasonics Ferroelectrics and Frequency Control, vol. 47, no. 2, pp. 317-332, 2000.
41. C. Hausleitner, R. Steindl, and A. Pohl, et. al, "Low cost radio interrogation systems for passive SAW sensors and transponders," Proceedings of IEEE International Symposium on Applications of Ferroelectrics, vol. 2, pp. 847-850, 2000.
42. C. C. W. Ruppel and T. A. Fjeldly, Advances in Surface Acoustic Wave Technology, Systems and Applications, World Scientific Co., Singapore, vol. 1 & 2, 2001.
43. W.-E. Bulst, G. Fischerauer, and L. Reindl, "State of the art in wireless sensing with surface acoustic waves," IEEE Transactions on Industrial Electronics, vol. 48, no. 2, pp. 265-271, 2001.
44. L. M. Reindl, A. Pohl, and G. Scholl, et. al, "SAW-based radio sensor systems," IEEE Sensors Journal, vol. 1, no. 1, pp. 69-78, 2001.
45. V. Kalinin, "Passive wireless strain and temperature sensors based on SAW devices," Proceedings of IEEE Radio and Wireless Conference, pp. 187-190, 2004.
46. B. De Geeter, O. Nys, and M. Chevroulet, et. al, "A wireless tyre pressure and temperature monitoring system," Proceedings Sensor Expo Conference, pp. 61–63, 1996.
47. W. Buff, M. Rusko, and E. Goroll, "Universal pressure and temperature SAW sensor for wireless applications," Proceedings of IEEE Ultrasonics Symposium, vol. 1, pp. 359-362, 1997.
48. A. Pahl, G. Ostermayer, and L. Reindl, et. al, "Monitor the tire pressure at cars using passive SAW sensors," Proceedings of IEEE Ultrasonics Symposium, pp. 471-474, 1997.
49. A. Pohl and F. Seifert, "Wirelessly interrogable surface acoustic wave sensors for vehicular applications," IEEE Transactions on Instrumentation and Measurement, vol. 46, no. 4, pp. 1031-1038, 1997.
50. W. Buff and S. Klelt, "Passive remote sensing for temperature and pressure using SAW resonator devices," IEEE Transactions on Ultrasonics, Ferroelectrics, and Frequency Control, vol. 45, no. 5, pp. 1388-1392, 1998.

51. A. Pohl and L. Reindl, "Measurement of physical parameters of car tires using passive SAW sensors," Proceedings of the 2nd Conference on Advanced Microsystems for Automotive Applications, pp. 249-262, 1998.
52. R. Grossmann, "Quartz crystals as remote sensors for tire pressure," Proceedings of IEEE Instrumentation and Measurement Technology Conference, vol. 3, pp. 1745-1749, 1999.
53. A. Pohl, R. Steindl, and L. Reindl, "The 'intelligent tire' utilizing passive SAW sensors measurement of tire friction," IEEE Transactions on Instrumentation and Measurement, vol. 48, no. 6, pp. 1041-1046, 1999.
54. G. Schimetta, F. Dollinger, and G. Scholl, "Wireless pressure and temperature measurement using a SAW hybrid sensor," Proceedings of IEEE Ultrasonics Symposium, vol. 1, pp. 445-448, 2000.
55. H. Wunderlich, G. Hettich, and M. Klein, et. al, "Concepts and steps in the development of wireless sensors and actuators for automotive applications," Proceedings of Sensor 99, pp. 157–162, 1999.
56. J. D. Turner and L. Austin, "Sensors for automotive telematics," Measurement Science and Technology, vol. 11, pp. R58–R79, 2000.
57. G. Schimetta, F. Dollinger, and R. Weigel, "A wireless pressure measurement system using a SAW hybrid sensor," IEEE Transactions on Microwave Theory and Techniques, vol. MTT-48, no. 12, pp. 2730–2735, 2000.
58. A. Stelzer, G. Schimetta, and L. Reindl, et. al, "Wireless SAW-sensors for surface and subsurface sensing applications," Proceedings of SPIE's 46th Annual Meeting, International Conference on Subsurface Sensing Technologies and Applications III, vol. 4491, 2001.
59. X. Zhang, F. Wang, and Z. Wang, "Intelligent tires based on wireless passive surface acoustic wave sensors," Proceedings of IEEE Conference on Intelligent Transportation Systems, pp. 960-964, 2004.
60. E. Tonuk and Y. S. Unlusoy, "Prediction of automobile tire cornering force characteristics by finite element modeling and analysis," Computer and Structures, vol. 79, no. 13, pp. 1219–1232, 2001.
61. R. J. Pinnington, "A wave model of a circular tyre. Part 1: belt modelling," Journal of Sound and Vibration, 2005.
62. R. J. Pinnington, "A wave model of a circular tyre. Part 2: side-wall and force transmission modelling," Journal of Sound and Vibration, 2005.
63. S. Basat, K. Lim, and I. Kim, et. al, "Design and development of a miniaturized embedded UHF RFID Tag for automotive tire applications," Proceedings of IEEE Electronic Components and Technology Conference, pp. 867-870, 2005.
64. W. D. Suh, K. A. Jose, and P. B Xavier, et. al, "Design optimization and experimental verification of wireless IDT based micro temperature sensor," Smart Materials and Structures, vol. 9, pp. 890–897, 2000.
65. R. B. Ward, "Temperature coefficients of SAW delay and velocity for Y-Cut and rotated LiNb03," IEEE Transactions on Ultrasonics, Ferroelectrics, and Frequency Control, vol. 31, no. 5, pp. 481-483, 1990.

66. R. M. Taziev and E. A. Kolosqvsky, "Deformation-sensitive cuts for surface acoustic waves in α -quartz," Proceedings of IEEE International Frequency Control Symposium, pp.660-664, 1993.
67. R. M. Taziev, E. A. Kolosovsky, and A. S. Kozlov, "Pressure-sensitive cuts for surface acoustic waves in α -quartz," IEEE Transactions on Ultrasonics, Ferroelectrics and Frequency Control, vol. 42, no. 5, pp. 845-849, 1995.
68. W. Ma and W. Shi, "Temperature-sensitive cuts far surface acoustic wave in Quartz," IEEE Transactions on Ultrasonics Ferroelectrics and Frequency Control, Vol. 48, no.1, pp. 333-335, 2001.
69. G. Schimetta, F. Dollinger, and G. Scholl, "Optimized design and fabrication of a wireless pressure and temperature sensor unit based on SAW transponder technology," IEEE MTT-S International Microwave Symposium Digest, vol. 1, pp. 355-358, 2001.
70. X. Zhang, F.-Y. Wang, and L. Li, "Optimal selection of Piezoelectric substrates and crystal cuts for SAW-based pressure and temperature sensors," to appear
71. P. Michelberger and D. Szoke, "Speed walking beam model, quarter car model dependent vertical vibrations of elastic vehicle bodies," International Journal of Vehicle Design, vol. 8, no. 1, pp. 87-95, 1985.
72. H. Mayer, "Model based detection of tyre deflation by estimation of a virtual transfer function," Proceedings of IEEE Conference on Control Applications, pp. 285-290, 1995.
73. T. Umeno, K. Asano, and H. Ohashi, et. al, "Pneumatic monitor for a vehicle based on the disturbance observer," Proceedings of IEEE Industry Applications Conference, Thirty-First IAS Annual Meeting, vol. 3, pp. 1687-1691, 1996.
74. T. Umeno, K. Asano, and H. Ohashi, et. al, "Observer based estimation of parameter variations and its application to tyre pressure diagnosis," Control Engineering Practice, vol. 9, no. 6, pp. 639-645, 2001.
75. C. Halfmann, M. Ayoubi and H. Holzmann, "Supervision of vehicles' tyre pressures by measurement of body accelerations," Control Engineering Practice, vol. 5, no. 8, pp. 1151-1159, 1997.
76. M. Yonetani and K. Ohashi, "Development of tire pressure monitoring system using wheel-speed sensor signal," Proceedings of the 16th International Technical Conference on the Enhanced Safety of Vehicles, pp. 492-496, 1998.
77. N. Persson, F. Gustafsson, and M. Drevo, "Indirect tire pressure monitoring using sensor fusion," SAE#2002-01-1250, 2002.
78. L. Li, F.-Y. Wang, and G. Shan, "Automatic tire pressure fault monitor using wavelet-based probability density estimation," Proceedings of IEEE Intelligent Vehicles Symposium, pp. 80-84, 2003.
79. H.-G. Jun, T. R. Way, and B. Lofgren, et. al, "Dynamic load and inflation pressure effects on contact pressures of a forestry forwarder tire," Journal of Terramechanics, vol. 41, no. 4, pp. 209-222, 2004.

80. U. D. Perdok and W. B. M. Arts, "The performance of agricultural tyres in soft soil conditions," Soil and Tillage Research, vol. 10, no. 4, pp. 319-330, 1987,
81. C. W. Fervers, "Improved FEM simulation model for tire–soil interaction," Journal of Terramechanics, vol. 41, no. 2-3, pp. 87-100, 2004.
82. V. Magori, V. R. Magori, and N. Seitz, "On-line determination of tyre deformation, a novel sensor principle," Proceedings of IEEE Ultrasonics Symposium, vol. 1, pp. 485-488, 1998.
83. M. Sergio, N. Manaresi, and M. Tartagni, et. al, "On road tire deformation measurement system using a capacitive-resistive sensor," Proceedings of IEEE Sensors, vol. 2, pp. 1059-1063, 2003.
84. R. Matsuzaki and A. Todoroki, "Wireless strain monitoring of tires using electrical capacitance changes with an oscillating circuit," Sensors and Actuators A: Physical, vol. 119, no. 2, pp. 323-331, 2005.
85. R. Marsili, "Measurement of the dynamic normal pressure between tire and ground using PVDF piezoelectric films," IEEE Transactions on Instrumentation and Measurement, vol. 49, no. 4, pp. 736-740, 2000
86. S. Yamamoto, O. Nakao, and H. Nishimura, "Touch mode capacitive pressure sensor for passive tire monitoring system," Proceedings of IEEE Sensors, vol. 2, pp. 1582-1586, 2002.
87. R. Bornefeld, W. Schreiber-Prillwitz, and O. Stover, et. al, "Low-Cost, Single-Chip Amplified Pressure Sensor in a Moulded Package for Tire Pressure Measurement and Motor Management," Advanced Microsystems for Automotive Applications, V. Jurgen and G. Wolfgang, eds., 2004.
88. M. Brandt, V. Bachmann, and A. Vogt, "Highly sensitive AlGaAs/GaAs position sensors for measurement of tyre tread deformation," Electronics Letters, vol. 34, no. 8, pp. 760-762, 1998.
89. O. Yilmazoglu, M. Brandt, and J. Sigmund, et. al, "Integrated InAs/GaSb 3D magnetic field sensors for 'the intelligent tire'," Sensors and Actuators A: Physical, vol. 94, no. 1-2, pp.59-63, 2001.
90. S. Gruber, M. Semsch, and T. Strothjohann, et. al, "Elements of a mechatronic vehicle corner," Mechatronics, vol. 12, no. 8, pp. 1069-1080, 2002.
91. H. Mayer, "Comparative diagnosis of tyre pressures," Proceedings of IEEE Conference on Control Applications, vol.1, pp. 627-632, 1994.
92. M. Durresi, A. Durresi, and L. Barolli, "Sensor inter-vehicle communication for safer highways," Proceedings of International Conference on Advanced Information Networking and Applications, vol. 2, pp. 599-604, 2005.
93. TKurashige, et al, "Low tire pressure warning device," SAE#820458, 1982.
94. H. Nishimura, "The Development of two simple tire warning devices," SAE#840067, 1984.

95. T. Tsukamoto, N. Matsuura, and T. Yamazaki, et. al, "Development of tire pressure monitor system," Society of Automotive Engineers of Japan, vol. 51, no. 11, 1997.
96. L. Reindl and W. Ruile, "Programmable reflectors for SAW-ID-tags," Proceedings of IEEE Ultrasonics Symposium, pp. 125-130, 1993.
97. L. Li, J. Song, and F.-Y. Wang, et. al, "New developments and research trends for intelligent vehicles," IEEE Intelligent Systems, vol. 20, no. 4, pp. 10-14, 2005.

INDEX

Printed in the United States of America.